REINFORCED CONCRETE

MAXWELL PRESS

REINFORCED CONCRETE

Albert Wells Buel
& Charles S. Hill

MAXWELL PRESS

Chennai Trichy New Delhi

MAXWELL PRESS

This edition has been published in india by arrangement with Carson Books, UK

ISBN 978-93-5528-004-6

Maxwell Press
No. 44, Nallathambi Street,
Triplicane, Chennai 600 005

Printed and bound in India

MJP 1296

Publisher : C. Janarthanan

Publisher's Note

The legacy of a country is in its varied cultural heritage, historical literature, developments in the field of economy and science. The top nations in the world are competing in the field of science, economy and literature. This vast legacy has to be conserved and documented so that it can be bestowed to the future generation. The knowledge of this legacy is slowly getting perished in the present generation due to lack of documentation.

Keeping this in mind, the concern with retrospective acquiring of rare books has been accented recently by the burgeoning reprint industry. Maxwell Press is gratified to retrieve the rare collections with a view to bring back those books that were landmarks in their time.

In this effort, a series of rare books would be republished under the banner, "Maxwell Press". The books in the reprint series have been carefully selected for their contemporary usefulness as well as their historical importance within the intellectual. We reconstruct the book with slight enhancements made for better presentation, without affecting the contents of the original edition.

Most of the works selected for republishing covers a huge range of subjects, from history to anthropology. We believe this reprint edition will be a service to the numerous researchers and practitioners active in this fascinating field. We allow readers to experience the wonder of peering into a scholarly work of the highest order and seminal significance.

MAXWELL PRESS

PREFACE.

In preparing this book the authors have had in mind a treatise for designing and constructing engineers following American practice and governed by the conditions which prevail in America. Theoretical discussions have been omitted, and in their place have been supplied practical working formulas, examples of representative structures, and records of actual practice in the selection of materials and of methods of workmanship and construction. For convenience of classification the book is divided into three parts or sections. In Part I are given working formulas for the calculation of all classes of structures in reinforced concrete and such facts about the properties of concrete and steel as are necessary in developing economical engineering designs. Part II contains illustrations and descriptions of a large number of representative structures of reinforced concrete. These record actual practice in design and show the adaptability of the new material to various types and forms of structures. Materials, workmanship, and methods of construction are considered in Part III and are illustrated by numerous examples from actual practice. In this part especial attention is given to the construction of centers and forms for concrete work and to methods of facing and finishing exposed concrete surfaces. In gathering materials for this book the authors have drawn freely upon the engineering journals of Europe and America, on the proceedings of technical societies, and on personal experience in designing and constructing works in reinforced concrete. Special acknowledgment of indebtedness is here made to "Beton und Eisen," "Le Génie Civil," "Ciment," "Annales des Ponts et Chaussées," "Engineering Record," and "Engineering News," and to the Proceedings of the Institution of Civil Engineers and the Transactions of the American Society of Civil Engineers.

A. W. B.
C. S. H.

New York City, September 1, 1904.

CONTENTS.

PART I. METHODS OF CALCULATION.

PART II. REPRESENTATIVE EXAMPLES.

PART III. METHODS OF CONSTRUCTION.

PART I.

METHODS OF CALCULATION.

REINFORCED CONCRETE.

CHAPTER I.—ECONOMIC USE AND PROPERTIES OF REINFORCED CONCRETE.

CONCRETE alone, considered as a building material, is nothing more nor less than a kind of masonry. The distinguishing features between rubble masonry and concrete are really confined to the methods of mixing and placing the materials. The results obtained with rubble masonry made of very small stone and with concrete made of large stone would be practically identical. The old Roman concrete was made with large stones, and may be classified either with rubble or concrete masonry. The value of either rubble or concrete as a material for construction depends largely on the quality of the cement used and the care exercised in the mixing and placing. Examples of masonry structures composed of large stones reinforced or tied together with iron rods and bars are found in the works of all periods, but usually only in connection with cut-stone masonry. The cost of such reinforcement was very great compared with the additional strength secured, and with rubble masonry the mechanical difficulties involved and the comparative cost render it impracticable.

Reinforced Concrete.—With the advent of modern concrete the facility with which reinforcing rods or bars of metal may be embedded anywhere in the mass of the masonry was soon seen and taken advantage of. The compressive resistance of concrete is about ten times its tensile resistance, while steel has about the same strength in tension as in compression. Volume for volume steel costs about fifty times as much as concrete. For the same sectional areas steel will support in compression thirty times more load than concrete, and in tension three hundred times the load that concrete will carry. Therefore, for duty under compression only, concrete will carry a given load at six-tenths of the cost required to support it with steel. On the other hand, to support a given load by concrete in tension would cost about six times as much as to support it with steel. These economic ratios are the *raison d'être* of reinforced concrete. If the various members of a structure are so designed that all of the compressive stresses are resisted by concrete and steel is introduced to resist the tensile stresses, each material will be serving the purpose for which it is the cheapest and best adapted and one of the principles of economic design will be fulfilled.

Other important advantages secured in the combination of concrete and embedded steel are that the protection of the metal elements from corrosion is practically perfect; that, with properly selected ingredients, the fire and heat resisting qualities are very high, perhaps surpassed by no other building material except fire-brick; and that, in many cases, the substantial appearance of a masonry structure is obtained at about the cost of a more or less temporary unprotected steel structure. When intelligently reinforced with steel, concrete becomes a material suitable and economical for beams, floors, and long columns, tanks, reservoirs, conduits, and sewers; admirably adapted to arch construction, and often economical for dams and retaining-walls. Even in concrete that is not subjected to tension or flexure it is often desirable to introduce steel reinforcement to prevent the occurrence of cracks due to shock or settlement or other causes.

Properties of Concrete.—A knowledge of the properties of materials is the first requisite for safe and economic designing of structures. The properties of reinforced concrete comprise not only those of the concrete and of the steel elements considered separately, but may be said to include those properties or characteristics of the composite mass that control the distribution of stresses between the elements of the combination of units and determine the nature of their interrelation. Such properties as are required by the practical engineer or architect in intelligent designing are here assembled in concise form, with values assigned to them that are considered to be safe and conservative deductions from the most recent experiments accessible. The scope and purpose of this work does not permit of an elaborate exposition of all the recent experiments nor of an exhaustive discussion of the deductions to be drawn therefrom.

Portland-cement concretes only will be considered. Concretes made with natural, slag, or Puzzolanic cements, although adapted to many uses, do not possess the qualities desirable for reinforced concrete structures. The object of reinforcing concrete with steel is to secure greater strength or safety, or both, than can be attained with concrete alone; and excepting a few special cases where the concrete is used principally for a filling or to add mass to the construction, the concrete made with Portland cement will be found to be the most economical for equal strength and safety.

The properties of concretes vary with their age and with the proportions and quality of the ingredients. The values given here are for concretes made with (1) true Portland cements having a tensile strength per square inch neat, in 7 days of 450 to 650 lbs., and in 28 days of 540 to 750 lbs.; (2) silica sand, not necessarily sharp nor coarse, but absolutely clean, and preferably a mixture of fine and coarse; and (3) good, hard, screened broken stone or clean gravel. The proportions of cement to sand generally used in the mortar or matrix, and for which there are reliable experimental data, vary from 1 of cement to 1 of sand up to from 1 of cement to 6 of sand; and the proportion of mortar or matrix to the aggregate (broken stone or gravel) is from 100 to 110 per cent. of the voids in the latter.

This method of specifying the proportions, by cement to sand in the mortar or matrix and by mortar or matrix to voids in the aggregate, is here adopted because it is believed that the ratio of matrix to aggregate, where the latter is good clean material, does not affect the strength of the concrete except in so far as sufficient matrix should be provided to fill the voids in the aggregate. Other things being equal, the strength of the concrete will be proportional to the strength of the mortar, and the maximum strength for a given matrix or mortar will be attained when all voids are filled. In practice this requires a volume of matrix about 10 per cent. in excess of the voids in the aggregate. Thus if by mixing several sizes of broken stone or gravel the proportion of voids to be filled is reduced from 45 per cent. or 50 per cent. down to 30 per cent., the proportion of matrix, cement and sand, to aggregate may be considerably reduced without reducing the strength of the concrete or affecting its properties. Where cement or sand are dear and stone and gravel are cheap, advantage may be taken of this method to reduce the cost of the concrete very materially.

The values here given are for concretes seven days, and one, three, and six months old. Those values should be used which correspond to the age at which the structure may be subject to its full load.

Compressive Strength.—The compressive strength of concrete is more often used than any other of its properties, and this follows from the fact that it is more economical and has heretofore been considered more reliable under compressive strains than transverse or tensile strains. Until very recent years engineers and architects hardly gave serious consideration to the value of concrete as a material to resist bending or tensile stresses, but at the present time comparatively few hesitate to use it in beams and similar situations where it is partly subjected to tensile stress, and a considerable number of eminent members of both professions have constructed works where the tensile strength of the concrete is taken advantage of. The best practice, where any tensile strains can occur, is to reinforce the section with steel. The two chief factors that determine the compressive strength of a concrete are its age and the proportion of sand to cement in the matrix. The quality of the cement, sand, and aggregate have more or less influence on the resulting concrete, but with any good brand of modern high-burned Portland cement, clean sand, and clean, hard stone, substantially the same results will be secured. Factors of far greater weight are the manner and conditions of mixing and placing and the personal equations of the operator. On this account it is extremely difficult to harmonize or draw conclusions from the large number of isolated tests that have been made by independent investigators under widely varying conditions and often with different objects in view.

A set of experiments made at the Watertown Arsenal for Mr. George A. Kimball, Chief Engineer of the Boston Elevated R.R., in 1899, are the most homogeneous and systematic set of tests that have as yet been published, and are given in Table I.

From these tests Mr. Edwin Thacher has deduced formulas for the ultimate strength of concretes. They give results that agree with the average of the experiments and can be entirely relied upon for concretes carefully made from good materials. They are as follows: The ultimate compressive strength in pounds per square inch of concrete:

$$\text{7 days old} = 1{,}800 - 200\left(\frac{\text{volume of sand}}{\text{volume of cement}}\right),$$

$$\text{1 month old} = 3{,}100 - 350\,(\quad\text{do.}\quad),$$

$$\text{3 months ``} = 3{,}820 - 460\,(\quad\text{do.}\quad),$$

$$\text{6 `` ``} = 4{,}900 - 600\,(\quad\text{do.}\quad).$$

These formulas give the results shown in Table II.

Tensile Strength.—The tensile strength may be safely placed at one-tenth of the compressive strength, and the modulus of transverse rupture, $\frac{M}{S}$, at about $1\frac{6}{10}$ that of the tensile strength. Tetmajer gives the ratio as follows for Portland-cement mortars consisting of 1 of cement to 3 of sand by weight:

$$\text{Tensile strength} = \left(\frac{\text{compressive strength}}{8.64 + 1.8 \text{ log. of age in months}}\right).$$

TABLE I.—SHOWING COMPRESSIVE STRENGTH OF CONCRETE AS DETERMINED BY TESTS MADE AT WATERTOWN ARSENAL IN 1899.

Brand of Cement.	Compressive Strength, Pounds per Square Inch.			
	7 Days.	1 Month.	3 Months.	6 Months.
MIXTURE 1:2:4.				
Atlas	1,387	2,428	2,966	3,953
Alpha	904	2,420	3,123	4,411
Germania	2,219	2,642	3,082	3,643
Alsen	1,592	2,269	2,608	3,612
Average	1,525	2,440	2,944	3,904
MIXTURE 1:3:6.				
Atlas	1,050	1,816	2,538	3,170
Alpha	892	2,120	2,355	2,750
Germania	1,550	2,174	2,486	2,930
Alsen	1,438	2,114	2,349	3,026
Average	1,232	2,063	2,432	2,969
MIXTURE 1:6:12.				
Atlas	594	1,090	1,201	1,583
Alpha	564	1,218	1,257	1,532
Germania	759	987	963	815
Alsen	417	873	844	1,323
Average	583	1,042	1,066	1,313

TABLE II.—SHOWING ULTIMATE COMPRESSIVE STRENGTH OF CONCRETE AS DETERMINED BY THACHER'S FORMULAS.

Mixture.	Age.			
	7 Days.	1 Month.	3 Months.	6 Months.
1:1 :3	1,600	2,750	3,360	4,300
1:2 :4	1,400	2,400	2,900	3,700
1:2½:5	1,300	2,225	2,670	3,400
1:3 :6	1,200	2,050	2,440	3,100
1:3½:7	1,100	1,875	2,210	2,800
1:4 :8	1,000	1,700	1,980	2,500
1:5 :10	800	1,350	1,520	1,900
1:6 :12	600	1,000	1,060	1,300

Modulus of Elasticity.—It has been said that no property of materials of construction is as uniform and reliable as the modulus of elasticity. This may be true of the modulus of elasticity of concrete and probably is, but the great variation in its value as determined by the experiments heretofore published left the matter very much in the dark. Its value has been stated all the way from 750,000 to 5,000,000. This was a discouraging condition with conservative constructors, and, no doubt, has greatly retarded the introduction of reinforced concrete in important works. The Watertown Arsenal tests of 1899 give values for the modulus of elasticity E of concretes as shown in Table III.

TABLE III.—SHOWING MODULUS OF ELASTICITY OF CONCRETE AS DETERMINED BY TESTS MADE AT WATERTOWN ARSENAL IN 1899.

Brand of Cement.	Modulus of Elasticity between Loads of 100 to 600.			
	7 Days.	1 Month.	3 Months.	6 Months.
	MIXTURE 1:2:4.			
Atlas	2,778,000	3,125,000	4,167,000	3,125,000
Alpha	....	2,083,000	4,167,000	3,125,000
Germania	2,500,000	....	3,571,000	4,167,000
Alsen	2,500,000	2,778,000	2,778,000	4,167,000
Average	2,592,000	2.662,000	3,670,000	3,646,000
	MIXTURE 1:3:6.			
Atlas	1,667,000	3,125,000	2,778,000	3,571,000
Alpha	....	2,083,000	3,571,000	4,167,000
Germania	2,273,000	2,273,000	2,778,000	3,125,000
Alsen	1,667.000	2,273,000	2,778,000	3,571,000
Average	1,869,000	2,438,000	2,976,000	3,608,000
	MIXTURE 1:6:12.			
Atlas	....	1,316,000	1,136,000	1,786,000
Alpha	....	1,667.000	1,786,000	1,923,000
Germania	....	961,000	2,083,000	1,786,000
Alsen	....	1,562,000	1,562,000	1,786,000
Average	....	1,376,000	1,642,000	1,820,000

From Table III the following formulas have been deduced, giving values very close to the averages of the experiments and sufficiently exact for all practical purposes. For concrete:

$$7 \text{ days old}, \quad E = 2{,}600{,}000 - 700{,}000 \left(\frac{\text{volume of sand}}{\text{volume of cement}} - 2\right),$$
$$1 \text{ month old}, \quad E = 2{,}900{,}000 - 300{,}000 \left(\quad \text{do.} \quad - 1\right),$$
$$3 \text{ months ''} \quad E = 3{,}600{,}000 - 500{,}000 \left(\quad \text{do.} \quad - 2\right),$$
$$6 \text{ '' ''} \quad E = 3{,}600{,}000 - 600{,}000 \left(\quad \text{do.} \quad - 3\right).$$

If the term $\left(\frac{\text{volume of sand}}{\text{volume of cement}} - c\right)$ is zero or less than zero (negative), the entire term is to be considered zero. In other words, all negative values must be considered as zero. Table IV shows the moduli of elasticity as determined by the above formulas. These values are sufficiently reliable for all ordinary purposes, and are probably as near to the truth as any that can be deduced from the experiments at present available. A large number of carefully executed experiments will be required to determine these values with greater precision.

TABLE IV.—SHOWING MODULI OF ELASTICITY OF CONCRETE AS DETERMINED BY FORMULAS.

Mixture.	Age.			
	7 Days.	1 Month.	3 Months.	6 Months.
1:1 :3	2,600,000	2,900,000	3,600,000	3,600,000
1:2 :4	2,600,000	2,600,000	3,600,000	3,600,000
1:2½:5	2,250,000	2,450,000	3,350,000	3,600,000
1:3 :6	1,900,000	2,300,000	3,100,000	3,360,000
1:3½:7	1,550,000	2,150,000	2,850,000	3,300,000
1:4 :8	1,200,000	2,000,000	2,600,000	3,000,000
1:5 :10	500,000	1,700,000	2,100,000	2,400,000
1:6 :12	....	1,400,000	1,600,000	1,800,000

Mr. W. H. Henby has given forty-eight determinations of the modulus of elasticity under tensile stress and eighteen under compressive stress, but the conditions were varied so that they can only be compared in groups of two or three tests with constant conditions, and as would naturally be expected the results were very erratic and are not conclusive. Prof. Wm. H. Burr concludes that the same values may safely be used for the modulus of elasticity in tension as in compression.

The values of E are only given for loads between 100 and 600, since these limits include the practical range of safe working stresses per square inch. For purposes of computing the ultimate strength, which would be for loads from 600 to 4,000 lbs., E would have considerably lower values. For loads between 1,000 and 2,000 lbs. the values would be from one-half to two-thirds of those given for loads between 100 and 600 lbs. For loads over 2,000 lbs. satisfactory data are not known to the writer.

Professors Boeck and Melan found a value of E at about 750,000 in connection with the Austrian experiments, where a number of arches were tested to destruction. In calculations of ultimate strength by formulas, assumed values of E ranging from 1,500,000 to 750,000, according to the mixture, age, and the ultimate load per square inch, would seem to agree more nearly with the average of previous experiments than values of E, corresponding to loads much less than the ultimate strength.

Two important points to be noted in connection with this subject are that the elastic limit of concrete, so far as it has been determined, is very close to the ultimate strength, and that its modulus of elasticity is a curve, instead of being practically a straight line, as it is with steel inside of the elastic limit. The nature of this curve cannot be determined from the limited number of determinations that have been published.

Working Loads.—In Table V are given what are considered safe working loads, in pounds per square inch, and properties for concretes in which the mortar or matrix is 1 of cement to 2 of sand and 1 of cement to 3 of sand, and in which all the voids in the aggregate are filled. According to present practice these mixtures will about cover the range for reinforced concretes.

Properties of Steel.—The following properties of steel for use in computing reinforced concrete sections, with the values assigned to them, will be used herein. These values are believed to be safe, but may be varied as conditions require according to the judgment of the designer.

Ultimate strength, 58,000 to 66,000 lbs. per square inch.

Elastic limit, 55 per cent. of the ultimate strength.

Modulus of elasticity, 29,000,000.

Working stress, factor of 4, 15,000 lbs. per square inch.
" " " " 5, 12,000 " " " "

Rate of expansion per degree Fahrenheit, 0.00000648 to 0.0000686.

Relations between Concrete and Steel.—The character of the relations that exist between the concrete and steel elements of reinforced concrete combinations depends first on the design of the section. If the two elements act independently in resisting the stresses, so that either the one or the other might carry all the load, it may be called a *composite design.*

TABLE V.—SHOWING SAFE WORKING LOADS FOR CONCRETE.

Mixture.	1 to 2 Matrix.				1 to 3 Matrix.			
Age.	1 Month.		6 Months.		1 Month.		6 Months.	
Safety factor	6	5	6	5	6	5	6	5
Compression, lbs.	400	500	600	700	340	400	500	600
Tension, lbs.	40	50	60	70	35	40	50	60
$f=\frac{M}{S}$	64	80	96	112	56	64	80	96
E	2,600,000		3,600,000		2,300,000		3,360,000	

Rate of expansion per degree Fahrenheit	(Clark)	.00000795
	(Rae and Dougherty)	.00000655 for 1:3:5 concrete
	(" " ")	.00000561 " 1:2 mortar
Adhesion to iron or steel metallic surface, ultimate	(Bauschinger)	570 to 640 pounds per square inch
	(Hatt)	636 to 756 " " " "
Safe working adhesion		60 to 100 " " " "

Note.—Prof. Hatt also found that the friction of smooth round rods embedded in concrete after they started to slip was from 50 per cent. to 70 per cent. of the adhesion. For concrete not reinforced with steel, use two-thirds the values given in the table for tension and $f = M \div S$.

If some of the forces are resisted entirely by the steel and other forces resisted entirely by the concrete, so that if the element resisting one force failed the entire section would fail, it may be called a *combination design.*

If the disposition of the steel and the concrete in the section is such that the two elements act as a single unit, all stresses being divided between the concrete and the steel, where the latter occurs, and that the entire omission of the steel would only result in reducing the strength of the section, it may be called a true *monolithic design.*

While many composite designs have been loosely classed with concrete-steel, they really have little in common with the combination and monolithic designs. Since the concrete and the steel are independent of each other, and either one may carry all the load, it is clear that each element should be calculated independently and like an all-concrete or an all-steel section, as the case may be. This is not to imply that the concrete may not stiffen the steel and prevent it from buckling, but as they do not act together as a combination or unit, and as the steel does not reinforce the concrete except in the manner that any additional and independent section may reinforce another, designs of this type should scarcely be classed with concrete-steel or reinforced concrete.

The illustrations, Figs. 1, 2, 3, and 4, of the Roebling, Rapp, and Youngstown systems of floor construction show examples of this type of designs. In none of these three is the metal either embedded in or protected by the concrete, and it can scarcely be considered as more

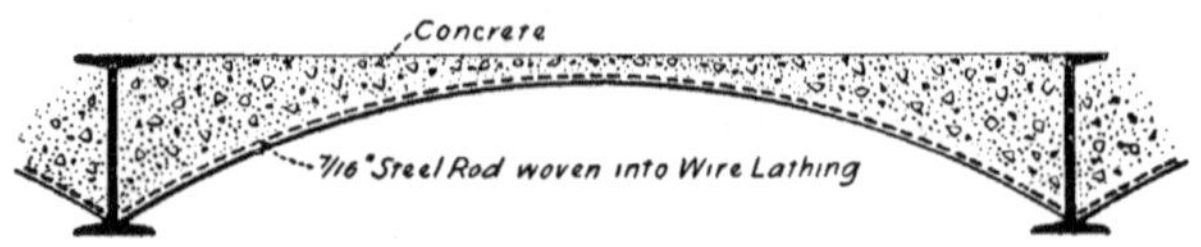

FIG. 1.—Roebling Arch Floor.

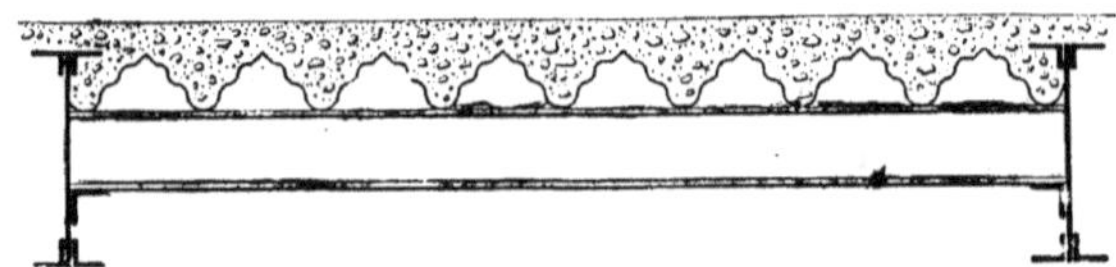

FIG. 2.—Youngstown Iron and Steel Roofing Company's Floor.

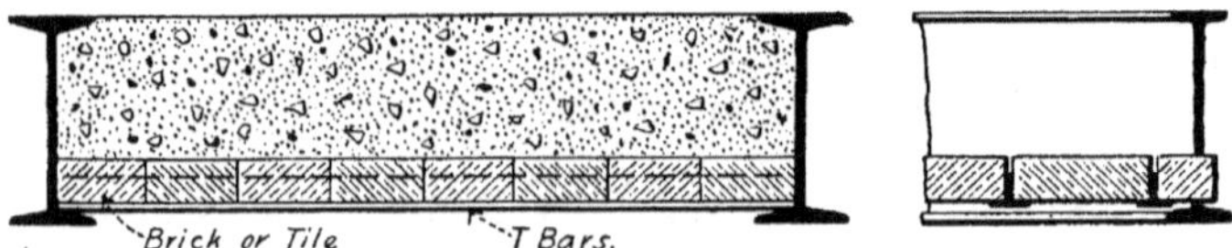

FIG. 3.—Rapp Floor with Flat Ceiling.

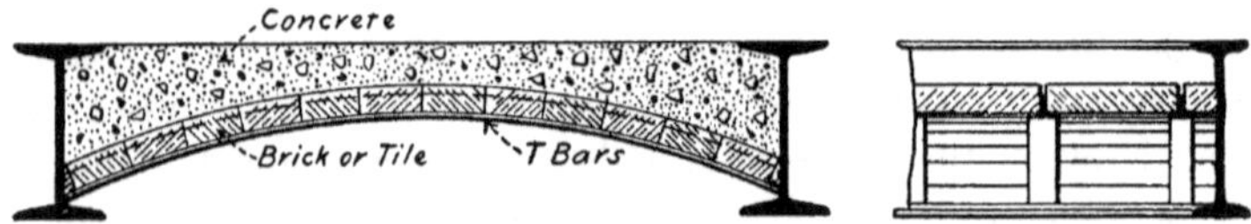

FIG. 4.—Rapp Floor with Arched Ceiling.

than a permanent centering or form on which to build the concrete structure. For temporary purposes it may be permissible to take into account the strength of the metal element acting alone, or, as in the case of the Roebling arch, held against buckling by the concrete.

Combination designs include concrete-steel beams after the concrete on the tension side has been strained beyond the point of rupture, which will occur in a well-designed beam long before the ultimate strength of the beam is reached. Concrete beams reinforced with steel, under loads that produce maximum tensile stresses in the concrete less than the ultimate strength, act as a single unit and may be classed as monolithic.

The most important characteristics or properties required to determine the distribution of stresses between the concrete and steel are the relations existing between the following:

A_c = area of the section of the concrete.
A_s = " " " " " " steel.
E_c = the modulus of elasticity of the conciete.
E_s = " " " " " " steel.

Under direct compression or tension the stresses will be distributed between the two elements in the proportion of $F_c:F_s::A_cE_c:A_sE_s$, where

$$F_c = \text{the total stress in the concrete}$$

and

$$F_s = \text{the total stress in the steel.}$$

From this is derived the equation

$$F_s = F_c \frac{A_s E_s}{A_c E_c}, \quad \ldots \ldots \ldots \quad (1)$$

and if $f_c = \frac{F_c}{A_c}$ = the stress per square inch in the concrete and $f_s = \frac{F_s}{A_s}$ = the stress per square inch in the steel we have

$$f_s = f_c \frac{E_s}{E_c}, \quad \ldots \ldots \ldots \quad (2)$$

which is to say that the stress per square inch in the two elements is directly proportional to their respective moduli of elasticity. This is derived directly from the definition of the modulus of elasticity which is the ratio of the stress per unit of section to the deformation. When the modulus of elasticity for steel is stated to be 29,000,000, it means that one pound per square inch tension or compression will stretch or compress the section an amount equal to its length divided by 29,000,000, and if E_c, for the concrete, is 2,900,000, one pound per square inch will stretch or compress it an amount equal to its length divided by 2,900,000. If $\frac{E_s}{E_c} = \frac{29,000,000}{2,900,000} = 10$, and if the same intensity of stress per square inch exists in both the concrete and the steel, the concrete will be deformed ten times as much per unit of length as the steel, or in the ratio $\frac{E_s}{E_c}$. If, however, the stress per square inch in the steel is ten times that in the concrete, or in the ratio of $E_s:E_c$, then the deformation will be the same per unit of length in both. Unless this latter condition maintains in every part of a concrete and steel structure of any description the surfaces of the two elements in contact will slide over each other or the concrete near the steel element will be strained beyond its elastic limit or its ultimate resistance.

While it is the invariable practice to meet this condition in the design of arches, columns, etc.. concrete-steel beams are quite generally designed on the theory that the steel does all the work on the tensile side of the neutral axis. There is no doubt whatever that the concrete on the tensile side of a well-designed reinforced concrete beam will fail long before the ultimate strength of the beam is reached, since most all of the tests to destruction have demonstrated it to be so. This theory will be treated at some length in the chapter on beams.

Coefficient of Expansion.—The thermal changes in reinforced concrete have ceased to be a matter for discussion from a practical view-point, and have been relegated to the laboratories for the determination of the last decimal in the rates of expansion. Some of the most recent and reliable determinations made by Rae and Dougherty at Columbia University and by Prof. W. D. Pence at Purdue University give the rate of expansion for Portland-cement concrete with various proportions of sand and stone or gravel, such as are generally used in practice, at 0.00000545 to 0.00000655 per degree Fahrenheit. The latter value by Rae and Dougherty is perhaps the more reliable, as the experiments were conducted with great care. Clark gives the rate at 0.00000795, which averaged with the mean of Prof. Pence's determination, 0.00000545, gives 0.00000670. This is less than 2½ per cent. greater than the value given by Rae and Dougherty.

The rate of expansion per degree Fahrenheit for wrought iron and steel is given by *K*ent at 0.00000648 to 0.00000686, and by U. S. Reports on Iron and Steel at 0.00000617 to 0.00000676. The mean of these is about 0.00000657. From this it appears that the difference in the rate for concrete and for steel is only a fraction of 1 per cent.

Aside from this the large number of reinforced concrete structures that have been exposed to the weather in severe climates for years without any indication of injurious effect from thermal changes is a sufficient proof that if there is any difference in the rate for the two materials, it is not enough to be of consequence.

Adhesion between Concrete and Steel.—Next in importance to the ratio between the stress per square inch and the moduli of elasticity is the adhesion between the concrete and the steel. Table V gives the ultimate and safe working values of this property in pounds per square inch. In the design of any combination or monolithic member of reinforced concrete the bond between the two elements is of vital importance. In the majority of cases met in practice the relation between the elements is such that the entire stress in the steel must be transmitted by this bond or adhesion. When the shear per foot run between the steel and concrete exceeds the safe working adhesion, resort must be had to a mechanical bond. Various devices have been used to obtain an effective bond, such as corrugating or twisting square or flat rods or bars, driving rivets in flat bars, the projecting heads of which serve the purpose, and deforming round rods so that they are made up of alternate round and flat sections but with the same sectional area at every point.

Some engineers have objected to the use of square or flat sections on the ground that the sharp re-entering angles formed in the concrete weaken the latter and induce cracks to start from the angle when subjected to loads or shocks. In cast iron, a material that has several properties similar to those of concrete, re-entering angles greatly weaken the sections, and therefore castings are generally boldly filleted at such angles. The writers do not know of any tests that throw light on this question, but notwithstanding the fact that considerable concrete has been reinforced with square and flat steel, it would seem to be safer and conservative practice to avoid all sharp re-entering angles in concrete. By far the larger part of all the reinforced concrete in Europe has been made with round rods or wires. In some cases steel angles, I beams, or T's have been used, but squares and flats if used at all do not seem to have met with general favor. The European methods and practices in reinforced concrete merit especial consideration, on account of the longer and greater experience from which they have been developed.

CHAPTER II.—BEAMS AND THE THEORIES OF FLEXURE.

NEARLY every writer on the subject has essayed to set up a new theory, or a variation of some other theory, of flexure in, and a new set of formulas for computing the strength of, reinforced concrete beams. While some of these theories are deduced from a few experiments, others are entirely theoretical, and none are fully demonstrated to be absolutely true. This condition is due to the fact that not enough experiments have been made finally to establish any theory. Any attempt to present the elements of the fifteen or twenty most prominent theories would not only expand this chapter to several hundred pages of theoretical and mathematical matter that few readers would care to wade through, but so many different formulas would be confusing to the average busy constructor who has not time to analyze them all and make a selection. Therefore, after briefly stating the basic principles of several of the most rational theories, and selecting those that most nearly conform to the experiments that have been made, a few of the most practical formulas covering the several conditions met in practice will be given in a form that can readily be applied.

Theories of Flexure.—Several authors have advanced the hypothesis that the deformation of a plane section of a reinforced concrete beam under transverse loading is in a parabolic or some other curve. The common theory of flexure is based on the assumption that the rate of strain or deformation of any fiber is directly proportional to its distance from the neutral axis. These two theories are illustrated in Figs. 5 and 6, where *mn* is the position of the plane section before flexure and *m′n′* its position after flexure.

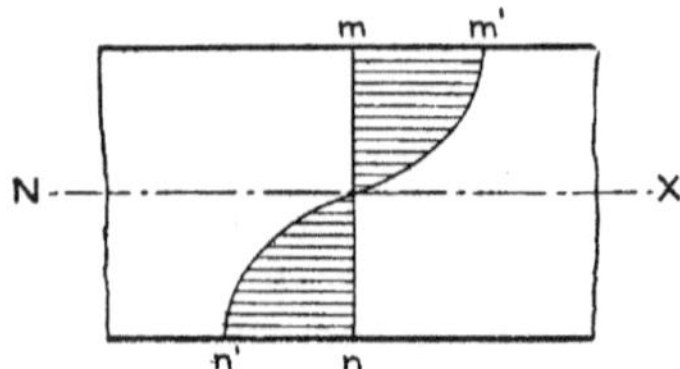

FIG. 5.—Diagram Illustrating Modified Theory of Flexure.

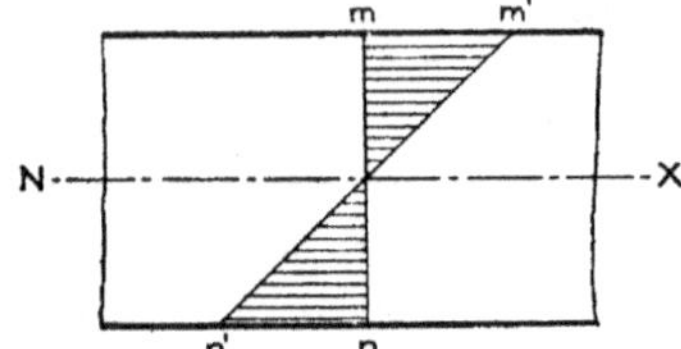

FIG. 6.—Diagram Illustrating Common Theory of Flexure.

While it is probable that the condition shown in Fig. 6 is not absolutely true, that shown in Fig. 5 has not yet been satisfactorily established by experiments. Formulas derived from the common theory of flexure, with empirical constants determined experimentally, give results sufficiently exact for all practical purposes, as will be seen by comparing them with the results of experiments.

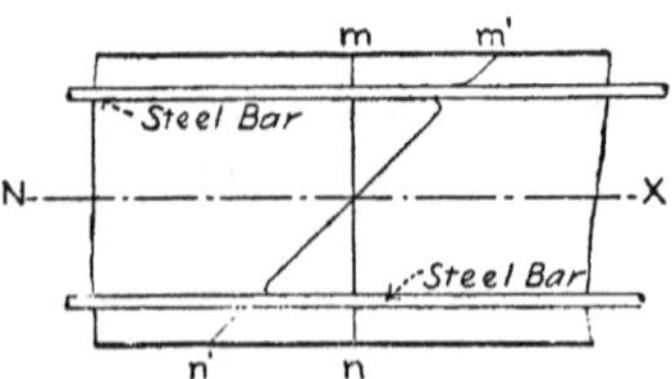

FIG. 7.—Diagram Illustrating Presumable Strains in Reinforced Concrete.

The fibers or particles of concrete in contact with or adjacent to the steel reinforcement are probably strained at a different rate from those fibers more distant from the steel, as indicated in Fig. 7.

Just how much this variation of strain is and how much it affects the strength of the beam has not been demonstrated, but enough is known from the experiments that have already been made to justify the neglect of this variation in practical designing.

Many writers have accepted the theory of M. Considère, that the concrete in tension in a reinforced concrete beam can be strained without breaking far beyond its elastic limit or even far beyond the strain that would break the same concrete if not reinforced. At the same time those who have accepted this theory state that the strain in the concrete beyond its elastic limit, or at least beyond the breaking strain of concrete not reinforced, is produced without any appreciable amount of work.

M. Considère made some experiments, from which he deduced the above theory. He found that a reinforced concrete beam could be deflected by transverse loading far beyond the amount at which a concrete beam not reinforced would break without causing any apparent failure in the concrete in the tension side.

Suppose a beam, *ABCD*, Fig. 8, supported at *R* and *R'*, and deflected by a load or system of loads to the position *EFGHp'K*. The

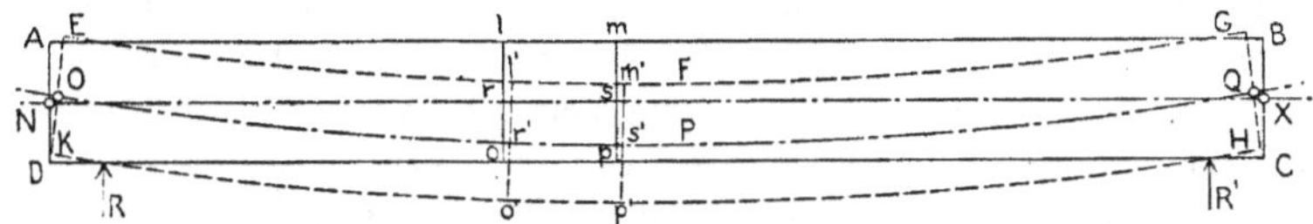

FIG. 8.—Diagram Illustrating Considère's Theory of Flexure.

neutral axis *NX* cannot change its length since there are no bending strains in the fibers on that line, and after flexure it will have a position similar to *OPQ*, and the parallel plane sections *lo* and *mp* will take the positions *l'o'* and *m'p'*, perpendicular to the new position of the neutral axis, and since this latter is a curved line, *l'o'* and *m'p'* will no longer be parallel.

The fibers *lm* will be compressed to a length *l'm'* and the fibers *op* will be stretched to a length *o'p'*. M. Considère claims that the fibers *op* were stretched to an amount greater than the breaking strain of concrete, since he observed a deflection of the beams corresponding to such an excessive stretch of the lower fibers, and did not discover any cracks or failure in the concrete. He then cut out pieces of concrete from the lower side of beams that had been thus subjected to transverse loading corresponding to strains in the concrete exceeding its ultimate tensile strength, and upon testing them found the concrete uninjured. From these results he concluded that the presence of the steel reinforcement so changes the action of concrete subjected to the tensile stresses of flexure that it can be stretched without breaking many times the amount it would stand without the steel reinforcement. It should be noted that these comparisons refer only to the strain or stretch in the concrete, and do not refer directly to the load or bending moment. An explanation of this observed phenomenon is given in the last edition of Prof. W. H. Burr's "Elasticity and Resistance of Materials."

Not only is the number of experiments on which this theory is based very limited, but a more recent series of tests made at the University of Wisconsin by Prof. F. E. Turneaure tends to discredit it entirely. Prof. Turneaure's beams had been kept under water until they were tested, and as they deflected under loadings, moisture appeared on the surface at some places on the tensile side, apparently coming through cracks. These places were marked, but an examination failed to disclose any caracks even under a magnifying-glass. Test pieces were then cut out of the concrete, some between the marks and some including the marks. Those that included a mark fell apart at the place where the moisture had appeared without the application of stress. Those that were cut out between the marks, like Considère's specimens, had not been injured by the flexure of the beams and showed good results on being tested. This proves that the lines where moisture appeared were real cracks, which were only prevented from opening sufficiently to be visible by the steel reinforcement. It also goes to prove that M. Considère did not happen to include a crack in the test pieces he cut out of his beams, or did not record any broken pieces. His failure to observe the microscopic cracks was probably due to the concrete being dry. But whether the concrete in M. Considère's beams cracked or not, Prof. Turneaure has proved beyond doubt that it cracks in some cases if not in all.

As between these somewhat contradictory experiments it seems far more rational to accept Prof. Turneaure's results since they conform to the previously accepted ideas of the action of elastic materials under stress, and do not require any contortions of the theory of elasticity or the properties of matter to explain them.

Accepting the phenomena observed by Prof. Turneaure as normal for reinforced concrete beams, the concrete and steel elements will work together up to the ultimate tensile strength of the concrete. After the ultimate strength of the concrete in tension is reached, invisible cracks will start at intervals on the tension side, gradually extending to the neutral axis, which will at the same time gradually recede to a new position nearer to the compression side. These cracks relieve the concrete of further stress, and between the cracks it is not injured.

Perhaps the most important point to note in connection with these results of Prof. Turneaure is that the cracks, however small, let moisture pass out, and if they let it pass out they would let it pass in. When exposed to the weather in Northern climates this would be a condition conducive to deterioration. In view of these facts, for permanent works in exposed locations the maximum tensile fiber stress in the concrete under transverse loading should never exceed the values for f given in Table V of the previous chapter. The general custom of many prominent builders of reinforced concrete is to ignore the tensile strength of the concrete. This is on the side of economy and at the present time is considered good practice where the work is not exposed to the weather, or in Southern climates, but it cannot be regarded as conservative for exposed works in severe climates. When this method is followed the properties of the beam are calculated for its ultimate resistance to transverse loading, and the safe loads are taken as some factor of the breaking load.

Formulas for Strength of Beams.—Before giving the formulas for the ultimate strength of reinforced concrete beams, the formulas that apply to the conditions existing before the concrete fails in tension will be developed. The notation to be used throughout this chapter will be as follows:

M = bending moment in inch-pounds.
l = length of beam in inches, center to center of supports.
h = depth of beam in inches, out to out of concrete.
b = breadth of beam in inches.
A = total area of cross-section of beam $= hb$.
A_s = area of steel reinforcement in tension.
A_s' = " " " " " compression.
$A_c = A - (A_s + A_s')$ = area of concrete.
a = area of steel in tension per unit of width $= A_s \div b$.
a' = area of steel in compression per unit of width $= A_s' \div b$.
x = distance from neutral axis to outer compression fiber of concrete.
y = " " " " " " tension fiber of concrete.
u = " " " " " " compression fiber of steel.
z = " " " " " " tension fiber of steel.
t = " " " " " center of steel sections in compres'n.
v = " " " " " " " " " " tension.
W = total load on beam, exclusive of weight of beam.
w = load per lineal foot of beam exclusive of weight of beam.
w' = weight of beam per lineal foot.
M = maximum bending moment in inch-pounds.
$z + d = y,\ u + d' = x,\ h' + d + d' = x + y = h.$

I = moment of inertia of reinforced beam about its neutral axis $= I_s + I_c$.
I_s = " " " " steel about neutral axis of beam.
I_c = " " " " concrete about neutral axis of beam.
E_s = modulus of elasticity of the steel.
E_c = " " " " " concrete.
e $= E_s \div E_c$
f_s = maximum intensity of tensile stress in the steel.
f'_s = " " " compressive stress in the steel.
f_c = " " " tensile stress in the concrete.
f'_c = " " " compressive stress in the concrete.
$N-X$ = neutral axis.
L_s = load sustained by concrete.
L_c = " " " steel.
s = proportion of load sustained by steel.
c = " " " " " concrete.

Fig. 9 is the cross-section of a beam showing the dimensions according to the notation given above. The steel reinforcement may con-

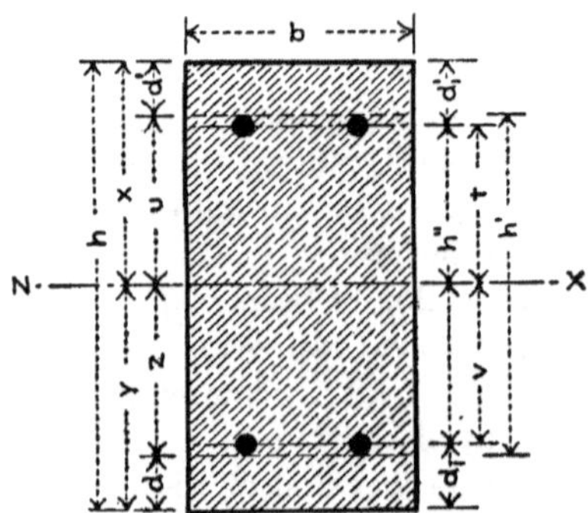

FIG. 9.—Cross-section of Beam Illustrating Standard Notation.

sist of one or any number of bars of any cross-section and on both sides of the neutral axis or only on the tensile side.

For a beam supported at both ends and loaded at the center with a load, W. . . . . $M = Wl \div 4$

For a beam supported at both ends and uniformly loaded . . . . $M = wl^2 \div 8$

For a beam fixed at one end and loaded at the other with a load, W. . . . . $M = Wl$

For a beam fixed at one end, unsupported at the other, and uniformly loaded . . . . $M = wl^2 \div 2$

For a beam fixed at one end and supported at the other and uniformly loaded. . . . . $M=wl^2\div 8$

For a beam fixed at one end and supported at the other and loaded at the center . . . . $M=3Wl\div 16$

For a beam fixed at both ends and loaded uniformly. . . . . $M=wl^2\div 12$

For a beam fixed at both ends and loaded at the center. . . . . $M=Wl\div 8$

For a beam supported at both ends and loaded with a single load, W, distant m inches from one end and n inches from the other end . . . . $M=W\cdot m\cdot n\cdot \div l$

Any two loads, W_1 and W_2, m inches apart, center to center, W_1 being equal to or greater than W_2, will produce the maximum bending moment when the distance from W_2 to the nearest support is n inches$=\dfrac{l-m}{2}-\dfrac{W_1 m}{2(W_1+W_2)}$, and the reaction at the opposite support will be $R=[W_2\cdot n+W_1(m+n)]\div l$, and the maximum moment $M=R[l-(m+n)]$.

$$y=\left\{\frac{bh^2}{2}+[A_s d_1{}^2+A_s'(h-d_1')^2](e-1)\right\}+[A+(A_s+A_s')(e-1)]; \quad (1)$$

$$x=h-y; \quad (2)$$

$$z=y-d; \quad (3)$$

$$u=x-d'; \quad (4)$$

$$v=y-d_1,\ \text{where } d_1=d+\text{radius of the steel}; \quad (5)$$

$$t=x-d_1',\ \text{where } d_1'=d'+\text{radius of the steel}; \quad (6)$$

$I_s=A_s v^2+A_s' t^2+$ (the moment of inertia of the steel bars about their individual neutral axes, which may be neglected for small rods without appreciable error); . . . . (7)

$$I_c=\frac{bh^3}{12}-(A_s v^2+A_s' t^2). \quad (8)$$

The distribution of the load between the concrete and the steel will be

$$\frac{L_c}{L_s}=\frac{E_c}{E_s}\frac{I_c}{I_s}=\frac{I_c}{eI_s}=g; \quad (9)$$

$$c=\frac{g}{1+g}\quad\text{and}\quad s=\frac{1}{1+g}; \quad (10)$$

$$f_s=(Mzs)\div I_s,\quad M=\frac{I_s\cdot t_s}{z\ s}\quad\text{and}\quad \frac{sM}{f_s}=\frac{I_s}{z}; \quad (11)$$

$$f_s'=(Mus)\div I_s,\quad M=\frac{I_s}{u}\frac{f_s'}{s}\quad\text{and}\quad\frac{sM}{f_s'}=\frac{I_s}{u};\quad .\ .\ (12)$$

$$f_c=(Myc)\div I_c,\quad M=\frac{I_c}{y}\frac{f_c}{c}\quad\text{and}\quad\frac{cM}{f_c}=\frac{I_c}{y};\quad .\ .\ .\ (13)$$

$$f_c'=(Mxc)\div I_c,\quad M=\frac{I_c}{x}\frac{f_c'}{c}\quad\text{and}\quad\frac{cM}{f_c'}=\frac{I_c}{x}.\quad .\ .\ .\ (14)$$

If $f_c=0$, that is if the concrete does not resist tensile stresses—the condition that exists after the concrete has failed in tension but before the ultimate resistance of the reinforced beam is reached—we have:

$$x=-\frac{e}{b}(A_s+A_s')+\sqrt{\frac{e^2(A_s+A_s')^2}{b^2}+\frac{2e}{b}[(A_s'd_1'+A_s(h''+d_1')]}.\quad (15)$$

$$I_c=\frac{bx^3}{3},\text{ nearly, if } A_s \text{ and } A_s' \text{ are small.}\quad .\ .\ .\ (16)$$

$$I_s=A_sv^2+A_s't^2,\text{ nearly.}\quad .\ .\ .\ .\ .\ .\ (17)$$

If $A_s'=0$, that is if the steel reinforcement is on the tension side of the neutral axis only,

$$x=-\frac{e}{b}A_s+\sqrt{\frac{e^2A_s^2}{b^2}+\frac{2e}{b}A_s(h''+d_1')}.\quad .\ .\ .\ .\ (18)$$

$$I_c=\frac{bx^3}{3},\text{ nearly, if } A_s \text{ is small.}\quad .\ .\ .\ .\ .\ (19)$$

$$I_s=A_sv^2,\text{ nearly.}\quad .\ .\ .\ .\ .\ .\ .\ (20)$$

The formulas given above cover nearly all cases met in practice and can be readily applied. They are all extremely simple, except perhaps (15) and (18), and even these two are very easily and quickly solved by ordinary arithemtic, when the values required for any particular case are substituted for the symbols.

Examples of Application.—What is the safe bending moment for a reinforced concrete beam 16 ins. deep and 4 ins. wide, having one steel rod $\frac{7}{8}$ in. diameter on the tensile side only, so placed that the center of the rod is $1\frac{1}{2}$ ins. above the bottom of the concrete, concrete to be made of 1 to 2 matrix and at six months old to have a safety factor of 5.

First Case.—Concrete to resist part of the tensile stresses without exceeding allowable fiber stress. From Table V of the previous chapter $f_c=112$, $f_c'=700$, $E_c=3{,}600{,}000$, and $E_s=29{,}000{,}000$.

From equation (1),

$$y=\frac{4\times 256}{2}+(0.6\times 2.25)7.06\div(64+4.236)=\frac{532.121}{68.236}=7.8\text{ ins.}$$

From equations (5), (7), and (8),

$$I_s = 0.6 \times 6.3^2 = 23.814 \quad \text{and} \quad I_c = \frac{4 \times 16^3}{12} - 23.814 = 1,341.52.$$

From equation (9),

$$g = \frac{1,341.52}{191.94} = 6.99,$$

and from equation (10),

$$c = 0.875 \quad \text{and} \quad s = 0.125.$$

From equation (13),

$$M = \frac{1,341.52 \times 112}{0.875 \times 7.8} = 22,014.7 \text{ in.-lbs.}$$

Second Case.—Resistance of the concrete to tensile stresses neglected.

From Table V of the previous chapter, $f_c' = 700$, $E_c = 3,600,000$, and $E_s = 29,000,000$.

Then from equation (18) we have

$$x = -\frac{8.06}{4.0} \times 0.6 + \sqrt{\frac{64.89 \times 0.36}{16} + \frac{16.11}{4} \, 0.6 \times 14.5};$$

$$x = -1.209 + \sqrt{1.46 + 35.04} = 4.833.$$

From equation (19) we get

$$I_c = \frac{4 \times 112.89}{3} = 150.52,$$

and from equation 20,

$$I_s = 93.45 \times 0.6 = 56.07.$$

Since all the tensile stresses are carried by the steel and all the compressive stresses by the concrete, and since the tensile stresses must balance the compressive stresses to fulfill the conditions of stability, we have $c = s = 0.5$, and from equation (14) we obtain

$$M = \frac{150.52 \times 700}{4.833 \times 0.5} = 43,601.9 \text{ inch-pounds.}$$

A comparison of Case 2 with Case 1 shows that for the conditions assumed there is a great economy in neglecting the concrete on the tensile side, and if the beams are designed so that the assumed loads will not cause the concrete to fail on the tensile side, nearly double the amount of material will be required with beams of the same depth that is required if the concrete is not considered to resist tensile stresses.

When the fiber stresses produced by specified loads are required instead of the bending moments that the beams will safely carry, the first part of equation 11 to 14 should be used, instead of the second part, as was done in the above examples. In other respects the operation is identical.

When the ultimate strength of concrete-steel beams is required, the formulas used in the second case of the example given above could be used if the modulus of elasticity of concrete for stresses near the breaking load were known. Table VI gives values of the modulus of elasticity for stresses up to 2,000 lbs. per square inch as determined at the Watertown Arsenal in the series of tests made for Mr. Geo. A. Kimball, Chief Engineer of the Boston Elevated Railroad, in 1899. These determinations show that the modulus of elasticity is very much less at stresses between 1,000 and 2,000 lbs. per square inch than between 100 and 600 and 1,000 lbs. per square inch, but they are not sufficiently comprehensive to form the basis of any satisfactory rule or formula for the ratio of the modulus of elasticity to the stress per square inch.

TABLE VI.—SHOWING REDUCTION IN VALUE OF E_c WITH INCREASING LOADS. VALUES GIVEN ARE THE MEAN OF THOSE FOR SEVERAL EXPERIMENTS WITH SEVERAL STANDARD BRANDS OF PORTLAND CEMENT.

Age.	Concrete 1 : 2 : 4.			Concrete 1 : 3 : 6.		
	100–600	100–1.000	1,000–2,000	100–600	100–1,000	1,000–2,000
7 days .	2,592,000	2,053,000	1,351,000	1,869,000	1,529,000	
1 mo. ..	2,662,000	2,444,000	1,462,000	2,438,000	2,135,000	1,219,000
3 mos. .	3,670,000	3,170,000	2,157,000	2,976,000	2,656,000	1,805,000
6 mos. .	3,646,000	3,567,000	2,581,000	3,608,000	3,503,000	1,868,000

The average compressive strength of 1:2:4 concrete six months old is given by Table I of Chapter I at 3,904 lbs. per square inch, but the values of the modulus of elasticity for loads over 2,000 lbs. per square inch have not as yet been satisfactorily determined, and without them the ultimate strength of a concrete-steel beam made with 1:2:4 concrete six months old cannot be correctly and rationally calculated. A number of formulas have been published for this case, and they have been compared with the few available tests to destruction of concrete-steel beams to the satisfaction of their authors, but until they are established by a long series of experiments, and until all the properties of concrete are better understood, it can hardly be recommended as good practice to rely on them exclusively in designing important structures. For these reasons the more rational method to pursue in designing concrete-steel beams would seem to be that of calculating the safe load from safe working stresses, as has been done in the example given above. The necessary experimental data for this are at hand. When the properties of concrete, and particularly its modulus of elasticity, have been more completely determined, the proper values may be inserted in the equations (1), (15) and (18), and the ultimate strength then be calculated precisely as was done in our example for the safe load.

Thacher's Empirical Formulas.—As a considerable number of concrete-steel designers are using empirical formulas based on ultimate safe loads, those developed and published by Mr. Edwin Thacher will be given here, because they agree more closely with experiments than any others and are practical and not difficult to apply.

Table VII, given by Mr. Thacher, shows the comparisons between the breaking loads by his formulas and the actual results of thirty tests to destruction. It should be noted that the Kirkaldy tests were made in 1876, at which date Portland cements were quite different from those now manufactured. Since Mr. Thacher used a value for $e=E_s \div E_c=20$ in all cases, it is evident that he assumed the modulus of elasticity to be the same for concretes made of Portland cement in 1876 as at the present time. This may or may not be true, but it is certain that the average ultimate strength of Portland cements have been increased considerably during the past 25 years.

TABLE VII.—SHOWING BREAKING LOADS OF BEAMS AS DETERMINED BY THACHER'S FORMULAS AND BY ACTUAL TESTS.

Series.	Number of Tests.	Size of Beam Tested.			Variation between Actual and Estimated Strength.	
		Length, Feet.	Depth from Top to Center of Rod, Inches.	Width.	Maximum, Per Cent.	Mean, Per Cent.
A	3	11.5	10.0	5.0	1.3	0.0
B	3	6.0	3.25	10.0	3.9	+0.6
C	9	5.0	10.0	12.0	14.0	−1.2
D	7	11.0	10.0	8.0	13.1	−1.4
E	8	6.67	6 and 7	8.0	16.0	+2.3
Mean from 30 tests						−0.013

Note.—Series *A* and *B*, made at Zanesville, Ohio, by Onward Bates and Edwin Thacher. Series *C*, made at London, England, by Prof. David Kirkaldy. Series *D*, made at Massachusetts Institute of Technology. Series *E*, made at Purdue University by Prof. W. K. Hatt.

Notwithstanding the arbitrary assumptions used in the comparisons between the formulas and actual results of tests, Table VII shows that the agreement is remarkably close, and for all ordinary cases Mr. Thacher's formulas can be relied upon to give safe results. The safety factor to be used with them will depend upon the conditions of loading and the judgment of the designer. The notation used is the same as before except as follows:

S = moment of resistance in foot-pounds.
M = bending moment in foot-pounds.
l = length of beam in feet.
W = load at center, including weight of beam for beam 1 in. wide.
w = uniform load per lineal foot, including weight of beam for 1 in. wide.
$w' = 12w$ = uniform load per square foot.
$q = h - d_1$ = effective depth of beam.

First Case.—Rectangular beams, steel in tension side only and concrete to take no stress in tension.

In order that a section before bending shall be a plane after bending, which is an assumption of the common theory of flexure, we have

$$x:\frac{f_c'}{E_c}::v:\frac{f_s}{E_s}; \text{ therefore } x=\frac{f_c'}{f_s}ev, \quad . \quad . \quad . \quad . \quad (21)$$

or

$$f_c' = f_s x \div ev. \quad . \quad . \quad . \quad . \quad . \quad . \quad . \quad . \quad (22)$$

$$q = x + v = (f_c' ev \div f_s) + v,$$

therefore

$$v = q \div \left(e\frac{f_c'}{f_s} + l\right), \quad . \quad . \quad . \quad . \quad . \quad . \quad . \quad (23)$$

and

$$x = q - v. \quad . \quad . \quad . \quad . \quad . \quad . \quad . \quad . \quad . \quad (24)$$

As the total compression must be equal to the total tension, we have

$$\frac{f_c' x}{2} = af_s, \quad \text{or} \quad x = 2a\frac{f_s}{f_c'}. \quad . \quad . \quad . \quad . \quad . \quad . \quad (25)$$

Substituting the value of f_c' found in equation (22), in equation (25) we have

$$(f_s \div e)(x^2 \div v) = 2af_s, \quad \text{or} \quad v = x^2 \div 2ea.$$

Substituting value of x from equation (25),

$$v = \left[\left(\frac{f_s}{f_c'}\right)^2 \left(\frac{2}{e}\right)\right]a,$$

$$q = x + v = 2\left[\frac{f_s}{f_c'} + \frac{1}{e}\left(\frac{f_s}{f_c'}\right)^2\right]a.$$

Therefore

$$a=q\div 2\left[\frac{f_s}{f_c'}+\frac{l}{e}\left(\frac{f_s}{f_c'}\right)^2\right]. \quad . \quad . \quad . \quad . \quad . \quad . \quad (26)$$

$$q=x+v=x+(x^2\div 2ae)$$

or

$$x^2+2aex=2aeq,$$

$$x=\sqrt{2aeq+(ea)^2}-ea. \quad . \quad . \quad . \quad . \quad . \quad . \quad . \quad (27)$$

For the concrete

$$S=\frac{f_c'x^2}{36}.$$

For the steel

$$S=\frac{f_s va}{12};$$

therefore

$$M=\frac{f_c'x^2}{36}+\frac{f_s va}{12}.$$

Substituting for f_c' its value from equation (22), we have

$$M=\frac{f_s}{36}\left[\frac{x^3}{ev}+3av\right]. \quad . \quad . \quad . \quad . \quad . \quad . \quad . \quad (28)$$

For a beam supported at both ends and loaded at the center,

$$W=\frac{4M}{l};$$

therefore

$$W=\frac{f_s}{9l}\left(\frac{x^3}{ev}+3av\right). \quad . \quad . \quad . \quad . \quad . \quad . \quad . \quad (29)$$

For a beam supported at both ends and uniformly loaded

$$w=\frac{8M}{l^2};$$

therefore

$$w=\frac{f_s}{4.5l^2}\left(\frac{x^3}{ev}+3av\right). \quad . \quad . \quad . \quad . \quad . \quad . \quad . \quad (30)$$

To design a beam, values must be assigned to f_s, f_c', E_s, E_c, h, d_1, q, and l. The length l will generally be fixed by the conditions of the problem, and the depth h, if not limited by such conditions, may be selected by trial to give the most economic design. The dimension d_1 will depend to some extent on the size and shape of the steel reinforcement, but should be large enough to leave from one to two inches of concrete outside of the steel, and had better be not less than the largest dimension of the aggregate plus the radius of the steel.

Values for E_s and E_c are given in Table V. Mr. Thacher uses a value $e=E_s \div E_c$ of 20. A comparison of Tables I and VI seems to indicate that this is rather high for concrete six months old, but on the other hand Table VII shows that it can be used with safety. The proper values to use for f_s and f_c' for any given conditions must be determined largely by the judgment of the designer. Those given in the foregoing chapter are safe for all ordinary conditions except where the live load is accompanied by considerable impact, in which case a higher safety factor may be adopted.

Equation (23) gives

$$v=q\div\left(\frac{f_c'}{f_s}e+1\right) \quad \text{and} \quad x=q-v.$$

Then from equation (26),

$$a=q\div 2\left[\frac{f_s}{f_c'}+\left(\frac{f_s}{f_c'}\right)^2\times\frac{1}{e}\right].$$

The distance center to center of rods will be the area of each rod divided by a.

M and W or w may be found directly from equations (28), (29), and (30).

To review a beam, find or assume the proper values for f_s, E_s, E_c, h, d_1, q, and l as before. Equation (27) gives the value of x, and equations (28) and (29) or (30) will then give M and W or w as required. Then if the value of f_c', found from equation (22), exceeds the probable ultimate strength of the concrete in compression (see Table I), the values of M and W or w found by equations (28) and (29) or (30) should be reduced in the ratio that the probable ultimate compressive strength of the concrete bears to f_c' as found by equation (22).

Second Case.—Steel in both tension and compression sides; concrete to take no stress in tension.

By the common theory of flexure we have the proportion

$$x:\frac{f_c'}{E_c}::v:\frac{f_s}{E_s}::t:\frac{f_s'}{E_s},$$

from which

$$x=\frac{f_c'}{f_s}e\cdot v \quad \text{and} \quad f_c'=(f_s\cdot x)\div(e\cdot v) \quad \text{and} \quad f_s'=tf_c'e\div x.$$

Substituting the value of f_c' above in the last equation we have

$$f_s'=tf_s\div v. \quad . \quad . \quad . \quad . \quad . \quad . \quad . \quad . \quad . \quad (31)$$

$$q=x+v=ev\frac{f_c'}{f_s}+v,$$

or

$$v=q\div\left(e\frac{f_c'}{f_s}+1\right) \quad \text{and} \quad x=q-v.$$

As the total tension must equal the total compression,

$$\frac{f_c'x}{2}+a'f_s'=af_s \quad \text{or} \quad x=\frac{2af_s-2a'f_s'}{f_c'}. \quad . \quad . \quad . \quad . \quad . \quad (32)$$

Let $n=a'\div a$ and $a'=na$, and substituting the value of f_s' from equation (31), we have

$$a=f_c'xv\div 2f_s(v-nt). \quad . \quad . \quad . \quad . \quad . \quad . \quad . \quad (33)$$

Substituting the value of f_c' and writing for q its value $x+v$ we have

$$q=(x^2\div 2ae)+nt+x;$$

therefore

$$x=\sqrt{2ae(q+nd_1')+[ae(n+1)]^2}-[ae(n+1)]. \quad . \quad . \quad (34)$$

For the concrete, . $S=f_c'x^2\div 36.$
For the steel in compression, $S=f_s'a't\div 12=f_s'nat\div 12.$
" " " " tension, $S=f_sav\div 12;$

therefore

$$M=(f_c'x^2\div 36)+(f_s'nat\div 12)+(f_sav\div 12).$$

Substituting the value of f_s' from equation (31), and of $f_c'=f_sx\div ev$, we have

$$M=(f_s\div 36)[(x^3\div ev)+(3nat^2\div v)+3av]. \quad . \quad . \quad . \quad (35)$$

For a beam supported at both ends and loaded at the center, $W=4M\div l$; therefore

$$W=(f_s\div 9\cdot l)[(x^3\div ev)+(3nat^2\div v)+3av]. \quad . \quad . \quad . \quad (36)$$

For a beam supported at both ends and uniformly loaded, $w=\frac{8M}{l^2}$;

therefore

$$w=(f_s\div 4.5l^2)[(x^3\div ev)+(3nat^2\div v)+3av]. \quad . \quad . \quad . \quad (37)$$

To Design a Beam.—Find or assign values for f_s, f_c', E_s, E_c, h, d_1, q, and l as in the first case. Equation (23) gives $v=q\div\left(e\frac{f_c'}{f_s}+1\right)$ and $x=q-v$.

Equation (33) gives $a=f_c'vx\div[2f_s(v-nt)]$.

Equation (31) gives $f_s'=f_st\div v$. Equations (35), (36), and (37) give the values of M and W or w as required.

To Review a Beam.—Find or assign values for f_s, E_s, E_c, h, d_1, q, and l as before.

Then from equation (34) we find

$$x=\sqrt{2ae(q+nd_1')+[ae(n+1)]^2}-[ae(n+1)] \text{ and } v=q-x.$$

a=total area of bars in tension side divided by the width of the beam in inches and $a'=na$.

Equation (31) gives $f_s'=tf_s\div v$, and equation (22) gives $f_c'=(f_sx)\div ev$. If this value of f_c' exceeds the probable ultimate resistance of the concrete to compression, the values of M and W or w, found by equations (35), (36), and (37), should be reduced in the ratio that the probable compressive strength of the concrete bears to the value found by equation (22).

Third Case.—A solution for a reinforced concrete T section, composed of a floor slab and joist, following Mr. Thacher's analysis, is given in the Third Case, as follows:

The slab is first designed as a beam with a span equal to the distance center to center of joists and with the reinforcing bars of the slab at right angles to the direction of the length of the joists. The spacing of the joists should be such as to give the greatest economy for the entire floor system when not limited by special conditions.

To insure that the concrete in the slab will act as a part of the joist, and not shear on the line of the lower surface of the slab extended through or across the joist, it should be placed at the same time as the concrete of the joist is placed and monolithic with it, or else steel bars should be so disposed as to connect the slab and the part of the joist below the slab to resist the longitudinal shear. The notation is the same as given for the First Case, except as follows, referring to Fig. 10.

b' =width of slab=center to center of joists, in inches.

r =thickness of slab, in inches.

A_s'=in this case the area of steel which would offer the same resistance to shortening as do the wings of the T, no steel being actually used in the compression side.

d_1' is taken by Mr. Thacher at one-half of r, the thickness of the slab.

To Design a Beam, Ends Supported.—Having determined b_1', r, and d_1' as indicated above, assign values to f_s, f_c, E_s, E_c, h, and

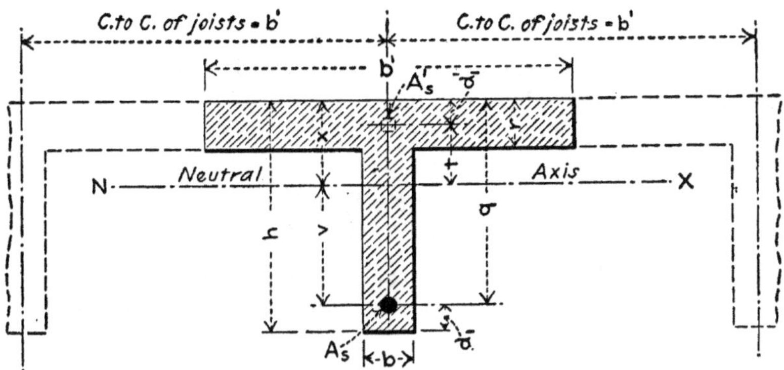

FIG. 10.—Cross-section of T-shaped Beam Illustrating Standard Notation.

d_1, as in the previous cases, noting that $A_s' \div A_s = n$, $h - d_1 = q$, and $A_s \div b = a$. Then

$$A_s' = (b'-b)r \div e; \quad v = q \div [e(f_c' \div f_s) + 1].$$

$$x = q - v; \quad f_s' = tf_s \div v. \quad \ldots \quad (38)$$

$$A_s = (f_c' bx + 2A_s' f_s') \div 2f_s. \quad \ldots \quad (39)$$

$$M = \tfrac{1}{12}(\tfrac{1}{3}f_c' bx^2 + f_s' A_s' t + f_s A_s v). \quad \ldots \quad (40)$$

$$W = \tfrac{1}{3} \cdot l(\tfrac{1}{3}f_c' bx^2 + f_s' A_s' t + f_s A_s v). \quad \ldots \quad (41)$$

$$w' = \tfrac{1}{8}b'l^2(\tfrac{1}{3}f_c' bx^2 + f_s' A_s' t + f_s A_s v). \quad \ldots \quad (42)$$

A_s can be reduced until M and W or w' have the required values.

To Review a Beam, Ends Supported, assign values to f_s, E_s, and E_c as before.

Then $A_s' = (b'-b)r \div e$; A_s = area of the steel.

$$x = \sqrt{2ae(q + nd_1') + [ae(n+1)]^2} - [ae(n+1)]. \quad \ldots \quad (43)$$

$$v = q - x; \quad f_s' = tf_s \div v; \quad f_c' = f_s x \div ev. \quad \ldots \quad (44)$$

Christophe's Formulas.—M. Christophe gives the following formulas for beams of T section:

Case I.—If the neutral axis is above the lower side of the slab,

$$x = -\frac{e(A_s' + A_s)}{b'} + \sqrt{\frac{e^2(A_s' + A_s)^2}{(b')^2} + \frac{2e}{b'}[A_s' d_1' + A_s(h - d_1)]}. \quad (45)$$

$$(x^2 b' \div 2) + e(A_s' t - A_s v) = 0. \quad \ldots \quad (46)$$

$$f_c' = Mx \div I. \quad \ldots \quad (47)$$

$$I = \tfrac{1}{3}x^3b' + e(A_s't^2 + A_sv^2). \quad (48)$$
$$f_s = f_c'ev \div x = eMv \div I. \quad (49)$$
$$f_s' = f_c'et \div x = eMt \div I. \quad (50)$$
$$x + v = h - d_1 = q. \quad (51)$$
$$t = x - d_1'. \quad (52)$$

Case II.—If the neutral axis is level with the lower side of the slab, $x = r$ in the preceding formulas.

Case III.—If the neutral axis is below the lower side of the slab, formulas (46) and (48) become

$$(x^2b' \div 2) - \tfrac{1}{2}(x-r)^2(b'-b) + e(A_s't - A_sv) = 0; \quad (53)$$
$$I = \tfrac{1}{3}x^3b' - \tfrac{1}{3}(x-r)^3(b'-b) + e(A_s't^2 + A_sv^2). \quad (54)$$

Thacher's Constants.—The constants in Table VIII have been calculated by Mr. Thacher, using the following values for the constants in his formulas:

E_s=							30,000,000
f_s, ultimate,							64,000
E_c, for	1:2:4	concrete	one month			old,	1,460,000
f_c',	"	"	"	"	"	"	2,400
E_c,	"	"	"	six months		"	2,580,000
f_c',	"	"	"	"	"	"	3,700
E_c,	"	1:3:6	"	one month		"	1,220,000
f_c',	"	"	"	"	"	"	2,050
E_c,	"	"	"	six months		"	1,860,000
f_c',	"	"	"	"	"	"	3,100

TABLE VIII.—GIVING CONSTANTS FOR BEAMS CALCULATED BY MR. EDWIN THACHER.

Proportions of Concrete.	1 to 2 to 4.		1 to 3 to 6.	
Age of Concrete.	1 Month.	6 Months.	1 Month.	6 Months.
Area of steel, sq. ins. required for 1 in. width of beam = a.	$\frac{q}{142}$	$\frac{q}{100}$	$\frac{q}{165}$	$\frac{q}{109}$
Ultimate bending moment, ft.-lbs. for 1 in. width of beam = M	$35.62\,q^2$	$51.25\,q^2$	$30.62\,q^2$	$46.25\,q^2$
Breaking weight at center for 1 in. width of beam = W	$\frac{142.5\,q^2}{l}$	$\frac{205.0\,q^2}{l}$	$\frac{122.5\,q^2}{l}$	$\frac{185.0\,q^2}{l}$
Breaking weight per lineal foot uniformly distributed for 1 in. width of beam = w	$\frac{285.0\,q^2}{l^2}$	$\frac{410.0\,q^2}{l^2}$	$\frac{245.0\,q^2}{l^2}$	$\frac{370.0\,q^2}{l^2}$
Breaking weight per square foot uniformly distributed = w'	$\frac{3,420.0\,q^2}{l^2}$	$\frac{4,920.0\,q^2}{l^2}$	$\frac{2,940.0\,q^2}{l^2}$	$\frac{4,440.0\,q^2}{l^2}$
Effective depth of beam required for a uniform load of w' lbs. per sq. ft. = q	$\sqrt{\frac{l^2w'}{3,420}}$	$\sqrt{\frac{l^2w'}{4,920}}$	$\sqrt{\frac{l^2w'}{2,940}}$	$\sqrt{\frac{l^2w'}{4,440}}$

To apply Table VIII, divide the constants given therein by the safety factor required.

In order that the steel reinforcement shall be sufficient to develop the full value of the concrete when at an age of six months it has gained nearly its full strength, it is preferable to design reinforced concrete beams for their strength at six months with such a safety factor that they will still have a sufficient margin of safety at an age of one month. This only applies to beams designed under the assumption that the concrete takes no stress in tension. If they are designed according to the formulas 1 to 9 inclusive, the safety factor based on the ultimate strength will be about twice as great as that used for the concrete in tension, and will be ample to provide for loadings at one month old or even less, when designed for the strength at six months old.

The safety factor to be used when the resistance of the concrete in tension is considered may therefore be taken with safety much lower than when the tensile resistance of the concrete is neglected. If this is done the economy noted in the remarks following the example illustrating the use of formula 18, concrete in tension neglected, would be reduced and might entirely disappear.

The following examples will show how Table VIII and the preceding formulas may be applied to practical cases.

EXAMPLE 1.—To find depth of slab 6 ft. in length between supports that will sustain a uniform load, including its own weight of 480 lbs. per square foot with a factor of safety of 5 in six months, using 1–2–4 concrete.

$$q=\sqrt{\frac{l^2w'}{984}}=\sqrt{\frac{6.0^2\times 480}{984}}=4.19 \text{ in.}; \quad h=4.19+0.81=5.0 \text{ in.}$$

$$a=\frac{4.19}{100}=0.0419. \quad \text{For rods } \tfrac{5}{8}'' \text{ diam., dist. c. to c.} =\frac{.282}{0.0419}=6.7''.$$

EXAMPLE 2.—To find safe load per square foot on slab having a clear length of 12 ft. and a depth of 10 ins. from top of slab to center of rods, using 1–3–6 concrete with a factor of safety of 4 in six months.

$$w'=\frac{1,110q^2}{l^2}=\frac{1,111\times 10^2}{12^2}=770 \text{ lbs.} \quad a=\frac{10.0}{109}=0.092.$$

For 1 in. rods $\frac{.714}{.092}=7.75$ ins. center to center.

EXAMPLE 3.—To find width of joist having clear span of 15 ft. and effective depth of 12 ins. that will support a load of 400 lbs. per

TABLE IX.—SHOWING RESULTS OF TESTS OF REINFORCED

No. of Test.	Size of Specimen.				Concrete Mixture.	Reinforcement.					
	Span, in Feet.	Height h, Ins.	d_1, Ins.	Width b, Ins.		Bottom.		Top.		Kind of Bar.	Vertical Metal.
						No.	Size.	No.	Size.		
1	5	12	2	12	Crushed	1	2.5"	—	—	Bolts	—
2	"	"	"	"	brick	"	×¼"	—	—	or	—
3	"	"	"	"	1 to 2	"	"	—	—	dowels	—
4	"	"	"	"	"	"	"	—	—	3" to 15"	—
5	"	"	"	"	"	"	"	—	—	apart	—
6	"	"	"	"	"	"	"	—	—	"	—
7	"	"	"	"	"	"	"	—	—	"	—
8	"	"	"	"	"	"	"	—	—	"	—
9	"	"	"	"	"	"	"	—	—	"	—
10	4.15	4	¼	60	1 : 3 : 6	10	⅜"	—	—	Twisted	—
11	6	4	¼	10	1 : 3 : 6	2	⅜"	—	—	Thacher	—
12	"	"	"	"	1 : 2 . 4	"	"	—	—	"	—
13	"	"	"	"	"	"	"	—	—	"	—
14	11.5	11	1	5	1 : 3 : 6	3	"	—	—	"	—
15	"	"	"	"	1 : 2 : 4	"	"	—	—	"	—
16	"	"	"	"	"	"	"	—	—	"	—
17	20	24		12	1 : 2 : 2½	4	13/16	—	—	Twisted	—
18	"	"		"	"	"	1" rd.	—	—	Plain	1"×1/16"
19	"	"		"	"	2	13/16	2	13/16	Twisted	—
20	"	"		"	"	"	1" rd.	2	1" rd.	Plain	1"×1/16"
21	11	12	1.85	8	1 : 3 : 6	—	—	—	—	—	—
22	"	"	"	"	"	1	¼"	—	—	Twisted	—
23	11	12	1.85	8	1 : 3 : 6	1	½	—	—	Twisted	—
24	"	"	"	"	"	"	¾	—	—	"	—
25	"	"	"	"	"		⅞	—	—	"	—
26	"	"	"	"	"	"	1	—	—	"	—
27	"	"	"	"	"	"	1¼	—	—	"	—
28	"	"	"	"	"	2	1	—	—	"	—
29	"	"	"	"	"	"	1¼	—	—	"	—
30	"	"	"	"	"	"	"	1	½	"	—
31	"	"	"	"	"	"	"	2	¾	"	—
32	"	"	2.06	"	"	"	"	—	—	"	16
33	"	"	"	"	"	"	"	—	—	"	stirrups
34	"	"	"	"	"	"	"	2	¾	"	¼" sq.
35	"	"	1.85	"	"	1	1" sq.	—	—	Plain	—
36	"	"	"	"	"	1	1" sq.	—	—	Twisted	—
37	"	"	"	"	"	2	¾" sq.	—	—	Plain	—
38	"	"	"	"	"	2	"	—	—	Twisted	—
39	"	"	"	"	"	4	½" sq.	—	—	Plain	—
40	"	"	"	"	"	4	"	—	—	Twisted	—
41	"	"	"	"	"	1	¼" sq.	—	—	Plain	—
42	"	"	"	"	"	2	⅞" sq.	—	—	Twisted	—
43	"	"	"	"	"	2	"	—	—	Plain	—
44	"	"	"	"	"	4	⅝" sq.	—	—	Twisted	—
45	"	"	"	"	"	4	"	—	—	Plain	—
46	"	"	"	"	"	2	¾" sq.	—	—	Twisted	Expanded
47	"	"	"	"	"	1	1" sq.	—	—	"	Metal

CONCRETE BEAMS COMPILED FROM VARIOUS SOURCES.

No. of Test.	Age at Test, Days.	Weight of Beam, Pounds.	Kind of Load.	Load Producing Failure, Pounds.	M. in In.-lbs. at Fracture.	REMARKS. C. = concrete. R. = rod. Bk. = broke or broken.	Authority.
1	60		Ctr.	16,761		C. gave way, R. not Bk.	Kirkaldy
2	to		"	18,441		C. parted, R. not Bk.	for
3	90		"	19,703		R. Bk. at bolt-hole	Hyatt
4	"		"	18,459		do.	do.
5	"		"	17,485		do.	do.
6	"		"	19,399		do.	do.
7	"		"	19,559		do.	do.
8	"		"	21,967		do.	do.
9	"		"	19,241		C. crushed, R. not Bk.	
10	21?		"	21,220		C. failed	E. F. Miller
11	79		"	2,228		R. Bk.	E. Thacher
12	84		"	2,237		do.	do.
13	84		"	2,091		do.	do.
14	78		"	4,985		do.	do.
15	81		"	5,095		do.	do.
16	83		"	4,985		do.	do.
17	42		"	46,000			Boston
18	56		"	42,600		Stirrups 2″ c. to c.	Rapid
19	42		"	34,000			Transit
20	56		"	23,000		Stirrups 12″ c. to c.	8th Rep'rt
21	40	1,198	"	1,302	62,733	Vert. Bk. near center	21 to 47
22	40	1,200	"	1,300	62,700	Do. and R. Bk.	M. I. of T.
23	39	1,205	2–Pt.	10,095	241,973	Vert. Bk. near one load. R. Bk.	
24	38	1,160	C.	13,680	470,580	C. crushed on top	
25	50	1,290	"	14,710	506,715	Longitudinal shearing break at or a little above the reinforcing bar, and in addition a break extending diagonally (often at an inclination of 45 degrees) upward toward the center. The point where the diagonal break joined the top of the beam was sometimes at the point of application of the load and sometimes near it. This note also applies to tests Nos. 42 and 44 to 47 inclusive.	
26	50	1,204	"	15 796	541,134		
27	41	1,195	"	12,805	442,283		
28	41	1,240	"	18,760	639,540		
29	42	1,274	"	23,105	783,486		
30	42	1,279	"	21,105	717,569		
31	45	1,294	"	23,105	783,816		
32	30	1,282	2–Pt.	24,000	553,553		
33	31	1,292	"	29,200	663,718		
34	30	1,341	"	24,000	554,527		
35	53	1,292	"	15,250	356,818		
36	49	1,211	"	16,500	382,983		
37	43	1,271	"	15,950	370,700		
38	40	1,200	"	19,000	438,807		
39	35	1,261	"	17,500	378,065	Bk. at crack about 18″ from center	
40	33	1,213	"	20,000	433,329	Failed by longitudinal shear	
41	57	1,213	"	12,500	295,015	Vert. Bk. at load. R. pulled out	
42	54	1,248	"	22,250	510,092		
43	57	1,221	"	20,250	465,647	Handling crack. Failed like 40	
44	47	1,203	"	19,250	443,350		
45	50	1,192	"	15,250	355,168		
46	40	1,215	"	24.250	553,548	Sheet of expanded metal in vertical plane of center line	
47	49	1,222	"	21,750	498,688		

TABLE X.—CONCRETE STEEL BEAMS, STONE OR

(Concrete (1–2–4). Age 6 months. Factor of safety 5. Rods

wide (exclusive of weight of

Clear Span, Feet.	Depth of Beam in Inches (*h*).						
	10	12	14	16	18	20	22
10.0	52.5	82.5	113.4	155.3	205.0	245.0	305.0
11.0	41.6	65.9	91.4	125.8	165.7	199.0	248.0
12.0	33.2	53.4	74.4	103.3	136.2	163.4	205.0
13.0	26.8	43.6	61.3	85.3	113.2	136.2	171.0
14.0	21.6	35.9	50.7	71.1	95.2	114.7	144.3
15.0	17.5	29.6	42.3	59.9	80.2	97.2	123.1
16.0	14.1	24.5	35.4	50.6	68.4	83.0	105.1
17.0		20.2	29.8	43.0	58.5	71.2	90.6
18.0		16.7	25.0	36.5	50.2	61.2	78.3
19.0			20.9	31.0	43.2	52.7	68.1
20.0			17.4	26.4	37.1	45.6	59.1
21.0				22.4	31.8	39.4	51.5
22.0				18.9	27.4	34.1	44.8
23.0					23.5	29.4	39.1
24.0					20.0	25.3	34.1
25.0						21.7	29.6
26.0							25.6
27.0							22.1
28.0							
29.0							
30.0							
31.0							
32.0							
33.0							
34.0							
35.0							
36.0							
37.0							
38.0							
39.0							
40.0							
	Weight of Beam per Lineal Foot.						
	10.4	12.5	14.6	16.7	18.8	20.8	22.9
$q = h - d_1$	8.75	10.75	12.5	14.5	16.5	18.0	20.0
d_1	1.25	1.25	1.5	1.5	1.5	2.0	2.0
	Steel Bars; Area, Square Inch, a.						
	.0875	.1075	.125	.145	.165	.180	.200
	Quantities for Beam 1 Foot Long and 1 Inch Wide.						
	Concrete, Cubic Feet.						
	.070	.084	.097	.111	.125	.139	.153
	Steel, Pounds.						
	0.298	0.366	0.425	0.493	0.561	0.612	0.680

For factor of safety of 4, $w =$

GRAVEL CONCRETE, BY THACHER'S FORMULAS.

medium steel. Safe load per lineal foot for beam 1 in.

beam). $w=\frac{82q^2}{l^2}$; $a=q\div 100$.

Depth of Beam in Inches (h).								Clear Span, Feet.
24	27	30	36	42	48	54	60	
372.0	484.0	612.0	883.0					10.0
303.0	396.0	500.0	724.0					11.0
251.0	328.0	415.0	603.0	844.0				12.0
210.0	275.0	349.0	508.0	712.0	954.0			13.0
177.0	233.0	297.0	433.0	608.0	817.0			14.0
151.0	200.0	255.0	372.0	525.0	705.0	894.0		15.0
130.0	172.0	220.0	323.0	456.0	613.0	777.0	978.0	16.0
112.6	149.0	191.0	281.0	399.0	538.0	682.0	859.0	17.0
97.4	130.0	167.0	247.0	351.0	474.0	603.0	759.0	18.0
85.0	114.0	147.0	218.0	310.0	420.0	535.0	676.0	19.0
74.0	100.0	130.0	193.0	276.0	375.0	477.0	604.0	20.0
65.0	88.0	115.0	172.0	246.0	335.0	428.0	543.0	21.0
57.0	78.0	102.0	153.0	220.0	301.0	384.0	488.0	22.0
50.0	69.0	90.5	137.0	198.0	271.0	348.0	441.0	23.0
44.0	61.0	80.5	123.0	178.0	245.0	314.0	401.0	24.0
38.5	54.0	71.4	110.0	161.0	222.0	285.0	364.0	25.0
33.7	48.0	63.8	99.0	145.0	202.0	260.0	332.0	26.0
29.5	42.3	56.8	89.0	131.0	183.0	236.0	304.0	27.0
25.5	37.3	50.8	70.0	119.0	167.0	216.0	278.0	28.0
22.3	33.0	45.2	72.0	108.0	152.0	198.0	255.0	29.0
	29.0	40.2	64.7	98.0	139.0	181.0	234.0	30.0
		35.7	58.3	89.0	127.0	166.0	216.0	31.0
		31.5	52.5	81.0	116.0	152.0	198.0	32.0
			47.0	73.7	106.0	140.0	183.0	33.0
			42.1	66.8	97.0	129.0	168.0	34.0
			37.6	60.7	89.0	118.0	156.0	35.0
			33.5	54.9	81.0	109.0	144.0	36.0
				49.7	74.0	100.0	133.0	37.0
				44.8	68.0	92.0	123.0	38.0
					61.0	84.0	113.0	39.0
					56.0	77.0	104.0	40.0
Weight of Beam per Lineal Foot.								
25.0	28.1	31.2	37.5	43.8	50.0	56.2	62.5	
22.0	25.0	28.0	33.5	39.5	45.5	51.0	57.0	$q=h-d_1$
2.0	2.0	2.0	2.5	2.5	2.5	3.0	3.0	d_1
Steel Bars; Area, Square Inch, a.								
.220	.250	.280	.335	.395	.455	.510	.570	

Quantities for Beam 1 Foot Long and 1 Inch Wide.

Concrete, Cubic Feet.							
.167	.188	.208	.250	2.92	.333	.375	.417
Steel, Pounds.							
0.748	0.850	0.952	1.139	1.343	1.547	1.734	1.938

$\left(\text{load in table}+25\%+\frac{\text{weight per lin. ft. of beam}}{4}\right)$

TABLE XI.—CONCRETE STEEL BEAMS, STONE OR

(Concrete (1-3-6). Age 6 months. Factor of safety 5. Rods

wide (exclusive of weight of

Clear Span, Feet.	Depth of Beam in Inches (h).						
	10	12	14	16	18	20	22
10.0	46.3	73.3	101.2	139.0	183.0	219.0	274.0
11.0	36.5	58.3	81.0	112.0	147.0	177.0	222.0
12.0	29.0	47.0	65.6	91.3	121.2	146.0	183.0
13.0	23.1	38.1	53.8	75.3	100.3	121.2	152.0
14.0	18.5	31.1	44.4	62.7	83.9	101.4	128.0
15.0	14.8	25.5	36.8	52.5	70.7	85.8	108.6
16.0		20.9	30.6	44.1	60.0	73.0	91.0
17.0		17.1	25.4	37.2	50.9	62.2	79.5
18.0			21.1	31.3	43.4	53.2	68.5
19.0			17.4	26.4	37.0	45.6	59.1
20.0				22.1	31.5	39.2	51.1
21.0				18.6	26.9	33.6	44.2
22.0					22.8	28.8	38.3
23.0					19.3	24.6	33.1
24.0						20.8	28.6
25.0							24.5
26.0							
27.0							
28.0							
29.0							
30.0							
31.0							
32.0							
33.0							
34.0							
35.0							
36.0							
37.0							
38.0							
39.0							
40.0							
	Weight of Beam per Lineal Foot.						
	10.4	12.5	14.6	16.7	18.8	20.8	22.9
$q=h-d_1$	8.75	10.75	12.5	14.5	16.5	18.0	20.0
d_1	1.25	1.25	1.5	1.5	1.5	2.0	2.0
	Steel Bars; Area, Square Inch, a.						
	.0803	.0986	.1147	.1330	.1514	.1652	.1835
	Quantities for Beam 1 Foot Long and 1 Inch Wide.						
	Concrete, Cubic Feet.						
	.070	.084	.097	.111	.125	.139	.153
	Steel, Pounds.						
	0.273	0.335	0.390	0.453	0.515	0.562	0.642

For factor of safety of 4, $w=$

GRAVEL CONCRETE, BY THACHER'S FORMULAS.

medium steel. Safe load per lineal foot for beam 1 in.

beam). $w = \frac{74q^2}{l^2}$; $a = q \div 109$.

Depth of Beam in Inches (h).								Clear Span, Feet.
24	27	30	36	42	48	54	60	
333.0	435.0	549.0	794.0					10.0
271.0	355.0	449.0	650.0	911.0				11.0
224.0	293.0	373.0	540.0	758.0				12.0
187.0	246.0	313.0	455.0	640.0	859.0			13.0
157.0	208.0	265.0	387.0	546.0	732.0	925.0		14.0
134.0	178.0	227.0	332.0	469.0	632.0	800.0		15.0
115.0	153.0	195.0	287.0	407.0	549.0	696.0	878.0	16.0
99.0	132.0	170.0	250.0	356.0	480.0	610.0	770.0	17.0
85.4	114.4	148.0	219.0	313.0	424.0	539.0	681.0	18.0
74.0	100.0	129.3	193.0	276.0	375.0	477.0	604.0	19.0
64.5	87.5	113.6	170.0	245.0	333.0	426.0	539.0	20.0
56.2	76.5	100.0	151.0	218.0	297.0	381.0	484.0	21.0
49.0	67.4	88.8	134.0	195.0	267.0	342.0	434.0	22.0
42.6	59.3	78.8	120.0	174.7	240.0	308.0	392.0	23.0
37.2	52.1	69.4	106.5	156.8	216.0	278.0	355.0	24.0
32.3	45.9	61.5	95.5	143.2	196.0	252.0	323.0	25.0
28.0	40.4	54.8	85.3	127.2	177.0	229.0	294.0	26.0
	35.3	48.3	76.5	114.2	160.0	208.0	268.0	27.0
	30.9	42.8	68.5	103.4	146.0	190.0	245.0	28.0
		37.8	61.3	93.7	132.0	173.0	224.0	29.0
		33.3	54.8	84.4	120.0	158.0	206.0	30.0
			49.0	76.4	110.0	144.0	188.0	31.0
			43.5	69.0	100.0	132.0	172.0	32.0
			38.7	62.2	91.0	121.0	158.0	33.0
				56.2	83.0	111.0	146.0	34.0
				50.6	75.0	101.0	134.0	35.0
					68.0	92.3	123.0	36.0
					62.0	84.3	113.0	37.0
					56.0	77.2	104.0	38.0
					51.0	70.3	96.0	39.0
						64.3	88.0	40.0
Weight of Beam per Lineal Foot.								
25.0	28.1	31.2	37.5	43.8	50.0	56.2	62.5	
22.0	25.0	28.0	33.5	39.5	45.5	51.0	57.0	$q = h - d_1$
2.0	2.0	2.0	2.5	2.5	2.5	3.0	3.0	d_1
Steel Bars; Area, Square Inch, a.								
.2019	.2294	.2569	.3073	.3624	.4174	.4679	.5230	
Quantities for Beam 1 Foot Long and 1 Inch Wide.								
Concrete, Cubic Feet.								
.167	.188	.208	.250	.292	.333	.375	.417	
Steel, Pounds.								
0.687	0.780	0.874	1.045	1.232	1.419	1.591	1.778	

$\left(\text{load in table} + 25\% + \frac{\text{weight per lin. ft. of beam}}{4}\right)$.

lineal foot with a factor of safety of 5 in six months, using 1–2–4 concrete.

$$\text{For 1 in. wide } w=\frac{82q^2}{l^2}=\frac{82\times 12^2}{15^2}=52.5;\ \frac{400}{52.5}=7.6 \text{ in. wide;}$$

$$\text{steel}=\frac{12.0\times 7.6}{100}=0.91 \text{ sq. in.}$$

EXAMPLE 4.—To find breaking load at center of beam 8 ft. long, 8 ins. wide, and 10 ins. deep in which $q=9$ ins., for 1–2–4 concrete one month old.

$$W=\frac{142.5\times b\times q^2}{l}=\frac{142.5\times 8\times 9^2}{8}=11,540 \text{ lbs.};$$

$$\text{steel}=\frac{9.0\times 8.0}{142}=0.51 \text{ sq. in.}$$

EXAMPLE 5.—To find the safe bending moment in foot-pounds that can be sustained by a beam 24 ins. deep ($q=22$ ins.) and 12 ins. wide, factor of safety 4, concrete 1–3–6 six months old.

$$M=\frac{46.25\times 12\times q^2}{4}=138.75\times 22^2=67,150 \text{ ft.-lbs.}$$

Longitudinal Shear between Steel and Concrete.—The longitudinal shear between the concrete and the steel may be found by the following method, as given by Prof. W. H. Burr in his "Resistance of Materials."

Let f_c= the intensity of stress in the concrete at the distance v from the neutral axis, Fig. 9.

p= the perimeter of the steel reinforcement.

k= intensity of the shear on the surface between the concrete and the steel.

F= total transverse shear at distance m from end of beam.

f_s= the intensity of the stress in the steel.

The variation of f_s for the infinitely small distance (dm) is (df_s).

The other notation is the same as given with Fig. 9. $M=f_c\cdot I\div v$, from which

$$dM=df_c\cdot I\div v. \qquad (55)$$

$$dM=F\cdot dm=df_c\cdot I\div v. \qquad (56)$$

Then $p\cdot dm\cdot k=A_s\cdot df_s$; therefore

$$dm=A_s\cdot df_s\div p\cdot k. \qquad (57)$$

Substituting the value of dm found in equation (57), in equation (56) we have

$$F\cdot A_s\cdot df_s\div p\cdot k=df_c\cdot I\div v;$$

and since $df_s \div df_c = e$, we have

$$k = e \cdot F \cdot A_s \cdot v \div p \cdot I. \quad \quad (58)$$

If k is greater than the safe working adhesive strength, resort should be had to some mechanical bond, such as the Thacher rods, corrugated or twisted bars, or any other device that will offer resistance to the tendency of the rods or other section of steel to pull out. Preference should be given to such devices as do not produce re-entering angles in the concrete.

Hatt's Formulas. — Professor W. K. Hatt gives the following formulas for flexure, following the method of M. Considère. These formulas have been recommended by a committee of the American Railway Engineering and Maintenance of Way Association.

hx = the distance from the compression face to the neutral axis.

hu = the distance from the compression face to the center of gravity of the reinforcement.

p = the ratio of the area of steel to that of the cross-section of the beam.

b = width of beam.

E_s, E_c, E_t = the moduli of elasticity of the steel, concrete in compression, and concrete in tension respectively.

$n = E_c \div E_t$.

$e = E_s \div E_t$.

f_s = stress in metal reinforcement.

f_c' = compressive stress in outer fiber of concrete.

f_c = tensional stress in outer fiber of concrete.

Reinforcement is supposed to be in the tension flange alone. E_c and E_t are measured at the stresses f_c' and f_c. The values of x, u, and p are ratios. p and u are at the control of the designer, while x depends on p, u, n, and e; n and e are fixed by the quality of the materials, and they change during flexure with the varying values of f_c', f_s, and f_c that is, the modulus of elasticity of the concrete varies with the stress at which it is measured. For practical purposes of computation, however, the constant values of n and e may be used appropriate to the loads or stress per square inch.

On the assumption of plane cross-sections during flexure, we may determine the ratio of f_s to f_c' and f_s to f_c as follows:

$$f_c' = f_c nx \div (1 - x). \quad \quad (59)$$

$$f_s = f_c e(u - x) \div (1 - x). \quad \quad (60)$$

Next, to locate the neutral axis, that is, to determine the value of x, we may equate the forces of tension and compression on the cross-section, assuming, as before, that the stress-strain diagrams are arcs of parabolas. Thus

$$\tfrac{2}{3}f_c'x=\tfrac{2}{3}f_c(1-x)+pf_s. \quad . \quad . \quad . \quad . \quad . \quad . \quad (61)$$

Inserting the values f_c' and f_s obtained above we obtain the following quadratic:

$$\tfrac{2}{3}x^2n=\tfrac{2}{3}(1-x)^2+pe(u-x).$$

Solving the quadratic we have

$$x=\frac{-\tfrac{1}{2}(4+3pe)+\sqrt{4n+9/4p^2e^2+p[6e\{u(n-l)+l\}]}}{2(n-1)}. \quad . \quad . \quad (62)$$

Having obtained x we may compute f_c' and f_s, and finally obtain the moment of resistance of the section. Taking moments about the neutral axis, we have

$$M=f_cbh^2\left\{\frac{5}{12}(1-x)^2+\frac{5nx^3}{12(1-x)}+p\frac{(u-x)^2}{1-x}e\right\} \quad . \quad . \quad . \quad (63)$$

No useful development will result from the substitution in equation (63) of the value of x obtained in general terms from equation (62). In practical computations, n, e, u, and p are given; x is then computed from 62; f_c' and f_s computed from equations (59) and (60); finally the moment of resistance is computed from equation (63).

At the load corresponding to the cracking of the concrete in the tension face these equations should be modified to correspond with the fact that the stress-strain diagram for the concrete in tension is more nearly a rectangle than a parabola. The difference, however, between the results, at the time of the appearance of the crack, due to the assumption of the rectangle or a parabola is small. With proper values of n and e the equations may be allowed to stand.

When, however, the crack, having formed itself, extends through the lower region of the cross-section, the equations must be modified by the omission of the effect of the tensional forces due to the resistance of the concrete under tension.

We have then

$$\tfrac{2}{3}f_c'x=pf_s \quad \text{or} \quad p\frac{E_s}{E_c}(u-x)=\tfrac{2}{3}x^2, \quad . \quad . \quad . \quad . \quad (64)$$

which serves to locate the neutral axis. When f_s is assumed to be the elastic limit of the reinforcing metal f_c' may be computed.

In the stone concrete beams tested by Prof. Hatt, the elastic limit of the iron was reached before the concrete failed in compression.

The resisting moment of the section is

$$M=bh^2\left\{\frac{5}{12}f_c'x^2+pf_s(u-x)\right\}. \quad . \quad . \quad . \quad . \quad (65)$$

The quality of steel used in reinforcing concrete should be as carefully specified as for an all-steel structure doing the same duty. While some engineers advocate the use of high steel on account of its high elastic limit, such practice cannot be recommended and is certainly not conservative. There is little question but that the steel should be just as soft and ductile as would be required for structural-steel members under the same kind of loading. Considerable concrete has been reinforced with bars rolled from old rails. Such material, being hard and brittle, would fail with less warning than bars of soft or medium steel. Whatever extra strength is secured by using high steel is at the expense of safety.

Shearing Stresses in Beams.—Shearing stresses in reinforced concrete beams have not as yet been satisfactorily analyzed, and few if any experiments have been made that throw much light on the question. It is quite certain, however, that the problem is radically different from that of a steel plate girder. In the latter the web is usually very thin and the strength of the metal practically the same for tensile and compressive stresses. The mode of failure of the web of a plate girder is by the buckling of the web, because in the direction of the compressive stress the web acts as a long column with a length of sometimes 100 or 150 times its thickness. Therefore when reduced by a formula for flexure under compressive stress and with an allowance for the beneficial effect of the tensile stresses in holding the web in line, we find that it is very much weaker on the lines of the compressive stresses.

In concrete, on the other hand, the thickness of the sections are always such that the question of flexure in compression does not enter into the resistance of the shearing stresses by what may be called the web. But, on the other hand, the tensile strength of the concrete is only about one-tenth of its compressive strength. From this it would appear that the manner of failure of a concrete beam in excessive shearing stresses would probably be by tension, which is exactly opposite to the manner of failure of a plate-girder web under excessive shearing.

A plate-girder web is reinforced by stiffeners to prevent buckling. To reinforce a concrete beam for shearing stresses the natural method would seem to be to insert steel members connecting the lower and upper fibers and inclined at an angle of 45° toward the nearest support. In Table IX the writer has arranged 47 tests of reinforced concrete beams compiled from various sources as noted. It does not appear that any one of these beams has failed primarily under vertical shear. As most of them were subjected to a center load and the others to loads at two points (position not located) the shear for those loaded at the center would be only one-half of the end shear that would be produced by a uniformly distributed load giving the same bending moment. Concentrated loads near the points of support would produce relatively much greater end shear than bending moment. This may in part account for the fact that tests to destruction of concrete beams have generally shown failure by transverse rupture and not by vertical shear.

M. Mesnagen states that the shearing strength of concretes are from 1.2 to 1.3 times the tensile strength. Bauschinger gives the shearing strength of concrete four weeks old at 1.25 times the tensile strength and at two years old 1.5 times the tensile strength.

By reference to Tables I and V, it will be seen that the shearing strength of the beams given in Table IX should be from 270 lbs. to 380 lbs. per square inch ultimate, depending on the mixture and age. The safe working shearing strength of concrete may be taken at 50 lbs. to 75 lbs. per square inch. The latter figure may safely be used on all ordinary cases and the former for cases of severe loading, including impact.

Taking tests numbers 11 to 16, inclusive, of Table IX and multiplying the area of their cross-section by the probable ultimate shearing strength for the proportions of cement to sand, and for the age at which they were tested as given therein, it will be found that the probable ultimate shearing strength of number 11 was about 12,000 lbs., while the maximum shearing due to its ultimate strength was only 1,114 lbs. Numbers 12 and 13 had a probable ultimate shearing strength of 15,000 lbs., whereas they failed under loads producing maximum shears of only 1,119 lbs. and 1,046 lbs., respectively. Number 14 had a probable ultimate shearing strength of nearly 17,000 lbs., and failed under a load producing an end shear of 2,493 lbs. Numbers 15 and 16 had a probable ultimate shearing strength of 20,900 lbs. and failed under loads producing end shears of only 2,548 lbs. and 2,493 lbs., respectively. From this it will be seen that these should not have failed by shearing, and even if they had been tested to destruction under uniformly distributed loads the shear would only have been doubled and there would still have been a safety factor of 4 for shearing. Mr. Thacher states that these beams all failed by transverse rupture due to bending, with vertical cracks in the concrete under the point of loading.

Taking tests numbers 23 to 47, inclusive, we find that the probable ultimate shearing strength should have been from 26,000 lbs. to 29,000 lbs. These beams failed under loads producing maximum end shears of from 5,000 lbs. to nearly 15,000 lbs., and 23 out of 25 failed under loads producing shears of 7,000 lbs. to 12,000 lbs. Under uniform load producing the same moment, the maximum shear would have been on an average about two-thirds of the ultimate shearing strength of the beams. Twenty out of these twenty-five beams failed by a longitudinal shearing break at or near the reinforcing bar and in addition a break extending diagonally (often at an angle of 45°) upward toward the center. The point where the diagonal break joined the top of the beam was sometimes at the point of application of the load

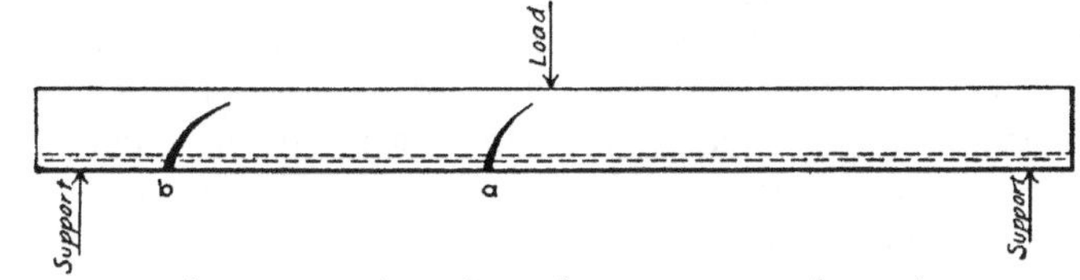

FIG. 11.—Diagram Showing Diagonal Break Frequently Attributed to Shear.

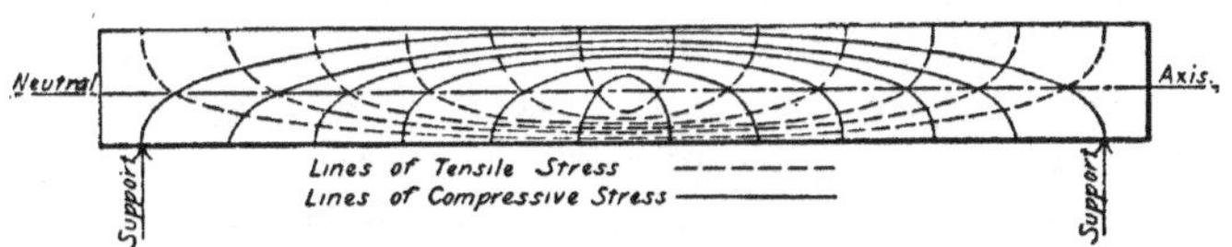

FIG. 12.—Diagrams Illustrating Lines of Stresses in Beams as Given by Rankine.

and sometimes near it. Fig. 11 illustrates the general characteristics of this manner of failure at the point "*a*."

Fig. 12 illustrates the lines of the principal stresses in a beam as given by Rankin. Comparing Figs. 11 and 12 and noting that the shear in a beam under a single concentrated load is the same at all sections when the dead weight of the beam is neglected, it is clear that the break *a* practically follows a line of compressive stress. The lines of compressive stress cross the lines of tensile stress at the neutral axis, making angles of 45° with the latter. The intensity of stress on these lines is maximum where they are horizontal, at the top of the beam for the compressive- and at the bottom for the tensile-stress lines, and are zero when they become vertical at the bottom or top. From this it appears that the crack *a*, Fig. 11, was practically started at the bottom by the cracking of the concrete due to the tension of flexure and was extended by the tensile stress of shear acting normal to the line of the crack and tending to pull the concrete apart. While the shearing stresses were less than the strength of the concrete and could not have caused these cracks primarily, in a material of this nature a crack once started by some other cause is often extended by stresses much less than the strength of the material.

If this explanation is the true one it should not be considered that these beams failed by shear, but that failure in the concrete having been caused by the tensile stress of flexure the cracks were extended by a shearing stress which otherwise would have been safe.

Again referring to Fig. 11, if a similar crack had occurred at a point nearer the support, as at *b*, it may be accounted for in a similar manner excepting that the primary crack in that case probably has been due to the longitudinal shearing between the reinforcing bars and the concrete, which is maximum toward the ends.

It is possible that similar cracks would result from a primary failure due to shear, and if the ultimate strength of the beam to resist shear was reached before its ultimate resistance to flexure it is likely that they would occur, but they would probably start at the point where tensile shear is maximum.

Taking test numbers 23 to 47 from Table IX, all except five of which failed as described, and noting that the shear at the points where failure occurred was less than at the ends by the amount of the shear due to the weight of the beam itself, it seems to be clearly demonstrated that these cracks are not due primarily to shear.

Beams numbers 32, 33, and 34 were reinforced with 16 stirrups of ¼ in. square rods, but there are no identical tests without stirrups of the same age and manner of loading with which to compare them and deduce the probable value of the stirrups. Comparing them by bending moments producing fracture we see that 32 and 33 compared with 29 show an apparent loss of 22.5 per cent. and 34 compared with 31 shows an apparent loss of 29.2 per cent. This would indicate that the stirrups are a disadvantage, but certainly no definite conclusions can be drawn from these comparisons.

Incidentally it is worth while to notice that 29 compared with 30 and 31 shows no benefit whatever derived from embedding reinforcing metal on the compression side of the neutral axis.

Tests numbers 18 and 20 with stirrups and 17 and 19 without stirrups were made for the purpose of determining the advantage of using stirrups. They are not entirely satisfactory comparisons because the beams with stirrups were reinforced with plain rods, while those without stirrups were reinforced with twisted rods. Fig. 13 shows the arrangement of the reinforcing metal in these four beams. The two beams with stirrups had an advantage over those without stirrups of fourteen days in age and also a slight advantage in the area of the steel reinforcement. Nothwithstanding this advantage number 18 shows a loss of 7.4 per cent. in strength as compared with number 17 and number 20 shows a loss of 32.3 per cent. as compared with number 19. As far as this series goes it tends to discredit the use of stirrups of that kind and arrangement.

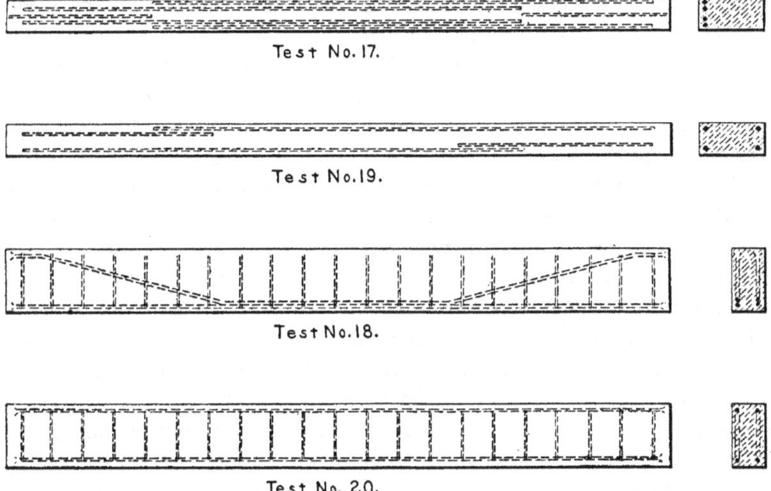

FIG. 13.—Construction of Beams for Tests 17 to 20 in Table IX.

Of tests 24 to 47, inclusive, most of them contained an excess of metal. Beams economically designed like those of tests 11 to 16, inclusive, where the metal is not in excess would rarely require reinforcement for shearing stresses.

No tests are known to have been published with beams reinforced with steel members running from the bottom to the top and inclined at the top at an angle of 45° toward the nearest support. A series of this kind would be very interesting if scientifically made. The inclined reinforcing rods should not be farther apart than the depth of the beam. An indication of what may be expected from tests of such beams is given by comparing numbers 46 and 47 with 38 and 36, respectively; 46 and 47, in addition to the twisted square rods, were reinforced with expanded metal extending throughout their length and depth, but were otherwise identical with 38 and 36.

Test number 46 shows a gain in ultimate strength of 27.6 per cent., and 47 a gain of 32 per cent., or an average gain for the two beams so reinforced of 29.8 per cent. There is no other explanation for this gain in ultimate strength except that it was due to the expanded metal. The additional metal thus inserted seems to have been placed where it would do the most good, and this accounts for the remarkable showing. Any other kind of metal properly distributed and connected to the concrete would probably make an equally good showing.

To provide properly for the longitudinal shear at the reinforcing members, in addition to investigating the stress by formula 58, care should be taken that the total width of bars on any horizontal section does not exceed one-half the width of beam on that section, and it would be better if it did not exceed one-third the total width of beam; also that the distance d_1 from center of reinforcing rods to edge of concrete should not be less than two diameters of the rods.

Care and judgment should be used in applying any formula involving the modulus of elasticity of concrete because of its very variable value. Considère found in two experiments, which were identical except as to the amount of water used, that between stresses of zero and 2,845 lbs. per square inch E_c=7,111,000 for the experiment containing the proper amount of water and E_c=2,845,000 for the experiment gauged with an excess of water. This shows a loss of 60 per cent. due to excess of water used in gaging. The concrete with an excess of water was soft and easily crumbled and did not acquire compactness. Where concrete may be subjected to severe stresses, particularly those of flexure, the greatest possible care should be used in the selection of the materials, mixing, and placing, to insure strength and compactness and a reasonably close agreement of the modulus of elasticity with the value assumed for it in designing.

Considère gives the economic proportion of steel to concrete in cross-section at 2.5 per cent. for static loads and 1.2 per cent. for moving loads, and points out that there is no economy and some danger in using more than 3 and 2 per cent., respectively, the danger being due to the excess of steel, which would cause the beam to fail first by compression of the concrete and therefore without warning.

There is not much to be gained by using steel in the compression side, except for beams continuous over three or more supports. In the latter case, if the supports are practically unyielding, as they should be, multiply the value of a from equation (26) by 0.8 and use the same amount of steel in the compression side.

While the formulas and examples here given are thought to be sufficient to meet the requirements of designers and constructors for

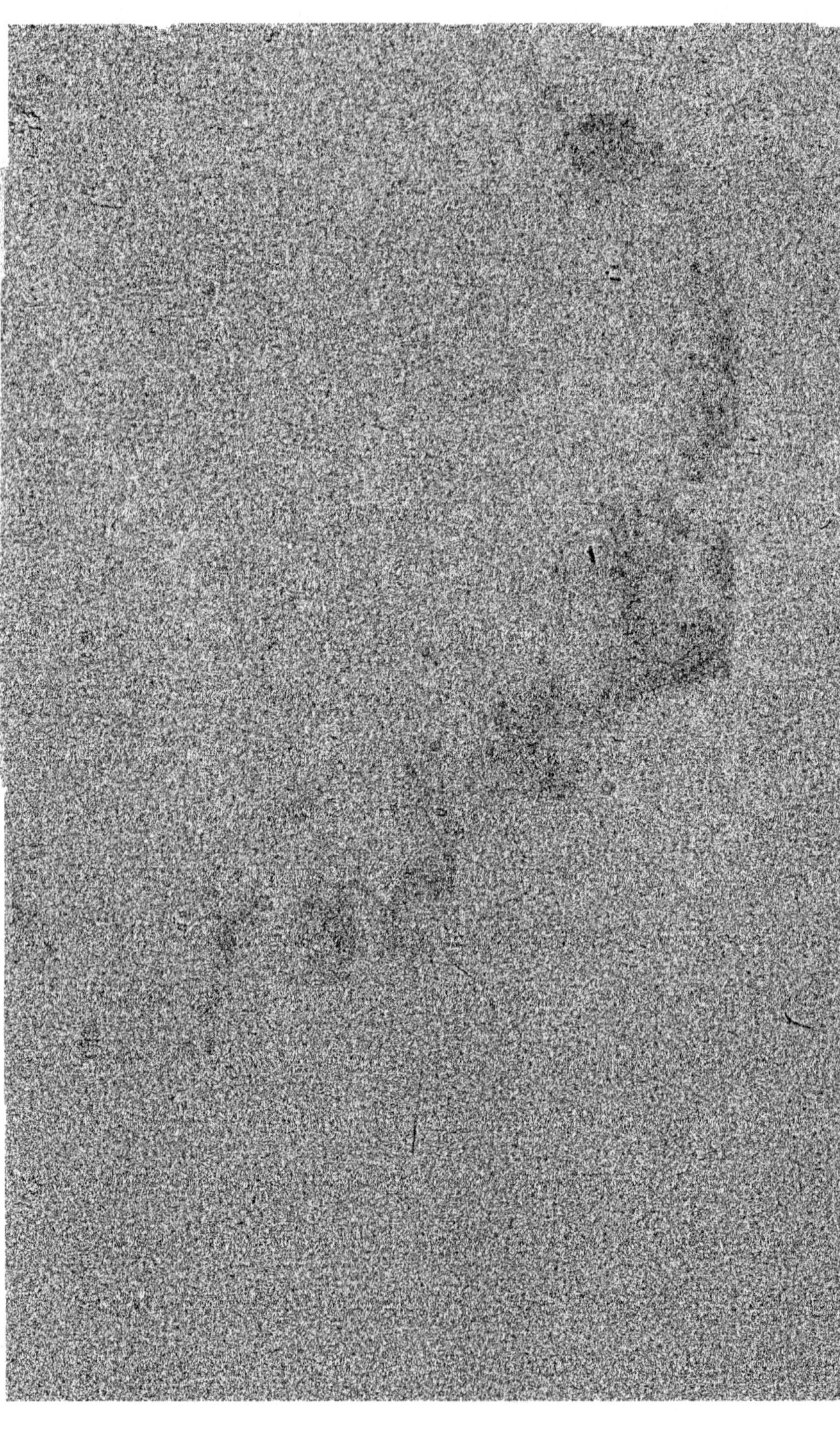

proportioning and calculating reinforced concrete beams, it is not at all intended that they should be considered as a complete exposition of the theory of flexure in reinforced concrete. The purpose of this book does not permit of a treatment of this subject even approximately complete, and those readers who desire to pursue further the theoretical discussion are referred to standard text-books and professional papers, such as those of Considère, Christophe, Lefort, Harel de la Noë, Resal, Von Emperger, "Resistance of Materials," by W. H. Burr, J. B. Johnson, Edwin Thacher, L. C. Wason, A. L. Johnson, and others. All of the above have been reviewed in the preparation of this chapter.

Tables X to XIII, inclusive, are taken from a paper by Mr. Edwin Thacher. They can be used with confidence, and will be found of great assistance in fixing trial sections even by those who prefer to use other formulas.

CHAPTER III.—COLUMNS.

THE results of a series of tests on reinforced concrete columns made at the Massachusetts Institute of Technology are given in Table XIV. The columns were 8×8 ins. and 10×10 ins. square and reinforced with either one or four rods parallel with their axes. Where one rod was used it was placed in the center of the cross-section, and where four rods were used they were so disposed as to form a square 4×4 ins. for the 8×8-in. columns and 5×5 ins. for the 10×10-in. columns, and parallel to that of the column section. The concrete was proportioned 1 of cement to 3 of sand and 6 of stone, two-thirds of the latter being 1 in. and one-third ½ in. in size. The columns were cast in a vertical position and tested in a horizontal machine. The tensile strength of the plain bars or rods was from 56,000 lbs. to 63,000 lbs., and that of the twisted rods about 80,000 lbs. per square inch. The notation used is

$P=f(A_c+eA_s)$ = compressive strength of section;

f = crushing strength of concrete;

A_c = area of section of concrete;

A_s = area of section of steel;

$e=\frac{E_s}{E_c}$, where E_s = modulus of elasticity of the steel and E_c = modulus of elasticity of the concrete.

TABLE XIV.—SHOWING RESULTS OF TESTS OF REINFORCED CONCRETE COLUMNS.

Number.	Age in Days.	Area of Section in Square Inches.	Length in Feet.	Number of Rods.	Side of Square Rod in Inches.	Plain, "P" or Twisted, "T."	Actual Breaking Load in Pounds.	Value of P in the Formula $P=f(Ac+eAs)$.*	Manner of Failure.
1	30	64	17	1	1	P	107,000	125,000	[illegible]d at [illegible]nd
2	30	64	17	1	1	T	127,000	125,000	[illegible]d [illegible], [illegible]n [illegible]d at [illegible]nd
3	29	64	12	1	1	P	100,000	125,000	[illegible]d [illegible], [illegible]n [illegible]d at [illegible]nd
4	28	64	12	1	1	T	126,000	125,000	[illegible]d at [illegible], [illegible]ply [illegible], [illegible]d [illegible]n [illegible]t off, [illegible]e [illegible]st [illegible]e 156,000 [illegible]s. at 40 days
5	32	64	6	1	1	P	138,000	125,000	[illegible]d at [illegible]le, [illegible]n [illegible]d off along [illegible]e [illegible]d to [illegible]e [illegible]nd
6	31	64	6	1	1	T	133,000	125,000	[illegible]d at [illegible]nd
7	31	64	17	1	$1\frac{1}{4}$	P	136,000	132,000	[illegible]d at [illegible]d, [illegible]g [illegible]y
8	35	64	17	1	$1\frac{1}{4}$	T	154,000	132,000	[illegible]d at [illegible]d, [illegible]g [illegible]ff 3 [illegible]t
9	35	64	17	4	$\frac{3}{4}$	P	182,000	140,000	[illegible]d at [illegible]nd
10	34	64	17	4	$\frac{3}{4}$	T	167,000	[illegible]40,[illegible]00	[illegible]d at [illegible]nd, [illegible]e [illegible]r pr[illegible] [illegible]nd rough at [illegible]t [illegible]nd
11	31	64	12	4	$\frac{3}{4}$	T	147,[illegible]00	[illegible]40,[illegible]00	[illegible]d [illegible]nd [illegible]t [illegible]pn at [illegible]nd
12	32	64	12	4	$\frac{3}{4}$	P	153,[illegible]00	[illegible]40,[illegible]00	[illegible]d [illegible]nd [illegible]t [illegible]pn at [illegible]nd
13	29	64	6	4	1	T	158,[illegible]00	[illegible]62,[illegible]00	[illegible]d [illegible]nd [illegible]t [illegible]pn at [illegible]nd
14	31	64	6	4	1	P	244,[illegible]00	[illegible]62,[illegible]00	[illegible]d at [illegible]nd
15	35	100	17	1	1	P	215,[illegible]00	[illegible]88,[illegible]00	Broke off [illegible]n 3 or 4 [illegible]t at [illegible]nd
16	35	100	6	1	1	P	[illegible]00,[illegible]00	188,[illegible]00	[illegible]d [illegible]y at [illegible]nd
17	45	100	6	1	1	T	228,400	188,[illegible]00	Sheared [illegible]ly at [illegible]nd [illegible]nd broke [illegible]k for half the length
18	31	100	12	1	$1\frac{1}{4}$	T	[illegible]62,[illegible]00	[illegible]94,500	[illegible]d [illegible]ally [illegible]r [illegible]nd
19	29	100	12	1	$1\frac{1}{4}$	P	257,[illegible]00	[illegible]94,500	[illegible]d at [illegible]nd
20	28	100	12	4	$\frac{3}{4}$	T	300,[illegible]00	203,[illegible]00	Did [illegible]nt [illegible], [illegible]d [illegible]y of [illegible]e
21	29	100	12	4	$\frac{3}{4}$	P	[illegible]74,[illegible]00	203,[illegible]00	[illegible]d at [illegible]; wedge-shaped [illegible]e [illegible]d in [illegible]b-tw[illegible]en [illegible]ds

* This formula has been used more or less to compute the breaking loads of such columns.

The numerical value of the constants that have been used in computing the values of P by the formula are the averages of this series of tests, viz.:

$E_s=28,000,000$ and $E_c=3,400,000$, therefore $e=8.23$; $f=1,750$ lbs. per square inch.

The values of P have been computed in order that the reader may be enabled to compare them with the actual breaking loads.

The results of the experiments given in Table XIV have been compared for this work in Table XV, so as to show the ratio of length to diameter, the breaking load per square inch of cross-section and per square inch of the equivalent section where the area of the metal reinforcement is multiplied by the ratio of E_s divided by E_c or e. This will facilitate the comparison of these results with those of other experiments.

Attention is called to the notes on manner of failure in Table XIV, which indicate that only numbers 2 and 3 developed column flexure, and from Table XV these are seen to have a ratio of l/d of 25.5 and 18 respectively. Out of six columns with a ratio of 25.5 for l/d, only one developed flexure; and out of eleven columns with a ratio of 18 or over for l/d, only two developed flexure. No column with a ratio of less than 18 for l/d showed any flexure. This agrees very well with the experiments with iron and steel columns, which usually fail by direct compression or crushing for lengths less than twenty or twenty-five times the least radius of gyration, and by flexure for greater lengths.

In Table XVI these results have been grouped according to the ratio of the length to the diameter, with the averages of the breaking load per square inch of the cross-section and the per cent. variation from the average breaking load per square inch of all, and also with the averages of P divided by (A_c+eA_s) and its per cent. variation from the average of all. It is apparent from these results that the total variations as well as the maximum variations from the average are greater for the breaking loads per square inch of total cross-section than for the total breaking load divided by (A_c+eA_s). Without some identical tests of concrete columns not reinforced it is difficult to show the exact value of the reinforcement by these tests. If, however, those columns that have ratios of length to diameter of 7.2, 9.0, 20.4, and 25.5 are compared, it will be seen that the agreement with the formula $P=f(A_c+eA_s)$ is very good. Also that the average of columns with $l/d=14.4$ is 19.7 per cent. above the average, while for the columns with $l/d=18.0$ it is 19.9 below the average, and therefore the average of these two groups taken together is only 0.1 per cent. below the average of the 21 experiments. It is further evident from these results that for all practical purposes no formula for flexure is required for columns not longer than 25 diameters, and as no experiments are known to have been made with columns of a greater ratio of length to diameter, there are no data by which such a formula can be verified.

TABLE XV.—COMPARISON OF RESULTS OF EXPERIMENTS GIVEN IN TABLE XIV.

Column Number.	Area Cross-section.	A_s	eA_s	A_c+eA_s	P	$\frac{P}{A_c+A_s}$	$\frac{P}{A_c+eA_s}$	Ratio Length to Diameter.
1	64	1.00	8.23	71.23	107,000	1,670	1,500	25.5
2†	"	1.00	"	"	127,000	1,985	1,780	"
3†	"	1.00	"	"	100,000	1,560	1,400	18.0
4	"	1.00	"	"	126,000	1,970	1,770	"
5	"	1.00	"	"	138,000	2,160	1,940	9.0
6	"	1.00	"	"	133,000	2,080	1,870	9.0
7	"	1.56	12.84	75.28	136,000	2,125	1,810	25.5
8	"	1.56	12.84	75.28	154,000	2,410	2,040	"
9	"	2.25	18.52	80.27	182,000	2,840	2,270	"
10	"	"	"	"	167,000	2,610	2,080	"
11	"	"	"	"	147,000	2,300	1,830	18.0
12	"	2.25	18.52	80.27	153,000	2,390	1,910	"
13	"	4.00	32.92	92.92	158,000	2,470	1,700	9.0
14	"	4.00	32.92	92.92	244,000	3,810	2,620	"
15	100	1.00	8.23	107.23	215,000	2,150	2,000	20.4
16	"	1.00	8.23	107.23	200,000	2,000	1,870	7.2
17	"	1.00	8.23	107.23	228,400	2,284	2,130	"
18	"	1.56	12.84	111.28	262,000	2,620	2,360	14.4
19	"	1.56	12.84	111.28	257,000	2,570	2,310	"
20	"	2.25	18.52	116.27	300,000*	3,000*	2,580*	"
21	"	2.25	18.52	116.27	274,000	2,740	2,360	"
	Average of all.					2,369	2,006	

* Did not break; exceeded capacity of machine. † Buckled.

This does not seem to be a matter of much practical importance, since the cases that occur in practice requiring columns longer than 25 diameters are very rare, and until more is known concerning the subject it would be well to limit the lengths of all concrete columns, either with or without reinforcement, to 25 diameters, and for important works or in cases of severe loading to 20 diameters or even less. As the length of columns in classic architecture, including both base and capital, rarely if ever exceeds one-half of the above limits, it would not appear that a formula for flexure in concrete columns over 25 diameters would have much practical value.

Finally, for concrete columns with or without longitudinal reinforcement, but not hooped, and not exceeding 25 diameters in length, the ultimate strength may be taken at 2,000(A_c+eA_s).

TABLE XVI.—RESULTS OF TABLE XV ARRANGED ACCORDING TO RATIO OF LENGTH TO DIAMETER.

Ratio of Length to Diameter.	Column Numbers.	$P \div A_c + A_s$ (Average).	Per Cent. Variation from Mean.	$P \div (A_c + eA_s)$ (Average).	Per Cent. Variation from Mean.
7.	16, 17	2142.0	− 9.60	2,000	− 0.30
9.	5, 6, 13, 14	2630.0	+10.10	2,032	+ 1.30
14.2	18, 19, 20, 21	2732.5	+15.30	2,402	+19.70
18.0	3, 4, 11, 12	2055.0	−13.25	1,727	−19.90
20.4	15	2150.0	− 9.25	2,000	− 0.30
25.5	1, 2, 7, 8, 9, 10	2273.3	− 4.04	1,913	− 4.64
	All	2369.0		2,006	

f=2,006 is 14.6 per cent. higher than the value assumed for the calculation of the values of P in Table XIV, and only 2.76 lower than the value given in Table I.

As such columns fail suddenly and with little or no warning, a safety factor of 4 to 6 should be used, depending upon the conditions of loading and importance of the work. The safe working loads may then be placed at 500(A_c+eA_s) to say 350(A_c+eA_s).

It appears from the comparison of Tables XIV and XV, that a reinforcing rod in the center is just as efficient as any other distribution of longitudinal reinforcement, and it is much the most convenient in construction.

Table XVII shows the results of European experiments on concrete columns and prisms reinforced with wire hooping wound in the form of a spiral, and compared with other columns and prisms without the hooping, but otherwise identical. Those numbered 1 to 6 inclusive were made by M. Considère, 7 to 12 inclusive by MM. Considère and Hennebique, and 13 to 22 inclusive were made at Quimper. Number 1 was reinforced by wire hooping and 8 longitudinal wires or rods, while number 2 was identical with number 1, except that it was not reinforced in any direction. The results indicate that the hooping increased the resistance by more than four times, or from 2,240 lbs. per square inch for number 2 to 9,150 lbs. per square inch for number 1. Number 3 stood 6,260 lbs. per square inch, but M. Considère says the concrete was not so good.

Numbers 4, 5, 6, Table XVII, are experiments on hollow cylindrical sections, numbers 4 and 5 being hooped and number 6 without hooping.

TABLE XVII.—SHOWING RESULTS OF TESTS OF HOOPED CONCRETE COLUMNS.

No.	Length, Inches.	Shape of Cross-section.	Diameter, Inches.	Spirals.			Longitudinal Rods.		Total Sectional Area, Sq. Ins.	Area of Section of Core Inside of Spiral, Sq. Ins.	Ratio of Length to Diameter.	Load, Pounds.	Load per Sq. In. of Cross-section.
				Diam. of Wire, Inches.	Diam. of Winding Cylinder, Inches.	Pitch, Inches.	No.	Diam., Inches.					
1	19.7	Octagonal	4.3	0.17	3.75	0.7	8	0.17	15.5	11.16	4.58	142,000	9,150
2	19.7	"	4.3	......	....	......	....	......	15.5	....	"	34,700	2,240
3	19.7	"	12.6	0.39	10.6	1.46	8	0.59	131.0	88.50	1.56	820,600	6,260
4	23.6	Hollow cylinder	7.1–4.9	Dimensions of hooping not given			....	......			3.32		2,580
5	23.6	"	"				....	......			"		2,290
6	23.6	"	"	Not [illegible]d			....	......			"		2,404
7	35.4	Octagonal	5.9	Not [illegible]ced			....	......			6.00		1,050
8	35.4	"	"	$\frac{1}{4}$	5.5	1.18	....	......			"		5, [illegible]0
9	35.4	"	"	0.17	5.5	0.59	....	......			"		5, [illegible]0
10	51.2	"	"	$\frac{1}{4}$	5.5	1.18	8	$\frac{1}{4}$			8.67		4,550
11	51.2	"	"	0.17	5.5	0.59	8	$\frac{1}{4}$			"		4,700
*12	51.2	"	"	......	....	......	8	0.35			"		2, [illegible]0
13	Not given		1.579	Fine wire with probable elastic limit of 78,200									4,870
14	do.		"	do.									6,540
15	do.		"	do.									7,360
16	do.		"	do.									4,930
17	do.		"	do.									10,500
18	do.		"	No reinforcement									569
19	do.		"	do.									711
20	do.		"	do.									853
21	do.		"	do.									853
22	do.		"	do.									2,420

As might have been expected, it is seen that the hooping does not improve a hollow section, because it fails by crushing inwardly, unless, as M. Considère points out, the hollow core is very small.

Of the series of six, numbers 7 to 12, the first was not reinforced, and the last was reinforced by longitudinal rods tied together with bands of iron wire, 0.17 in. in diameter, and 3.15 ins. center to center of bands. Numbers 8 and 9 were hooped, while numbers 10 and 11 were reinforced both by hooping and longitudinal rods.

Number 7, not reinforced, broke suddenly at 1,050 lbs. per square inch and number 12 also broke suddenly at 2,420 lbs. per square inch. It appears probable, from a comparison with the other tests, that the wire bands had more effect on the ultimate resistance than the longitudinal rods.

Numbers 8 and 9 were reinforced with spiral hooping only, the former failing between the spirals at a pressure of 5,120 lbs. per square inch, and the latter resisting 5,400 lbs. per square inch without failure. As the pitch of the spiral of number 8 was 1.18 ins. and that of number 9 only 0.59 in., or $\frac{1}{5}$ and $\frac{1}{10}$ of the diameters, respectively, and as the pitch of the spirals of numbers 1 and 3 were about $\frac{1}{6}$ and 1/8.6 of the respective diameters of the prisms, it is clearly indicated, as far as these series of experiments go, that the pitch of the spirals should not exceed $\frac{1}{6}$ of the diameter of the column. Considère recommends a pitch of $\frac{1}{7}$ to $\frac{1}{10}$ the diameter.

Numbers 10 and 11 failed by "column pressure," according to Considère, at 4,550 and 4,700 lbs. per square inch, respectively. As these columns only had a length of 8.67 diameters, it is difficult to account for their failure in this manner, in view of the results of the American experiments given in Table XIV. Considère finds that hooping does not greatly increase the resistance to column flexure, and that the modulus of elasticity is quite different for the two sides of a column under flexure due to the conditions existing,—the compression being increased on one side and simultaneously relieved or "unloaded" on the other, the modulus for the latter condition being about double that for the former. This may be a sufficient cause for the failure by flexure of the two out of the three columns having a length of 8.67 diameters, and the third one not being hooped failed by crushing at little over half the load that caused failure by flexure in the other two. If this is a true explanation it is remarkable that only two out of the 21 tests of Table XIV having lengths of from 7.2 to 25.5 diameters developed flexure, and these two were 18 and 25.5 diameters in length.

Considère also records flexure in three other similar columns of the same dimensions, at from 5,000 to 7,000 lbs. per square inch, although he does not state that they were tested to destruction or that they failed by flexure. In view of conflicting results of the more extensive American experiments, these results and theories of M. Considère can hardly as yet be accepted as a basis for practical construction. Since in practice the safe working loads would never be half enough to induce flexure in a concrete column under 20 or even 25 diameters in length, the question is not a serious one in practical work.

The idea of hooping a concrete column was suggested by the great resistance of sand when confined. The familiar pyramidal form of failure in cubes of stone tested to destruction also point to the value of hoops, although recent experiments have shown that the pyramidal form is due to the friction on the plates through which the pressure is applied. When this friction is reduced to a minimum by lubrication, it is found that the cubes fail along surfaces parallel to the line of pressure.

Following out this idea the series of tests numbered 13 to 22 in Table XVII, made in 1901, show that wire hooping, constituting from 2 to 4 per cent. of the volume of the small mortar prisms, increased the resistance from four to nine times that of identical prisms not reinforced. Prisms 13, 14, 15, 18, 19, and 20 had 675 lbs. of cement per cubic yard of sand, and the others, 16, 17, 21, and 22, 730 lbs. of cement per cubic yard of sand. Numbers 13 and 18 were tested at 8 days old, 14 and 19 at 14 days, 15 and 20 at 22 days, 16 and 21 at 23 days, and numbers 17 and 22 at 100 days old.

These tests indicate that the hooping is of more relative advantage to the poorer mortars or concretes than to the richer ones, or that it is not necessary to use very rich concrete for hooped columns. This is a result that would naturally be expected and it has been verified by other tests and by theory.

Considère gives the value of the hooping at 2.4 times that of an equal amount of metal in longitudinal reinforcing rods, and his analysis is rational and agrees very well with the experimental results.

If A_h=the area of cross-section of an amount of metal equivalent to the hooping, arranged as longitudinal rods, then the entire strength of a reinforced concrete column with steel hooping may be written as

$$P=f(A_c+eA_s+2.4eA_h).$$

If Y=the elastic limit of the metal, we would have, according to Considère,

$$P=fA_c+Y(A_s+2.4A_h),$$

where f is the ultimate compressive strength of the concrete.

If the longitudinal rods have an elastic limit of Y, and the hooping an elastic limit of Y', there would result

$$P = fA_c + YA_s + 2.4Y'A_h.$$

The tests numbers 8 and 9 of Table XVII would seem to indicate that the longitudinal reinforcing rods might be omitted in hooped columns, since their value is only 40 per cent. of the value of an equivalent amount of metal used in hooping. But tests 10 and 11, with the longitudinal rods, did not crack or chip until the pressure per square inch amounted to 2,900 and 3,360 lbs., while 8 and 9 cracked at 1,730 and 2,480 lbs. per square inch, and soon after began to chip. It is therefore desirable to retain the longitudinal reinforcing rods in connection with the spirals of hooped columns, and especially so when the ratio of length to diameter is large. It is important, as before stated, to keep the pitch of the spiral below $\frac{1}{6}$ or $\frac{1}{7}$ of the column diameter.

As hooped columns do not fail suddenly, but crack and chip under loads of from $\frac{1}{3}$ to $\frac{2}{3}$ of their ultimate strength, it will be safe to use with them somewhat higher working stresses or lower safety factors. Safety factors of $3\frac{1}{2}$ to 5, giving working stresses of 900 to 1300 lbs. per square inch, will cover the range of conditions for well-made hooped concrete columns.

CHAPTER IV.—RETAINING-WALLS, DAMS, TANKS, CONDUITS, AND CHIMNEYS.

RETAINING-WALLS.

THE usual form of retaining-wall in which the earth pressure is resisted by the moment of the weight of the wall about the center of resistance of the base, called the righting moment of the wall, is dependent for stability on the weight of the wall and is called a gravity wall. In such a wall, whether built of brick, stone, or concrete, there would be nothing to gain in stability by reinforcing it with steel. In cases of very long walls, especially when built of concrete, cracks due to thermal changes and other secondary causes are apt to occur at intervals, and these can be prevented by reinforcing the wall with longitudinal steel bars or rods. The determination of the proper amount of reinforcement to use for this purpose and its disposition will be given in the succeeding section on Dams.

The principles involved in the economic application of reinforced concrete to the construction of retaining-walls are, first, the utilization of the capacity of the concrete-steel combination to resist bending moments due to transverse loading, acting as cantilever and supported beams, thus permitting the use of small sections with only a fraction of the amount of material required for a gravity wall, and, second, the substitution of a weight of earth on the base or footing, costing practically nothing, for the weight of expensive masonry omitted, to resist the overturning moment.

Plane Walls.—The most simple form of the reinforced concrete retaining-wall, illustrated in Fig. 14, consists of a base, *EFGH*, to

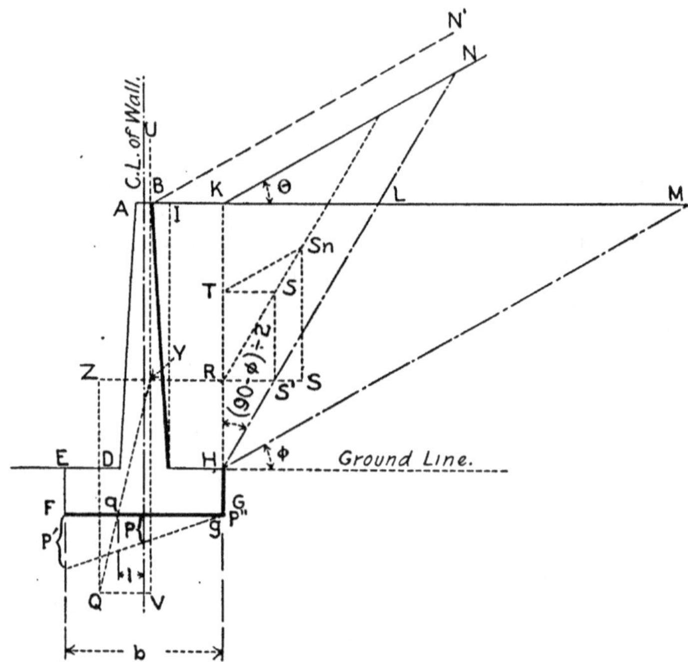

FIG. 14.—Diagram of Thrusts and Moments on Retaining-walls.

which is rigidly connected the vertical part of the wall *AB*CD, which latter acts as a cantilever beam fixed at *CD*.

The projecting portions of the base also act as cantilevers, the part projecting in front, or the toe, loaded by the upward pressure of the reaction and the portion projecting behind the wall, or the heel, loaded by the downward pressure of the weight of material filled in above it. Therefore the base will require transverse reinforcement both in the top and in the bottom, and the part *AB*CD will require vertical reinforcement in the back. As the moment is maximum at the plane DC and zero at AB, the section of the wall may taper to the mimimum practical thickness at the top and be finished with a coping when desired.

When the true angle of repose ϕ of the material behind the wall is known the determination of the thrust is a simple matter, but this angle of repose is not often known with any considerable degree of accuracy. It may have any value between zero for hydrostatic pressure to 90° for solid rock. A material in which ϕ is 30°, or even 45°, may be so changed by becoming saturated with water that ϕ would become zero or nearly so. Even if a wall were built against a bank of solid rock a thin vertical sheet of water collecting between the wall and the rock would produce hydrostatic pressure on the wall. For these reasons any retaining-wall to be successful must be provided with ample drainage to prevent the accumulation of water behind it and it must be designed to resist the pressure due to the minimum angle of repose that can obtain under any conditions in the material to be retained, or else it must be designed for hydrostatic pressure with ϕ=zero.

Let P_y= the thrust from the material behind the wall in a direction parallel to the surface of the bank, the horizontal component of the resultant of which acts at a height $x \div 3$ above the base. Let x= the height of the wall HK, Fig. 14, and w= the weight of a unit of volume (1 cu. ft.) of the material behind the wall. Then

$$P_y=\frac{wx^2}{2}\cos\theta\,\frac{\cos\theta-\sqrt{\cos^2\theta-\cos^2\phi}}{\cos\theta+\sqrt{\cos^2\theta-\cos^2\phi}}. \quad . \quad . \quad . \quad (1)$$

If the surface slopes at the angle of repose, that is, if the angle of surcharge is the angle of natural slope, $\theta=\phi$, and

$$P_y=\frac{wx^2}{2}\cos\phi. \quad . \quad . \quad . \quad . \quad . \quad . \quad . \quad . \quad (2)$$

If the surface is horizontal, θ=zero, and

$$P_y=\left(\frac{wx^2}{2}\right)\cdot\left(\frac{l-\sin\phi}{l+\sin\phi}\right). \quad . \quad . \quad . \quad . \quad . \quad . \quad (3)$$

The value of the functions used in equations (2) and (3), which apply to most cases met in practice, and for values of ϕ from zero to 60° are given in Table XVIII.

TABLE XVIII.—VALUE OF FUNCTIONS USED IN EQUATIONS (2) AND (3).

Slope =		1 on 6		1 on 2		2 on 3	1 on 1	
ϕ =	0	9° 28′	15°	26° 34′	30°	33° 42′	45°	60°
$\sin \phi$ =	0	0.164	0.259	0.447	0.500	0.555	0.707	0.866
$\cos \phi$ =	1	0.986	0.966	0.894	0.866	0.832	0.707	0.500
$\left(\frac{1-\sin \phi}{1+\sin \phi}\right)$ =	1	0.718	0.588	0.382	0.333	0.286	0.172	0.072

The horizontal component of the thrust P_y for a length unity is $RS'=P=P_y \cos \theta$. Therefore when $\theta=0$, $\cos \theta=1$, and $P=P_y$. When $\theta=\phi$, $P_y=P_x=$ the weight of the prism HKN.

It is clear that θ can never be greater than ϕ unless the slope is held in place by cribs or some equivalent device.

The line of action of P, the horizontal component of the resultant thrust, will intersect the vertical through H at R, where $HR=\frac{x}{3}$. Let the weight of a unit of length of $ABCD$, $EFGH$, and $BCHK=W=YV$, and find the horizontal distance, YR, of the vertical through their common center of gravity from the plane HK. Then the resultant on the foundation will be $\sqrt{P^2+W^2}$ and it will intersect the lower surface of the base at a distance $G-q=\overline{YR}+(\overline{RG}\cdot P\div W)$ in front of the plane GHK. Then the righting moment $(W\cdot \overline{vq})$ will be equal to the overturning moment $(P\cdot \overline{RG})$.

The overturning moment at the section DC will be $(P\cdot RH)$. The bending moment on the wall at DC will be $M=Px\div 3$. The downward moment on the heel of the base will be the weight of the prism $CIKH$ multiplied by one-half the length of the heel CH. It will be near enough and on the side of safety to design the heel to resist the downward moment, but to find the resultant moment on the heel the difference between the downward moment and the upward moment due to the part of the reaction that comes on the heel should be taken.

To find the intensities of the pressure on the base, let

$b=$ the width of the base.

$W'=$ the vertical component of the reaction acting through q.

$l=$ lever arm between q and the center of figure of the base $=\overline{Gq}-b/2$.

$p=$ the average intensity of pressure on the base.

$p'=$ the maximum intensity of pressure on the base.

$p''=$ the minimum intensity of pressure on the base.

Then for a length unity, the average intensity of pressure on the base will be $p = W' \div b$; the maximum intensity at the outer edge of the toe, F, will be

$$p' = (W' \div b) + \left(W'l \div \frac{b^2}{6}\right) \quad . \quad . \quad . \quad . \quad . \quad . \quad . \quad (4)$$

and the minimum intensity at the inner edge of the heel, G, will be

$$p'' = (W' \div b) - \left(W'l \div \frac{b^2}{6}\right). \quad . \quad . \quad . \quad . \quad . \quad . \quad (5)$$

When $l = b/6$, p' will equal $2p$ and p'' will be zero. When l is greater than $1/6$ of b, p'' will be negative and the heel will have a tendency to lift. In the latter case if the ultimate crushing resistance of the material under the toe is exceeded the wall will overturn.

Let the length of the lines p' and p'' in Fig. 14 represent these intensities of pressure and complete the figure $FfgG$, then will the length of the intercepts of the ordinates within that figure represent the intensities of pressure at every point, and the half sum of the intercepts at any two points will represent the average pressure between those points.

Find the horizontal distance of the center of gravity of that part of the figure $FfgG$ that is vertically under ED, from the vertical through D, and multiply it by the average pressure between E and D and by the length of the toe ED. This product will be the moment on the toe of the base. If the upward moment of the heel of the base is desired it may be found in the same manner.

By graphical statics the thrust is easily found as follows: Draw a vertical HK from the inner edge of the base EH and lay off HM at the natural angle of repose, ϕ. Bisect the angle of KHM by the line HL, which is the plane of fracture that satisfies the condition of maximum thrust and overturning moment, and the angle KHL will be $(90° - \phi) \div 2$. Lay off HR on HK, making $HR = \frac{1}{3}HK$, and draw RS through R and parallel to HL. The line RS will pass through the center of gravity of the triangle HKL. Consider a section of wall and of filling behind it of a length unity. The weight of the prism HKL will then be $W'' = \overline{HK} \times \overline{KL} \times w \div 2$.

With any convenient decimal scale lay off $RT = W''$ on the vertical line HK and draw the line TS through T, parallel to the surface KL, cutting RS at S; TS will then be the measure of the resultant conjugate pressure P_y, which acts through R. The horizontal projection of TS at RS' will be the measure of the horizontal component of the conjugate pressure, and the overturning moment in foot-pounds will be the amount of this horizontal component in pounds by the scale of forces multiplied by the lever arm HR in feet. When the surface is horizontal, $\theta=0$, and $RS'=TS$. Find the common center of gravity of the wall and of the material contained between the wall and the vertical HK, giving due account to the difference in the specific gravity of the two elements. Draw the vertical UV through this common center of gravity and intersect it at Y with a horizontal line drawn through R. On RY produced, lay off YZ equal to RS', and by the scale of forces lay off YV on the vertical UV equal to the weight of the wall and the material between the wall and the vertical plane HK. Complete the parallelogram of forces $YZQV$, and the diagonal YQ will then represent in amount and direction the resultant pressure on the base. This resultant should cut the lower side of the base within the middle third of its width or else the maximum intensity of the pressure on the toe should not exceed the safe compressive resistance of the material on which the wall is founded.

When the wall is surcharged with a bank sloping on the line KN, at an angle θ with the horizontal, the weight of the prism HKN, of length unity, must be used in laying off the load RT. In other respects the solution is identical. When $\theta=\phi$, KN will be equal to HK.

Having found the moments on the wall at the section CD, and on the base, the sections can be readily proportioned by the equations given in the chapter on beams. The shear at CD should be investigated and provided for. Equation (58) of the chapter on beams may be used to find the longitudinal shear on the vertical reinforcing rods or bars where they enter the base, and the latter should be made thick enough to prevent the rods from pulling out, allowance being made for the mechanical bond where deformed rods are used. The required width of the base, b, is generally found to be from one-quarter to one-half of the height of the wall, x, depending on the values of ϕ, θ, and w. In general railway work it is quite common practice to make $b=0.4x$, when the natural slope, ϕ, is 2 on 3, and to make $b=0.45x$ to $0.50x$ for unfavorable conditions or structures of great importance. For reinforced concrete walls, in which the weight of earth filling is substituted for the heavier masonry to make up a part of the moment of stability, the base should be somewhat wider, and for ordinary cases the value $b=0.50x$ may be selected for trial and verified or adjusted by either of the methods given above.

If the bank slopes from the top of the wall on the line BN', Fig. 14, the thrusts P_y and P, and consequently the bending moment and overturning moment, will be the same, while the righting moment will be increased by the weight of the material between *IK* and the slope *BN′*. This may be neglected, as it is on the side of safety.

EXAMPLE. — Required a wall to form the side of a coal-pocket 18 ft. high and surcharged at a slope of $\theta = \phi$, the angle of repose, ϕ, being 35° and the weight of the coal 55 lbs. per cubic foot. Cos $\phi = 0.81915$, and $x = 18$ ft. Then

$$P_y = (55 \times 18^2) \div 2 \times 0.81915 = 7{,}300 \text{ lbs.}$$
$$P = P_y \cos \theta = 7{,}300 \times 0.81915 = 5{,}980 \text{ lbs.}$$

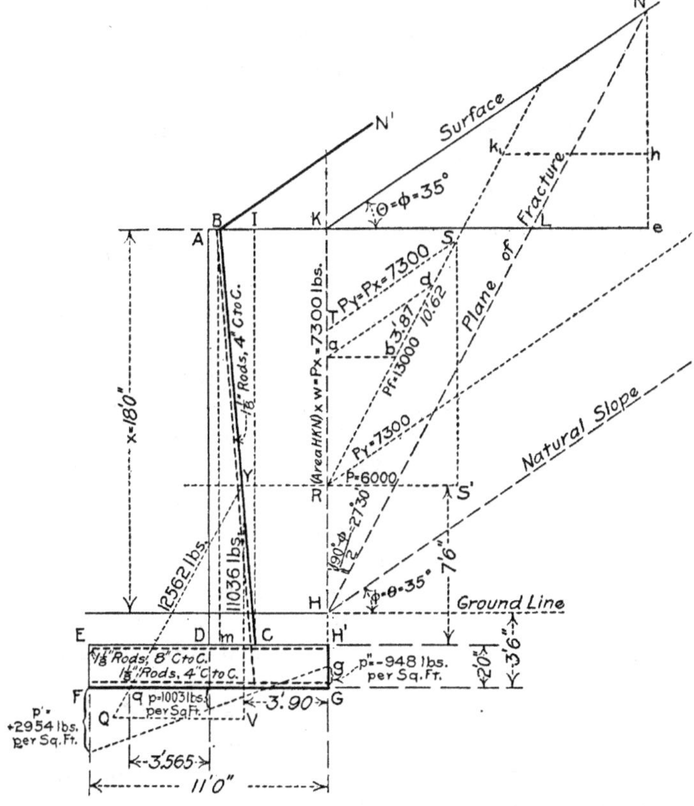

FIG. 15.—Graphical Calculation of Plain Retaining-wall.

Graphically, draw the vertical *HK*, in Fig. 15, through the inner edge of the heel and from *H* draw *HM*, making an angle of 35° with the horizontal. Bisect the angle *KHM* by the line *HN* and draw the slope line *KN* parallel to *HM*. Make $HR = \frac{1}{3}x = 6$ ft., and draw

RS parallel to HN, which will pass through the centers of gravity of the triangles HKL, KLN, and HKN.

Area of HLN	$=9.33 \times 10.33 \div 2=$	48.2
" " HKL	$=9.33 \times 18.00 \div 2=$	84.0
" " $HKN=$		132.2

As a check on the work, lay off $Ha=12=\frac{2}{3}x$ and draw the horizontal cutting RS at b. Lay off $eh=\frac{1}{3}eN$ and draw hk horizontal, cutting RS produced at k. Then will b be the center of gravity of the triangle HKL and k will be the center of gravity of the triangle KLN. Scale the distance bk at 10.62 ft. Then $10.62 \times 48.2 \div 132.2=3.87$ ft. Lay off $bd=3.87$ ft. and d will be the center of gravity of the triangle HKN.

The area HKN multiplied by 55 lbs. will be the weight of the prism 1 ft. long, and is $132.2 \times 55=7{,}260$ lbs. With the decimal scale of forces make RT vertical equal to 7,260 lbs. and draw TS parallel to KN, cutting RS at S. TS will then represent by the scale of forces the amount and direction of the resultant P_y, the center of action of which is through R. Drop a vertical from S, cutting the horizontal through R at S'; RS' will then be the resultant horizontal component of P_y, and scales 6,000 lbs.

Assume the foundation 3.5 ft. below the ground line and try a base 2 ft. thick. The top of the base will then be at EH', or 1.5 ft. below the floor and ground level at H and the lever arm of the resultant horizontal component will be $RH'=7.5$ ft. The bending moment on the wall ABCD at CD will be $6{,}000 \times 7.5=45{,}000$ foot-pounds. Table X gives the safe loads per lineal foot for beams 1 in. thick. Therefore for a span of 10 ft. the load per lineal foot for 1 in. wide that will give a moment of 45,000 foot-pounds for 1 ft. wide will be $w=(8 \times 45{,}000) \div (100 \times 12)=300$ lbs. By Table X we find that a depth h of 22 ins. is required. Make DC equal 24 ins. and $AB=6$ ins. and AD vertical. Try a base 11 ft. long with its center in the plane of the vertical face of the wall AD. Then $b=11$ ft., $ED=5.5$ ft., and $CH'=3.5$ ft. Assume the weight of concrete at 140 lbs. per cubic foot and calculate the weight and center of gravity of the wall, base, and filling over the heel as follows:

$CIKH'$,	19.5 ft. × 3.5 ft. × 55 lbs.	= 3,753 lbs. × 1.75 =	6,567.75
BCI,	19.5 " × 1.5 " ÷ 2 × 55 lbs. =	804 " × 4.00 =	3,216.00
BCm,	19.5 " × 1.5 " ÷ 2 × 140 "	= 2,034 " × 4.50 =	9,153.00
$ABmD$,	19.5 " × 0.5 " × 140 lbs.	= 1,365 " × 5.25 =	7,166.25
$EFGH'$,	2.0 " × 11.0 " × 140 "	= 3,080 " × 5.50 =	16,940.00
		11,036 lbs. × 3.90 =	43,043.00

Lay off $RY=3.9$ ft. and from Y lay off $YV=11,036$ lbs. Lay off VQ equal and parallel to $S'R$, then YQ will be the resultant on the foundation, and its amount can be found by scale or by the equation $YQ=\sqrt{11,036^2+6,000^2}=12,562$.

The resultant will intersect the base at q, where $Gq=3.90+(9.5\times6,000\div11,036)=3.90+5.165=9.065$, and $Dq=9.065-5.5=3.565$ ft. Then

$$p\ =11,036\div1.1=1,003+,$$

and

$$p'=\ 1,003+(11,036\times3.565\div20.17)=+2,954,$$

and

$$p''=\ 1,003-(11,036\times3.565\div20.17)=-\ \ 948,$$

from which it is seen that while the intensity of the uplift at the extreme end of the heel is less than 0.5 ton per square foot, the maximum intensity of pressure at the outer end of the toe is less than 1.5 tons per square foot, which is well within the allowable pressure on any satisfactory foundation. The wall will, therefore, be safe from overturning, and while the value of p' would be somewhat increased if there were nothing to resist the uplift p'', the frictional resistance of the material filled in behind the wall will be sufficient to balance the uplift when the latter is small. In this case the total uplift will be $(11\times948\div3,902)\times948\div2=1,267$ lbs., while a friction of 20 per cent. of P_y is 1,460 lbs. This, however, should not be depended upon to insure stability, and is not in this case, since without any resistance to the uplift, p' would be $11,036\div(11-9.065)\times\frac{3}{2}=3,802$ lbs., or less than two tons per square foot.

The resultant of the reaction on the toe would be $[(1,003\times5.5\times2.75)+(1,951\times5.5\times3.67\div2)]\div[(1,003\times5.5)+(1,951\times2.75)]=3.2$ ft. in front of the plane AD, and its amount is 10,882 lbs. The moment on the toe will then be 10,882 multiplied by 3.2 ft.$=34,822.4$ foot-pounds. If there is no resistance to the negative reaction, it would be nearly $11,036\times3.67=40,502$ foot-pounds. Using the larger value, we find w of Table X by the equation $w=(8\times40,502)\div(100\times12)=270$ lbs. per lineal foot, and Table X, by interpolation, gives 21 ins. as the depth of beam required.

The shear on the plane of CD is equal to P and is 6,000 lbs., which is $6,000\div(24\times12)=21$ lbs. per square inch, nearly, or only about one-third of the safe shearing strength.

The longitudinal shear on the vertical rods where they enter the base is found by equation (58) of Chapter II as follows:

$k = eFA_s v \div pI = 4{,}085{,}000 \div 55{,}400 = 73.6$ lbs. per square inch,

where $e = 20$,

$F = 6{,}000$ lbs.,

$p = 10.6$ sq. ins.,

$A_s = 3$ sq. ins.,

$v = 11.35$ ins.,

and $I = 5{,}527$ in ins.,

the rods being $1\frac{1}{8}$ ins. diameter and 4 ins. center to center. The values of v and I are easily found by equations (18), (19), (20), and (5) of Chapter II. The rods must extend into the base a distance equal to $DC = 24$ ins., as the longitudinal shear must be taken up within a length equal to the depth of the beam. As the longitudinal shear is near the limit of the safe working value of adhesion, the rods should be deformed or some mechanical bond should be used.

If the wall is very long and maximum economy is sought, the calculations may be adjusted for a base 22 ins. thick, but if this is done great care must be used to prevent any possibility of the vertical rods pulling out the base. The length of the heel and toe may also be adjusted so as to make the bending moments the same amount on each and thereby give the most economical proportions.

While the moment on the heel in this case will be much less than on the toe, for practical reasons it will generally be preferable to make the base of uniform thickness throughout.

The reinforcement in the base will consist of $1\frac{1}{8}$ ins. round rods $2\frac{1}{4}$ ins. above the bottom and below the top, spaced 4 ins. on centers in the bottom and 8 ins. on centers in the top, and all running transversely.

The weight of the wall itself will add to the compression per square inch on the concrete at D, caused by the bending moment, an amount proportional to its height. In this case it will be near enough to take this at $18 \times 140 \div 144$, or say 18 lbs. per square inch, which may be neglected. When the wall is very high and extreme accuracy is desired this compressive stress due to the weight of the wall may be considered.

Wall with Counterforts.—Another form of retaining-wall that is particularly well adapted for reinforced concrete is that with counterforts. By spacing the counterforts a considerable distance apart, and making the wall between the counterforts no thicker than required for strength and convenience of construction, very considerable reduction in cost may be effected, especially where the wall is high. In this case the overturning moment and the bending moment produced by the resultant horizontal thrust about an axis in the plane of the top of the base are resisted entirely by the counterforts, while the wall between the counterforts acts as a horizontal beam or slab supported by the counterforts and carrying a load acting horizontally. The load on the slab will decrease from the bottom to zero at the top, but will be uniformly distributed in the direction of the length of the wall. The base, if it has the same width between as at the counterforts, will act as a beam supported at the counterforts and loaded uniformly between them by the upward reaction. This load will be maximum at the outside edge or toe.

If the counterforts are n feet apart center to center, the thrust on the counterforts will be nP, the resultant horizontal thrust P having been found for a unit of length. The moment on the counterfort will be $\frac{1}{3}nPx$.

The wall between the counterforts may be considered in sections of unit height, that is, of x beams each 1 ft. = 12 ins. in breadth. The load per lineal foot on these beams will be $2Pr \div x^2$, where r is the distance from the center of the section or beam considered to the top of the wall.

These loads may be readily obtained graphically by a diagram constructed as shown in Fig. 16, which will be understood without further explanation.

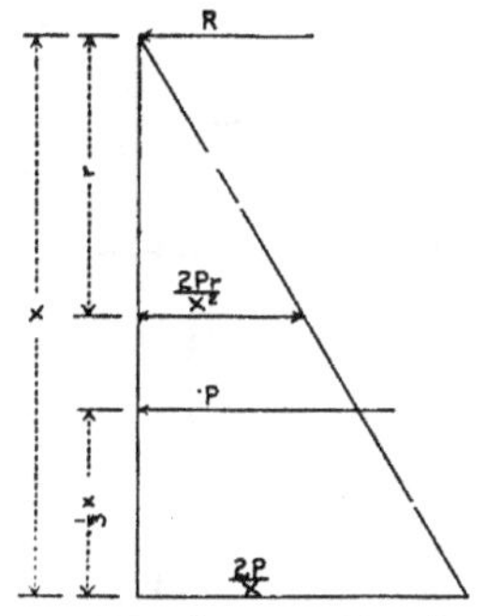

FIG. 16.—Diagram Showing Intensity of Horizontal Thrust at Any Depth Below Top of Wall.

EXAMPLE.—Taking the same data that were used for the previous example, where it was found that $P = 6,000$ lbs., and assuming the counterforts spaced 10 ft. apart on centers, the thrust on each counterfort will be 60,000 lbs. and the bending moment at CD will be 450,000 foot-pounds.

Then by Table X for a span of 16 ft. we find the load per lineal foot, w, for the width of b inches to give an equivalent bending moment must have the value $wb = 14,062.5$ lbs. If b is 23 ins. the load per lineal foot per inch of width will be 611.4 lbs., which corresponds to a depth of beam in Table X of 48 ins. The counterforts may be designed 24 ins. wide with a depth CD of 48 ins. (see Fig. 17).

The wall between the counterforts will then have a clear span of 8 ft., but it will be on the side of safety and possibly better practice to calculate the bending moment for a span of 10 ft., or the distance center to center of the counterforts. Where the maximum economy is desired it would be safe enough to use a span of say 9 ft., or say 1 ft. more than the clear span.

The load per square foot on the lowest 1-foot section will be $2 \times 6{,}000 \times 17.5 \div 324 = 648$ lbs. On the middle section it will be 333 lbs. and on the top section 19 lbs. From Table XII we find that the wall should taper from 10 ins. at the bottom to 4 ins. at the top, making some allowance for the fact that the table gives the safe loadings for the clear span.

The counterforts may taper from 48 ins. at *CD* to 12 ins. at *AB*, and will be reinforced with 5 rods $1\frac{3}{4}$ ins. diameter embedded in the concrete parallel to the inclined surface. To provide for the longitudinal shear they should be deformed to effect a mechanical bond with the concrete and thoroughly anchored in the base.

The reinforcement of the wall between the counterforts will consist of rods placed horizontally about 1 in. to $1\frac{1}{2}$ ins. inside of and parallel with the front face. These rods may all be $\frac{5}{8}$ in. diameter and spaced 3 ins. center to center at the bottom and 9 ins. center to center at the top, or larger rods may be used in the bottom and smaller ones in the top, so spaced as to give an equivalent area of metal. The shear between the intermediate sections of wall and the counterforts should be investigated and reinforcement introduced if necessary, and care should be exercised in construction to insure a good bond.

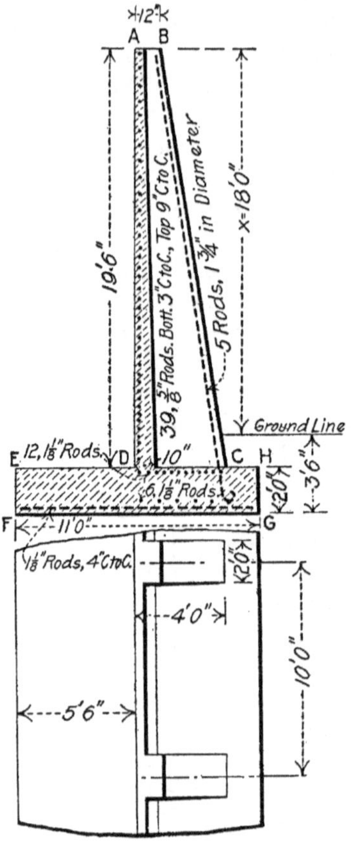

FIG. 17.—Retaining-wall with Counterforts.

In this case as the heel only projects 18 ins. beyond the counterforts, from *C* to *H′*, the transverse rods in the top of the base may be omitted and longitudinal rods in the bottom of the heel substituted. Six rods $1\frac{1}{8}$ ins. in diameter will be sufficient. Longitudinal rods in the top of the base will be required to resist the bending due to the upward reaction of the foundation between the counterforts.

The weight of 10 lin. ft. of the wall with counterforts will be $(1+4)$

$\div 2 \times 2 \times 19.5 = 97.5$ cu. ft.; $(0.33+0.83) \div 2 \times 8 \times 19.5 = 90.5$ cu. ft.; $97.5+90.5=188$ cu. ft.; $188 \times 140 = 26{,}320$ lbs.

The weight of the filling over the heel behind the wall will be $(5.5 \times 10 \times 19.5) - 188 = 884.5$ cu. ft., $884.5 \times 55 = 48{,}647.5$, say 48,650 lbs.

The weight of the base is $2 \times 11 \times 10 \times 140 = 30{,}800$ lbs. and the total weight on the foundation will be $26{,}320+48{,}650+30{,}800=105{,}770$ lbs. or 10,577 lbs. per lineal foot, against 11,036 lbs. in the previous example.

The difference being less than 5 per cent., it will not be necessary to recalculate the resultant pressures on the foundation.

Assuming that the reaction is transferred longitudinally to the counterforts by that part of the width of the base only which is under and covered by the counterforts, and dividing 10,577 by 48 ins., the width of the portion assumed to carry the load, we get 220 lbs. per lineal foot per inch width of beam. From Table X for a span of 10 ft. we find that a beam 19 ins. deep is required, but our base is 24 ins. deep, which gives a good margin of safety. The longitudinal reinforcement in the top of the base will then consist of twelve round rods $1\frac{1}{8}$ ins. in diameter.

The material required to build 10 lin. ft. of the wall with counterforts will be 15.111 cu. yds. of concrete and 2,691.1 lbs. of rods, while the wall without counterforts will require 17.176 cu. yds. of concrete and 3,488.4 lbs. of rods. This shows a saving in favor of the wall with counterforts of 2.065 cu. yds. of concrete and 797.3 lbs. of rods for each 10 ft. of wall. With concrete costing $8 per cubic yard and steel 5 cts. per pound in place, the saving per lineal foot of wall in favor of the design with counterforts would be $5.64, or about 18 per cent. of the total cost of the wall without counterforts.

Braced Retaining-walls.—The vault space under the sidewalks in front of a number of buildings have been enclosed with walls braced at the top by the beams carrying the sidewalks to the framework of the building. One great advantage of this use of reinforced concrete is that considerable valuable space is saved by keeping the wall thin. These braced vault walls have usually been built with beams filled in with concrete, as shown in Fig. 18, but they can be built just as well and more economically of concrete reinforced by rods. The I-beam wall requires too much metal and does not give as effective and economical distribution of the steel as can be attained in a concrete wall reinforced with rods.

The I-beam wall is usually computed on the assumption that the steel beams resist the transverse loading and the concrete acts only as a slab between the beams and as a protection to the latter.

Using the notation of Fig. 16, and letting W = the horizontal thrust on the part of the wall above the section considered, a distance

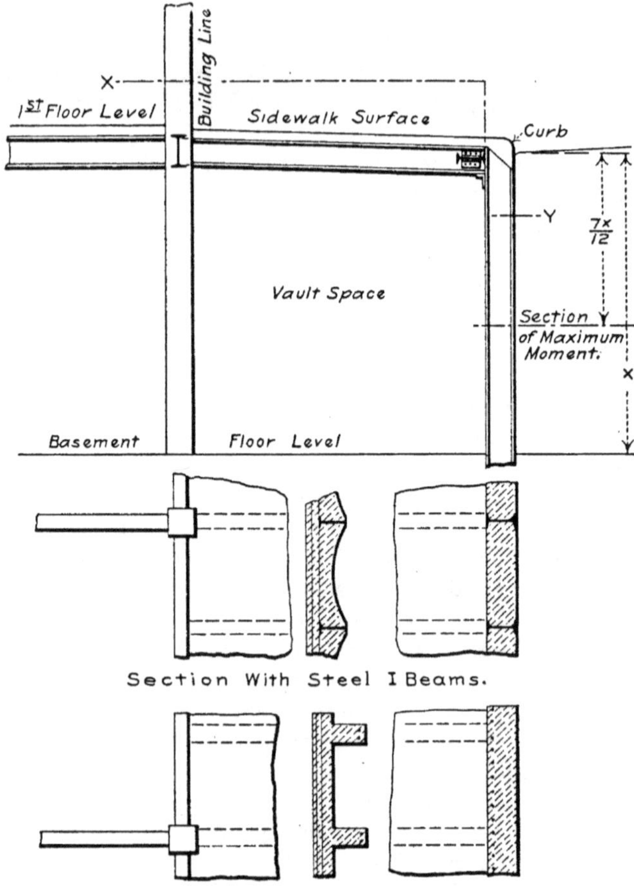

FIG. 18.—Designs for Braced Retaining-walls.

from the top designated by r, and R=the reaction at the top of the wall on the braces or sidewalk beams, we find that

$$M = rR - \tfrac{1}{3}rW, \quad \ldots \ldots \quad (6)$$

$$R = \tfrac{1}{3}P, \quad \ldots \ldots \quad (7)$$

and

$$W = Pr^2 \div x^2. \quad \ldots \ldots \quad (8)$$

Substituting the values of R and W from (7) and (8) in equation (6) we get

$$M = \frac{1}{3}Pr - (Pr^3 \div 3x^2). \quad . \quad . \quad . \quad . \quad . \quad (9)$$

M will be maximum when $r = \frac{7}{12}x$, which substituted in equation (9) gives

$$M = 7.8Px. \quad . \quad . \quad . \quad . \quad . \quad . \quad . \quad . \quad . \quad (10)$$

This is only $2\frac{1}{2}$ per cent. greater than the moment at the center from a uniformly distributed load, $M = Wl \div 8$, and it will be near enough to proportion the beams or slabs for a uniform load equal in total amount to P.

Unless the bracing to the top of the wall is continuous, the top

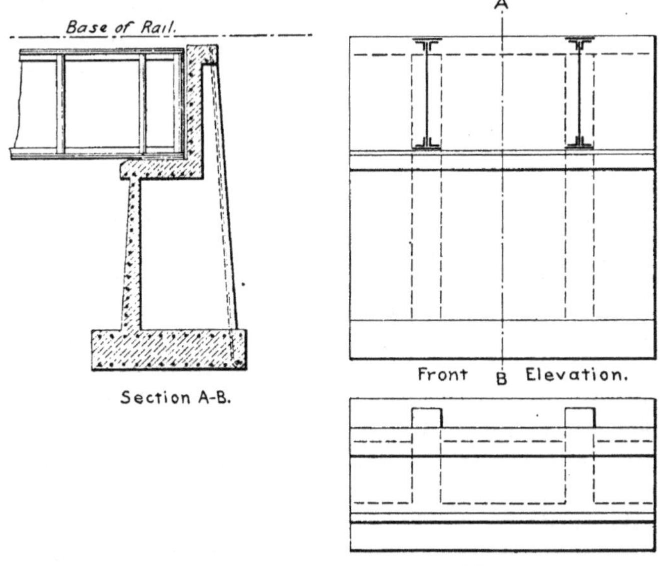

FIG. 19.—Design for Reinforced Concrete Bridge Abutment.

should be supported by a coping designed to resist a uniform load of R per lineal foot acting horizontally, with a span equal to the distance center to center of braces.

Bridge Abutments.—The reinforced concrete retaining-wall with counterforts has been successfuly applied to the construction of bridge abutments, the general form of which is shown in Fig. 19.

The only condition to be provided for which has not been covered in the methods and formulas that have been given are the loads on the bridge seats from the superstructure and moving load. These can best be taken care of by building a counterfort directly under each bridge seat or point of heavy concentrated loading and increasing the section of these counterforts so that the sum of the compression per square inch due to the direct loading and to bending in the outer fibers on the compression side shall not exceed the permissible unit stress of the concrete in compression.

Wing walls may be built at any angle with the face of the abutment, from zero degrees, giving a straight abutment, to 90 degrees, giving a U abutment, and they can be designed by the rules already given for retaining-walls.

If founded on piles, care should be taken that the horizontal component of the thrust in the plane of the foundation is properly provided for. When these points are intelligently considered, an abutment designed according to the methods given above for retaining-walls, and carefully constructed, can be relied upon to give satisfactory results.

DAMS.

Reinforced concrete has been used to some extent for the construction of dams, for which purpose it is often economical. It possesses the distinct advantage that the opening up of cracks can be guarded against by running the steel reinforcement in every direction desired.

Cracks and failures caused by insecure foundations, seepage, etc., will not be taken up here, as the discussion of these points may be found in works on the general subject of dams, the object of this chapter being only to cover some of the more important applications of reinforced concrete and the general methods that may be safely followed in the calculations and design.

Cracks caused by shrinkage in setting and by thermal contraction require special treatment, as upon these two causes depend in large measure the stresses in the steel reinforcement. Experience has shown that concrete walls not reinforced tend to crack and open up, and in the case of one very long wall these cracks appeared at intervals of about 50 ft. For concrete hardening and remaining in air such cracks may be due in about equal measure to shrinkage in hardening and to thermal changes.

The available experimental data bearing on this subject being practically limited to the coefficients of thermal expansion and changes of volume in setting and hardening, the following determination of the proper percentage of steel to use in the cross-section of any given wall to prevent cracks from opening up, under shrinkage and thermal stresses, is not claimed to be more than a reasonable theory, and may require revision in the light of future experiments and experience.

The change of length in specimens of neat cement and concretes during setting and hardening, deduced from a large number of experiments, are summed up by M. Considère as follows:

Neat cement,	hardening in water,	swelled	from	.001	to	.002
" "	" " air,	shrank	"	.0015	"	.002
Concrete poor in cement,	" " water,	swelled	"	.0002	"	.0005
" " " "	" " air,	shrank	"	.0003	"	.0005

Neat cement after hardening for two years in air, on subsequent immersion for three weeks in water swelled 0.00024 part in length. Mortar of one part of cement to three parts of sand, after hardening for fifteen months in water, shrank 0.0005 part in length during two months' exposure in air. From these experiments it appears that neat cements which have hardened in air and are subsequently immersed tend to swell only a fraction of the amount that they shrank during slow setting, and when hardened in water and then exposed in air, the shrinkage would be a fraction of the previous swelling. But with concretes poor in cement the swelling due to immersion after hardening in air or the shrinking caused by exposure in air after hardening in water may equal or exceed the amount of the shrinkage or swelling, respectively, due to the slow setting.

From these results M. Considère recommends that reinforced concrete works built and allowed to harden in air should be kept very wet if they are subsequently to be immersed in whole or in part, in order to reduce to a minimum the internal stresses caused by the swelling of the concrete after it has hardened. He points out that this is more important in the poorer concretes than in the rich ones, and has calculated that the stress on the steel reinforcement due to changes in volumes may be from 5,500 to 8,500 lbs. per square inch. While the changes after hardening are greater in the poorer mortars than in those rich in cement, the changes in volume accompanying slow setting are less in poorer than in the rich mortars. These properties of concrete are of especial importance in the construction of dams, and should be considered in fixing working unit stresses, both for the concrete and for the steel.

Assuming that "concrete poor in cement" is applicable to such mixtures as are generally used for walls, or one of cement to six to nine or more of sand and stone, the average shrinkage of concrete hardening in air may be taken at 0.0004 part per unit of length. While this is several times greater than concrete not reinforced will stand without cracking, reinforced concrete can be stretched from $2\frac{1}{2}$ to

3 times that amount, or from 0.001 to 0.0012 without at least showing any visible cracks. According to M. Considère reinforced concrete under tensile stresses exceeding 200 or 300 lbs. per square inch stretches a considerable amount without any increment of stress in the concrete, or in other words without the concrete doing any appreciable work. If this theory is correct, and it has very able advocates, the reinforcement not only prevents the cracks from opening up, but actually prevents them from occurring. But the stress in the steel reinforcement must at least be equal to the total stress less the ultimate tensile strength of the concrete, and if the concrete should crack from any cause whatever the steel would have to take the entire stress.

It follows that to ensure safety, sufficient steel should be used to take all of the tensile stresses within its elastic limit, which may be taken at 37,000 lbs. per square inch for medium structural steel and at 50,000 lbs. per square inch for high carbon steel. According to the theory of M. Considère, if the concrete does not crack at any point, the stress in the steel would depend only upon the coefficients of shrinkage and thermal contraction and its own modulus or coefficient of elasticity, and would be independent of the ratio of the area of the steel to that of the concrete.

Thermal Stresses.—In Chapter I the rate of change of length for 1° F. was given at 0.00000657 for steel and 0.00000655 for concrete, which are averages of several determinations, the differences between which were many times greater than the difference between these two mean values. This difference of 2 in the last place cannot make a difference of over 100 lbs. per square inch and the probable difference will not exceed 50 lbs. per square inch in the calculated maximum stress in the steel, which is far within the limits of accuracy of our present knowledge of the coefficients.

The range of extreme temperatures throughout the mass of concrete, since the material is a poor conductor of heat, will be materially less than that in the air or water surrounding it, and as the average temperature of the mass is that which will determine the amount of contraction or expansion, it will be safe in most localities to take 70° F. as the maximum range. If the sections used are very thin and exposed on all sides, a somewhat greater range may be used. In severe climates it may be desirable to provide for a considerably greater range of temperatures, but in the tropics thermal stresses may often be neglected with perfect safety.

Then for a range of temperature of 70° F. and the coefficient of 0.00000657 the change of length of an unrestrained body would be 0.0004599 part per unit of length, or say 0.00046 part.

If a wall or other mass of reinforced concrete were rigidly fixed at the ends and subjected to a fall of 70° F. in the average temperature throughout the mass, the thermal stress would be equal to that required to stretch it 0.00046 time its length, which would be 0.00046 time the modulus of elasticity. The thermal stress in the steel would then be $0.00046 \times 29,000,000 = 13,340$ lbs. per square inch. If the modulus of elasticity of the concrete is 3,600,000, the corresponding thermal stress would be $0.00046 \times 3,600,000 = 1,656$ lbs. per square inch, which is from four to eight times its ultimate tensile strength, and if not reinforced it would crack, since its ends are supposed to be rigidly fixed so that it cannot contract. The maximum stress that can come upon the steel is its own thermal stress of 13,340 lbs. per square inch plus the ultimate tensile strength of the concrete. If the latter is taken at 300 lbs. per square inch, the percentage of steel required in the cross-section of the reinforced wall, in order that the elastic limit shall not be exceeded, will be, for steel having an elastic limit of 37,000 lbs., $100 \div [(37,000 - 13,340) \div 300] = 30,000 \div 23,660 = 1.268$ per cent., and for steel with an elastic limit of 50,000 lbs., $30,000 \div 36,660 = 0.818$ per cent. Mr. W. W. Colpitts recommends the use of about one-half of the amount given in the last figure, or 0.6 sq. in. of steel per square foot of wall section.

The intermediate lengths of very long walls in which cracks are prevented from opening up by steel reinforcement are under exactly the same conditions of thermal stress as a wall supposed to be fixed rigidly at the ends, like that considered above.

As the maximum variations in temperature occur at approximate intervals of six months, the properties of concrete at that age should be used.

Shrinkage Stresses.—When concrete sets and hardens in air the shrinkage of 0.0004 part per unit of length will produce tensile stresses in the concrete unless the mass is free to contract. The intermediate lengths of long walls, when not relieved by the opening up of cracks, are stressed exactly as if they were rigidly held at the ends. When not reinforced with steel such walls do crack at intervals the length of which depends upon the amount and character of the friction on the foundation and surrounding material resisting movements, and on the strength of the wall. The greater the friction the shorter will be the intervals between cracks, and the stronger the material in the wall the longer will be the intervals The cracks will occur at the points of least resistance, either where the friction or strength of the wall or both are minimum.

No material is perfectly uniform in strength throughout, and even the best grades of steel when tested in long bars show the points of minimum strength by the reductions or drawing out of the section. A bar of very uniform steel will sometimes show five or six points of reduction of area before it breaks, its edges taking the form of a series of waves. Concrete is far less uniform than steel, and therefore while under the same stress some sections of a wall or other mass may be at the point of failure while other sections with the same area may still have a considerable reserve strength.

The longitudinal steel reinforcement, being fixed or immovable at the ends of the length of wall under consideration, cannot change its total length, and therefore the summation of the stresses in the steel caused by shrinkage of the concrete must be zero. Then at those sections where the concrete is weakest the steel will be in tension and where the concrete is strongest the steel will be in compression (provided the external friction were uniform) and the algebraic sum of these stresses in the steel will be zero.

The maximum tensile stress that can come upon the steel will be equal to the ultimate tensile strength of the concrete and the maximum compressive stress in the steel will be $0.0004 \times 29,000,000 = 11,600$ lbs. per square inch. The per cent. of steel reinforcement required for shrinkage stresses alone, the ultimate tensile strength of the concrete being 300 lbs. per square inch and the elastic limit of the steel 37,000 lbs., will be $30,000 \div 37,000 = 0.81$ per cent. With steel having an elastic limit of 50,000 lbs. it will be $30,000 \div 50,000 = 0.6$ per cent.

Thermal and Shrinkage Stresses Combined.—Under maximum thermal and shrinkage stresses combined it seems probable that the tensile thermal stress in the steel due to its own contraction neutralizes the compressive stresses caused by the shrinkage of the concrete. That the maximum tensile stress in the steel can never exceed its own thermal stress of 13,340 lbs. per square inch plus the ultimate tensile strength of the concrete seems clear because if it were greater the concrete would necessarily fail, or else, according to Considère's theory, it would elongate without doing appreciable work, and in either case it would not deliver any increment of stress to the steel. If this theory is correct, as it appears to be, the correct percentage of steel to use in the cross-section of a wall is the same as given for thermal stresses alone, or 1.27 per cent. for steel having an elastic limit of 37,000 lbs. and 0.82 per cent. for steel with an elastic limit of 50,000 lbs.

It thus appears that high carbon steel is the more economical for the duty of resisting thermal and shrinkage stresses, for which purpose it can be safely used, high ductile qualities not being required. For use in reinforced beams, columns, tanks, etc., the saving in weight by using high steel is not nearly so great, and in fact does not appear to be sufficient to compensate for the loss in reliability due to its inferior ductile qualities.

The basis of the theory given above for reinforcing concrete for thermal and shrinkage stresses is that the steel equalizes the strains between different sections along the length of the structure, and changes the mass from a material very deficient in uniformity as regards strain to an exceedingly uniform material.

Pressure on the Immersed Surface.—The graphical method of finding the pressure on the immersed surface of a dam, the overturning moment and the resultant on the base is illustrated in Fig. 20.

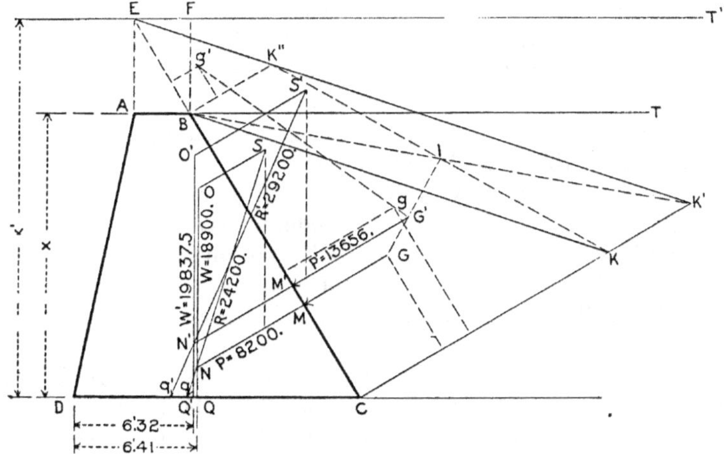

FIG. 20.—Graphical Solution of the Forces Acting on a Dam.

Let *BC* be the immersed surface, *DC* the base, *BT* the surface of the water, and x the depth of the water. Draw *CK* perpendicular to *BC* and equal to x, then will the weight of the water in the prism *BCK*, of a length unity, be the amount of the pressure on the immersed surface one unit in length. This pressure, *P*, will act normal to the surface *BC* and through the center of gravity of the triangle *BCK* at *G* and will intersect *BC* at *M*, *CM* being equal to one-third of *BC*.

Find the center of gravity of the section of the dam *DAB*C and its weight, *W*. From the intersection of *P* with the vertical through the center of gravity of the dam at *N*, lay off *NO* equal to the weight, *W*, by the scale of forces, and draw *OS* parallel to *NG*, making *OS* equal to *P* by the scale of forces. Then the resultant on the base, *R*, will be represented in direction and amount by the line *SN*, and its intersection with the base at *q* will be the center of reactions on the base.

If q is found within the middle third of the base the dam will be in stable equilibrium provided the maximum intensity of pressure on the foundation is not excessive. The intensity of pressure on any part of the base may be found by the method given for retaining-walls, using the vertical component of R acting through q. The intensity of pressure at any point of the immersed surface BC can be readily obtained by the method shown in Fig. 16.

When the immersed surface is vertical the methods given for finding the thrust on retaining-walls are applicable by making ϕ equal to zero.

When the crest is submerged the solution will be as follows. Produce the line BC to intersect the surface of the water at E and lay off CK' equal to x' and perpendicular to EBC. Draw BK'' perpendicular to BC from the crest of the dam and find the center of gravity of $BK''K'C$. This may be done by locating the centers of gravity of the triangles ECK', BCK, and EBK'' at g, G, and g' respectively and that of the parallelogram $BKK'K''$ by the tntersection of the diagonals at I, then the center of gravity of $BK''K'C$ will be at G' where GI intersects $g'g$ produced. Or if on $g'g$ produced, gG' is laid off equal to $g'g$ multiplied by the area of EBK'' and divided by the area of $BK''K'C$, G' will be the center of gravity sought. The area of $BK''K'C$ is the difference of the areas of the triangles ECK' and EBK''.

The center of gravity of a right angle triangle is at the inside corner of a rectangle the sides of which are one-third the length of the sides of the triangle adjacent to the right angle, the rectangle being inscribed in and including the right angle of the triangle.

A force equal to the weight of $BCK'K''$, for a length unity, acting through G' normal to BC at M' will be the resultant pressure P' on a unit of length of the surface BC.

Find the weight W' of a unit of length of the section of the dam and the prism of water $AEFB$, over the crest, and the intersection of the vertical through the common center of gravity with P' at N', and lay off $N'O'$ vertical and equal to W'. Draw $O'S'$ parallel and equal to P', then $S'N'$ will be the resultant, R', intersecting the base at q'.

The values of P, W, R, and the distance DQ and P', W', R', and the distance DQ', given in Fig. 20, were found by assuming the following dimensions and unit weights.

$x = 15$ ft.

$x' = 20$ ft.

The base $DC = 15$ ft.

The batter of $AD = 3$ ft. in its height.

The crest $AB = 3$ ft.

The average weight of the prism of the dam = 140 lbs. per cubic foot and the weight of water = 62.5 lbs. per cubic foot.

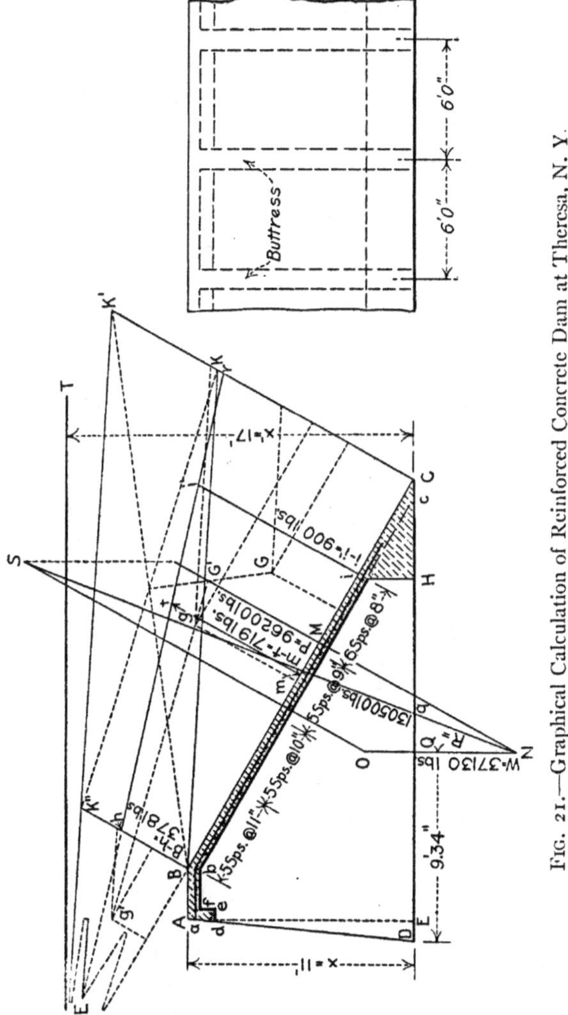

Fig. 21.—Graphical Calculation of Reinforced Concrete Dam at Theresa, N. Y.

For P, W, R, and DQ the water level is at ABT, and for P', W', R', and DQ' the surface of the water is at EFT'.

Example.—The reinforced concrete gravity dam recently built at Theresa, New York, is illustrated in Fig. 21. The highest water runs 6 ft. deep over the crest or about 17 ft. over the point of the tce, making the dam about 11 ft. high. The base is 22 ft. wide inside of the slab and the crest about 2 ft. 6 ins. wide. The dam is constructed with concrete buttresses, not reinforced, 12 ins. thick and spaced 6 ft. center to center, and having a batter of about 1 ft. in their height on the down-stream side. The inclined up-stream side of the buttresses supports a concrete slab 6 ins. thick and reinforced with ¾-in. Thacher rods and expanded metal. The toe is solid concrete with a width of 4 ft. on the base inside of the slab. The crest is strengthened by a reinforced concrete beam on its down-stream edge, which has a section of 6 ins.×8 ins. and two ½-in. Thacher rods. The reinforcing rods in the slab which forms the immersed surface run longitudinally and are spaced 8 ins. apart at the bottom near the toe, the spacing gradually increasing to 11 ins. at the top near the crest.

Assuming the weight of concrete at 140 lbs. per cubic foot, the weight of a section of the dam 6 ft. long, which is the length of the sections supported by each buttress, and the horizontal position of its center of gravity are found as follows, the letters referring to Fig. 21.

		Dimensions, Feet.	Weight per Cu. Ft., Lbs.	Weight.	Lever Arms about D.	Moment.
Buttress,	*DAE*....	11×1÷2	×140	= 770	×⅔	= 513.0
"	*ABFE*...	2.5×11	×140	= 3,850	×2.25	= 8662.5
"	*BFC*	19.5×11÷2	×140	=15,015	×10	=150150.0
Toe,	*HIc*......	4×2¼÷2×5	×140	= 3,150	×19⅓	= 60900.0
Dam,	*BCcb*....	21.75×½×5	×140	= 7,612	×13	= 98962.5
Crest,	*ABba*....	2½×½×5	×140	= 875	×2¼	= 1969.0
Nose,	*adef*.....	½×⅔×5	×140	= 223	×1¼	= 291.0
$W=$..................................				31,505	×(10.2)	=321448.0

321,448 *foot-pounds*÷31,505 *lbs.*=10.2 *ft.* from D to the vertical through the center of gravity.

Adding the weight of the water vertically over AB to the weight of the dam, we have 31,505+5,625=37,130 lbs. This shifts the center of gravity 0.86 ft. toward D, and Q will then be 10.2−0.86=9.34 ft. from D. Find the intersection of P with the vertical through Q at N just as was done in Fig. 20. The weight of $BCK'K''=P$, for a section 1 ft. in length, is $(BK''+CK')\div2\times(BC\times62.5)=(6+17)\div2\times22.3\times62.5=16{,}028$ lbs., and for a section 6 ft. in length to correspond with the unit section of the dam it is 96,200 lbs., nearly. Then laying off $NO=37{,}130$ lbs. and $OS=96{,}200$ lbs., we find $SN=R$

$=130{,}500$ lbs. R intersects the base of the buttress within about 4 ins. of its center, and as the average intensity of pressure on the base, which is founded on rock, is less than 3 tons per square foot the dam has a wide margin of safety against overturning.

The average intensity of pressure per square foot on the surface BC may be found by the method given in Fig. 16, P being taken equal to the weight of the prism ECK', with a length of 1 ft., and $x=EC$, or by the following method.

Taking $BC=22.3$ ft., the mean pressure per square foot will be $16{,}028 \div 22.3 = 719$ lbs., which is the intensity of pressure per square foot at the middle of $BC=m$ in Fig. 21. Erect a normal to BC, bisecting the latter at m, and lay off $mt=719$ lbs. From the apex of the triangle ECK' draw Ehr through the point t, then will the intercepts of lines parallel to mt or CK' between BC and hr be the intensities of pressure per square foot on BC. These are found to be 378. lbs. at B and 900 lbs. at i.

A 6-in. slab with $\frac{5}{8}$-in. round rods 5.7 ins. center to center, according to Table XII is good for 909 lbs. per square foot, with a safety factor of 5, and the area of the steel used, $\frac{3}{4}$-in. round rods, 8 ins. center to center, is about 4 per cent. greater than given in the table. In addition to the rods the slab is reinforced with expanded metal, which gives a further excess of strength, besides binding it together in every direction.

Other Forms of Dams.—That other forms of dams besides that shown in the example illustrated in Fig. 21 may often be advantageously constructed of reinforced concrete is so apparent that they need only to be mentioned. If another reinforced concrete slab is built on the down-stream or exposed faces of the buttresses, it is clear that the dam will consist of a series of pockets or coffers. These coffers may be filled with any cheap material, such as a concrete mixed in the proportions of 1 to 12 or 1 to 15 (1 part cement to 4 or 5 parts of sand and 8 or 10 of stone), or even with riprap stone or earth if the details of the design are well considered, and the added weight of this filling, increasing the resistance to the overturning moment, will permit a considerable reduction in the width of the base. In overflow or submerged dams or waste weirs the slab on the down-stream side, as seen in cross-section, may be given an ogee form or that of any curve desired. The discussion of the best form to use in particular cases will be found in the treatise by Clemens Herschel and other works on the subject.

Where a dam is to be constructed in a narrow gorge or cañon with solid rock walls, and an extreme width of not over 300 to 500 ft., an arched dam may be substituted for a gravity dam. The arched dam will be found to be more economical than the gravity dam up to a length that can be determined by comparative estimates for each case, and as the length is reduced below that at which the cost of the arched and the gravity dams are equal the saving due to the former will increase.

Consider a horizontal section or slice of unit height at the base and find the pressure on the immersed surface both in amount and direction. The stresses may then be computed by the elastic method given in the succeeding chapter on arches, and the section designed accordingly. In the same manner other sections at any convenient heights above the base, up to the crest, can be determined. It will generally be sufficient to compute two or three sections and interpolate between them. The horizontal components of the pressure on the immersed surface can be used for the loads on the divisions of the arch. This will facilitate the work and any error will be on the side of safety. If the immersed surface is vertical, the pressures will of course be horizontal and there will be no error by this assumption. Therefore it will be some advantage to make the immersed surface vertical or nearly so. The load on each division of the arch cannot be taken as acting perpendicular to the chord or span, but must be considered to act on the line of the radius of the extrados at the center of the division, and the load line must be laid off accordingly.

Core Walls.—Reinforced concrete should give very economical and satisfactory results in the construction of core walls for dams built of earth or stone. Such core walls can be built as thin as the practical limitations of construction will permit, as they can be supported at intervals by buttresses on one or both sides. They require reinforcing for shrinkage and thermal stresses only, and as they are generally deeply buried in the earthwork of the dam the thermal changes will be comparatively small and may often be neglected. When properly reinforced they should be found to be as free from cracks and consequent seepage as a core wall of steel plates, and they will not deteriorate or rust out, which cannot be claimed for the steel core. No more difficulty will be encountered in guarding against seepage under the base with the reinforced concrete construction than with any other.

TANKS.

Reinforced concrete has been extensively employed in Europe for the construction of tanks, bins, reservoirs, etc. Tanks for holding water and other fluids should be built with concrete of the maximum density, that is to say rich in cement. Mixtures of 1 to 4½ to 1 to 6 are the best for the purpose, as those richer than 1 to 4½ are liable to show hair cracks. Concrete water-tanks have frequently been lined with asphalt.

The general practice in designing the reinforcement is to consider that all of the tensile stresses are resisted by the steel. The concrete will certainly take a part of these stresses, but the additional margin of safety secured by neglecting the tensile strength of the concrete is obtained at a very small cost.

The main reinforcing rods must be thoroughly connected so as to form perfect hoops, and they should be placed either in the center of the concrete section or between the center and two diameters of the rods from the outer face. In addition to the hoops it will frequently be desirable to insert a few smaller vertical rods in the shell. In place of hoops made of rods or bars, sheets of wire netting or expanded metal may be used, but as by far the greater part of the metal reinforcement is required in the circular direction, rods or bars are more economical.

The intensity of the pressure on a unit of surface at a distance x below the surface of the contained fluid, whose weight per unit of volume is w, is $p=wx$. The tension on a section of the shell, the center of which is at a distance x below the surface of the fluid, and the height of which in the direction of x is unity, assumed to be entirely sustained by the steel, is $F=pD\div 2$, where D is the internal diameter of the tank.

The area of the steel reinforcement required in this section of the shell of a unit height is $A_s=F\div f$, where f is the permissible intensity of the working stress in the steel.

Example.—Let $x=15$ ft., $w=62.5$ lbs. (water), $f=12,000$ lbs., and $D=20$ ft. Then $p=937.5$ lbs., $F=9,375$ lbs., and $A_s=0.781$ sq. in. for 1 ft. height of shell at the section 15 ft. below the surface. Use two bars $2\times\frac{1}{4}$ ins. spaced $7\frac{1}{2}$ ins. center to center$=0.80$ sq. in. per foot height of shell.

Towards the top the spacing may be wider, or smaller sections of steel may be used, or both, thus gradually reducing the section of the steel with the pressure from bottom to top.

The bottom of the tank may be built of reinforced concrete beams and slabs, which can be readily proportioned by one of the methods given in Chapter II, or it may be constructed as a spherical invert. In the latter case proper provision must be made for the stresses in the lower section of the shell and between the invert and the shell.

The stresses in tanks for holding grain and other solids can be determined in a similar manner, finding P in the formulas given in the discussion of retaining-walls, and p by the equation $p=2P\div x$.

COLUMN FOOTINGS AND SPREAD FOUNDATIONS.

An excellent and economical substitute for steel I-beam grillage in foundations is offered by reinforced concrete. The methods, formulas, and examples given in the first part of this chapter for the determination of the base of retaining-walls are exactly applicable to the case of column and wall footings and foundations, and there should be no difficulty in applying them to these cases.

CONDUITS, SEWERS, AND CULVERTS.

The extensive and successful use of concrete without reinforcement for the construction of sewers and culverts, and the favor with which engineers have generally received this material for these uses, places the question of the advantages offered by reinforced concrete for building such works entirely on the ground of economy and increased stability. Generally speaking, the sections in reinforced concrete may be made about one-half of those which are customary with concrete not reinforced, but for sewers and culverts of small cross-section, up to about 8 ft. in diameter, many engineers prefer to adopt minimum sections which are sufficiently strong without any reinforcement.

In some cases, where there is danger of unequal settlement, the use of longitudinal reinforcing bars will prevent failure. It is very difficult to give a rule applicable to all cases for such longitudinal reinforcement, and the amount of metal used will depend largely on the judgment of the designer.

It is customary to build concrete sewers in monolithic lengths of not much over 50 ft. to avoid shrinkage cracks. When longitudinal reinforcement is used, the length of the monolithic sections may be made as long as practical conditions permit. The amount of steel required to prevent shrinkage cracks in long monolithic sections may be determined by the formula given for shrinkage stresses in the section on dams.

For the computation of the stresses, sections, and reinforcement for flat-top sewers and box culverts, the methods given in Chapter II for beams are directly applicable. It will be safe to assume the load equal to the weight of the material and moving loads that can come vertically over the clear span of the slab that forms the roof of the culvert or sewer, as this will give the maximum possible load. It is probable that in many cases the actual weight on the slab will be less than this, and in some conditions very much less, but no safe general rule is known by which the ratio of the actual load to the weight of the material vertically over the roof can be determined. If the fil over the top of the culvert or sewer is not very great something should be added to the weight of the moving loads for impact. The side walls should be treated as abutment retaining-walls for the earth pressure and load from the top.

An advantage that is common to concrete sewers with or without reinforcement is the high resistance that the material offers to attrition. Mr. W. B. Fuller observed in some sewers designed and built by him in Duluth, in which the inverts were constructed of granite blocks with one to one mortar in the joints and in which the velocity was 42 ft. per second, that the granite wore down below the joints, leaving the mortar of the joints projecting enough to make perceptible ripples.

Mr. Fuller states that in his experience he has secured better concrete where reinforcement was used than when it was not, due to the fact that the contractors and workmen take more care when embedded steel is used and that therefore the use of reinforcement has a tendency to produce better workmanship.

The method for determining the stresses in arched culverts and sewers will be given in the chapter on arches.

Reinforced concrete pipe sewers are entirely feasible and may be built either in place or can be cast in sections and laid in place in the same manner as tile. The author has no experimental data on reinforced-concrete pipe under external pressure, but in 1900 he made a series of three tests to destruction on concrete pipe without reinforcement. These pipes had an internal diameter of 30 ins. and were 4 ins. thick, making the external diameter 38 ins. They were 40 ins. long. When embedded in gravel up to the height of their horizontal diameter they carried a load of 8.7 tons distributed on their extrados, without showing any appreciable deflection. It not being practicable to increase the load the gravel was gradually removed from their sides and when raked out to a depth of 6 ins. below their horizontal diameter a deflection of $\frac{1}{64}$ to $\frac{1}{16}$ in. was observed. They collapsed when the gravel had been raked out to a depth of $9\frac{1}{2}$ ins. to 11 ins. below their horizontal diameter. The deduction to be drawn from these tests is the importance of thoroughly embedding the pipe, whether of plain concrete or reinforced concrete. When laid on a solid bed and well backed up to their horizontal diameter with good material, thoroughly tamped, they will carry any load that is liable to come on them in ordinary cases. When not thoroughly backed up they are liable to fail under a very small load.

Conduits under internal pressure should be treated in a way very similar to tanks. If p is the intensity of the internal pressure per square inch, and d the internal diameter of the conduit in inches, the tension on the shell per longitudinal inch is $F=pd\div 2$. Assuming that the entire tensile stress is resisted by the steel reinforcement, as is customary in the desinging of tanks, and that the allowed working stress per square inch in the steel is f, the area of steel required in each longitudinal foot of the conduit will be $A_s=12F\div f$.

In cases where unequal settlement may occur or where it is not practicable to obtain a uniform foundation or bed throughout the length of the culvert, longitudinal reinforcing members may be introduced. The circular reinforcing members (A_s above) should either be thoroughly connected at their ends to form perfect hoops or may be inserted as a continuous spiral. Wire netting or expanded metal may be used if preferred.

Conduits under internal pressure should be made with 1 to 6 concrete in order to give maximum density and avoid seepage.

CHIMNEYS AND STAND-PIPES.

Reinforced concrete is particularly well adapted to the construction of tall chimneys and stand-pipes. For the former purpose it will often effect a saving of 30 or 40 per cent. of the cost of a brick chimney and will often compete in cost with steel chimneys with the advantage as to durability in favor of the reinforced concrete. The following is a simple method for computing the stresses and sections of steel required in a reinforced-concrete chimney.

Compute the overturning moment of wind at the base and the moment of resistance of the steel reinforcement about the center of figure of the base. Then the maximum stress per square inch in the steel on the windward side will be $f_s=M\div S_s$, where S_s is the moment of resistance of the steel.

EXAMPLE.—Required a circular chimney 100 ft. high and 5 ft. internal diameter. The inside wall will be laterally supported by, but not connected with, the outside wall. The load per square inch on the foot of a concrete column of uniform cross-section due to its own weight alone is about 1 lb. per foot of height, which for 100 ft. is 100 lbs. per square inch, or about one-fifth of the safe load. Therefore the inside wall may have a uniform thickness of 4 ins. from bottom to top.

Assuming a 4-in. space between the inner and outer walls and the thickness of the latter to be 8 ins. at the bottom and 4 ins. at the top, we find the external diameter to be 7 ft. 0 in. at the top and 7 ft. 8 ins. at the bottom.

The weight of the exterior wall will be 128,400 lbs. and the wind pressure, at 50 lbs. per square foot on a normal plane, will be 18,325 lbs. on the windward side of the chimney. The center of action of the resultant of the wind pressure will be 49.26 ft. above the base and its moment about the plane of the base 902,506 foot-pounds, of 10,830,072 inch-pounds. The resultant would then intersect the plane of the base at a point over 7 ft. from the axis of the chimney or over 3 ft. outside of the figure of the base, and without reinforcement the chimney would fail.

Assume the principal reinforcing rods, which will be vertical, to be ½ in. thick and arranged in an annular space the inside diameter of which is 83 ins. and the outside diameter 84 ins. Calculate the moment of resistance of an annular shell ½ in. thick of the same diameters as the annular space in which the reinforcing rods are arranged, by the formula $S=0.0982 \times$ [(the cube of the external diameter) minus (the fourth power of the internal diameter divided by the external diameter)]. For this example the value of S_s is found to be 2,723 in inches. Dividing the overturning moment from the wind pressure in inch-pounds by the safe working stress in the steel we have 10,830,072 $\div$12,000=902.5, which is the value of S_s required. Then the section of steel required in each foot of circumference of the shell, measured on the center line of the reinforcement, will be 12 ins. multiplied by ½ in. (or 6 sq. ins.), multiplied by 902.5 and divided by 2,723, which gives 1.99 sq. ins.; say 2 sq. ins., and use 2-in. ×½-in. bars 6 ins. center to center.

The maximum intensity of compression in the concrete on the leeward side at the base is found in a similar way as follows:

$$S_c=0.0982[92^3-(76^4\div 92)]=40,859.$$

For one to nine concrete one month old Table V gives 2,300,000 for the value of E_c, and the value of e will then be 29,000,000 ÷ 2,300,000 = 12.6, from which we get $S=S_c+eS_s=40,859+12.6(2722.6\div 3)$ $=52,294$. Then the maximum intensity of the compression in the concrete due to wind only will be $f_c=M\div S=10,830,072\div 52,294$ $=207.1$ lbs. per square inch. Adding the weight of the shell distributed on the 8-in. annular base we have 207+75=282 lbs. per square inch maximum compression, while Table V shows that 340 lbs. per square inch is permissible with a factor of safety of six.

The rods must enter the base, or foundation, far enough to develop their full strength and the dimensions of the base must be such that the permissible maximum intensity of pressure for the material on which the structure is founded will not be exceeded. In some cases an additional saving in cost may be effected by using reinforcement in the base. The methods and formulas given in full for the bases of retaining-walls may be directly applied to find the stresses and sections for the base of a chimney. These points have been treated in sufficient detail in the section on retaining-walls and do not require further discussion here.

Stand-pipes may be analyzed for overturning moment due to wind in the same manner as given above for chimneys, and in addition to the vertical reinforcement they must be hooped with circular reinforcement. The method for computing the sections and spacing of the hoops or rings is identical with that given for tanks.

CHAPTER V.—TESTS AND DESIGN OF ARCHES.

Evolution and Advantages of Reinforced-concrete Arches.—The idea of reinforcing masonry arches with metal antedates by many years the advent of reinforced concrete as now understood. The voussoirs were banded or clamped together on the extrados by straps of iron, so that the joints could not open up, and tensile stresses on the side of the extrados were thus provided for. But, since the summation of the moments in the ring must be equal to zero, when there are negative moments producing tension on the side of the extrados there must be positive moments, equal in total amount, on the side of the intrados, and no provision was made for the latter. Therefore the advantage derived from the banding was not very great, as failure would take place by the opening up of joints at the intrados.

The general and commercially successful application of reinforced concrete in engineering and architectural works dates from the introduction of the system invented by Jean Monier in Paris, between 1876 and 1878. In Monier's United States patent, which was issued on July 29, 1884, the first and principal claim read as follows:

> As an article of manufacture, an integral element of construction composed of a metal skeleton comprising longitudinal bars or rods and transverse ribs secured by ligatures of metal, and a covering of cement, in which said metal skeleton is embedded, all constructed and arranged substantially as set forth.

From this it appears that the generally accepted belief that Monier's invention was confined to the use of WIRE NETTING for the reinforcing members is not correct, since he distinctly covers any form of metal.

skeleton comprising longitudinal and transverse bars or rods, and that most of the later so-called "systems" differ from Monier's only in detail or in the omission of the transverse ribs. The latter variation reduces the amount of metal required and thus cheapens the cost, but it is doubtful if the expedient is often advisable.

The first Monier arches were constructed with one wire netting only, embedded in the concrete near the intrados. This arrangement does not materially increase the resistance of the ring to the moments producing tensile stresses at the extrados, and as was the case with stone voussoir arches banded with iron on the extrados, a very small increase in strength is derived from the metal reinforcement. Monier arches were afterwards built with two nettings, one near the intrados and the other near the extrados. This arrangement gave much better results, as will be shown in the tables giving the results of experiments in which several Monier arches among a number of others were tested to destruction.

One objection to Monier arches as usually built is that the same quantity of metal is used in the transverse as in the longitudinal direction, whereas in an economical design the principal reinforcement should be in the latter direction, the transverse metal members either being much smaller in section or entirely omitted. Another objection that has been advanced is that it is difficult to embed a wire netting with 2-in. to 3-in. meshes near the surface of ordinary concrete and that consequently mortar or a very fine and expensive gravel concrete has generally been used. It is claimed by some that, unless very carefully done, the tamping is liable to displace or pull on the netting and thus disturb concrete that has already taken its first set. In 1894 there had been at least 200 Monier arches built in Europe.

Rolled shapes were first used for reinforcing concrete arches by R. Wünsch of Buda-Pesth, Hungary, in 1884. The arrangement of the steel members, which may be angles, tees, beams. or channels, either rolled or built, is clearly shown in Fig. 22. Between 1884 and 1894 Mr. Wünsch built six bridges of various spans from 13 ft. 4 ins. to 55 ft. 8 ins., the rise varying from 16 ins. to 45 ins. It will be noticed that these were all very flat arches. An arrangement of or variation from the Wünsch arch, which is believed to be original, is shown in Fig. 23. This permits the use of a curved instead of horizontal extrados, and thus gives to the system a wider range of application. The author does not know of any Wünsch arches in which this arrangement has been used, but there is no reason why it should not be entirely successful. With the addition of transverse steel members of lighter section, so as to tie the entire arch together in all directions and resist the stresses due to shrinkage and the thrust of the earth filling on the spandrel walls, it would be as effective as any sys-

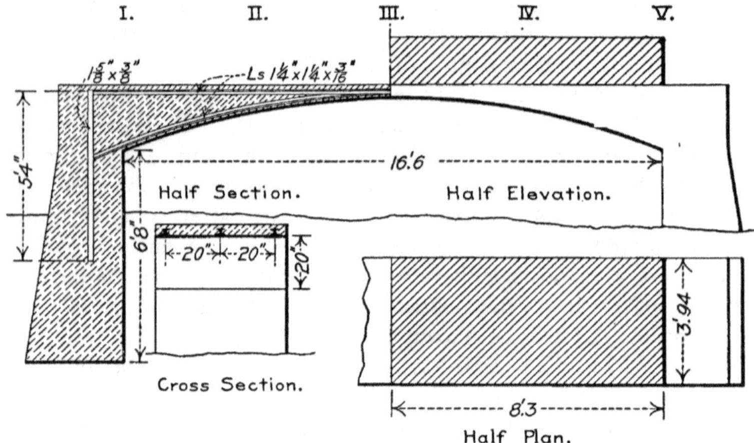

FIG. 22.—Wünsch Arch, Showing Manner of Loading and Points at which Deflections were Measured in Test.

tem yet devised for reinforcing concrete arches, and in some respects superior to those in general use.

The next system of reinforced concrete arches that attracted atten-

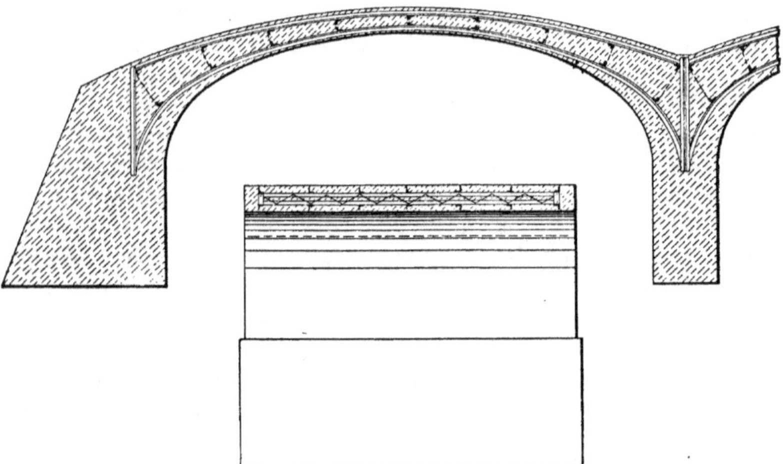

FIG. 23.—Buel-Wünsch System of Reinforcing Concrete Arches.

tion was that of Prof. J. Melan, which was patented in the United States on Sept. 12, 1893, and was described in the principal claim as

"A vault or arch consisting of abutments, beams, or girders, arched ribs rigidly connected with said abutments, beams, or girders, and a filling of concrete or the like between said ribs, substantially as described."

The earlier plans for Melan arches showed the "beam or girder" as described embedded in the concrete of the abutment and with the arched ribs riveted or bolted thereto. The function of the transverse reinforcement at the skew back was to prevent shrinkage cracks from opening up longitudinally along the soffit, caused by the shrinkage of the concrete in the ring, the older concrete of the abutment having already shrunk. M. Considère clearly points out the danger of shrinkage cracks occurring when fresh concrete is laid on that in which the shrinkage has already taken place in whole or in part, and advises that reinforcing members be introduced to prevent them. In most of the Melan arches built in America these "beams or girders" have been omitted and longitudinal cracks have occurred in some of them. Such cracks, however, do not usually indicate any danger of failure, but they are unsightly and may be the occasion of alarm to the public.

The Melan system of arch construction was introduced in the United States by Herr Fr. Von Emperger about ten years ago, and for several years it was the only one used in this country. About 34 Melan arch spans had been built in America in 1889, and in 1904 the number is considerably over 100. The length of these spans varies from 12 feet to 125 feet, and one of 136 feet is under contract.

Herr Von Emperger afterwards patented a system of reinforcement in which the rolled or built-up arched ribs of the Melan system were replaced by bars of metal embedded in the concrete near and parallel to the extrados and intrados, the bars being connected by lacing, tie bars, or struts, but the system has not come into general use.

The successful introduction of reinforced concrete arch bridges in the United States is due to the efforts of Mr. Edwin Thacher more than to those of any other one man. Mr. Thacher was the first engineer in America to design these arches according to the elastic theory and it was he who first worked out a comparatively simple method of summation by which this method of computing the stresses can readily be applied in practical work.

Almost at the beginning of his work in reinforced concrete Mr. Thacher perceived that the shearing stresses in an arch were practically negligible and that therefore the web of the Melan ribs could safely be omitted with a considerable saving in metal. He secured a patent embodying this principle, and at the present time more Thacher arches are being built in America than any other one kind.

The advantages gained by reinforcing concrete arches with steel are:

(1) In long spans the thickness of the ring can be made very much less, thus reducing the dead load and effecting a saving of 50 per cent. or more in the cost of the ring.

(2) The load being reduced, the thrust is correspondingly less, and the abutments may be very much smaller.

(3) Reinforced with steel, the working unit stresses in concrete can be considerably increased with perfect safety.

(4) The danger of cracks occurring from any cause can be prevented by the proper distribution of the reinforcing metal.

(5) Arch rings of such forms and proportions that the bending moments would render them impossible with any ordinary masonry can be, and have been, constructed of reinforced concrete both easily and successfully.

(6) Much longer spans are practical for any given set of conditions than could be seriously connsidered with any other class of masonry.

Tests of Reinforced Concrete Arches.—The following extracts from the reports of the extensive tests made in Austria ten or twelve years ago are taken from the "Engineering Magazine," no complete translation having been made:

The tables herewith presented are a part of the report of the Arch-tests Committee of the Austrian Society of Engineers and Architects. The report covers: (1) tests of 17 floor arches with spans of 4 ft. 4.3 ins., 8 ft. 10.2 ins., and 13 ft. 3.6 ins., representing all kinds of floor construction; (2) tests of two culverts of 32.8 ft. span and $\frac{1}{10}$ rise; (3) tests of four bridges of 75 ft. span and $\frac{1}{5}$ rise; (4) an exhaustive series of tests to determine the strength and elasticity of all materials used in the above arches; (5) a theoretical calculation based on the results attained; (6) conclusion from the result in regard to theory and construction. All the tests were made under the strict supervision of the most prominent engineers to insure conformity to usual practice, while the theoretical work was so divided up that each part fell to an expert. A full record of deflection and other details is given.

The conclusion derived from results indicated in Tables XIX and XX are: (1) concrete arches (1:4) show in both cases adequate dimensions; (2) the Monier arch, with one wire net, has failed to show the superiority claimed for it, which is the more significant because the Monier arch has been, up to date, the most favored type of fire-proof construction in Central Europe.

The report disapproves the use of a concrete floor with the Monier netting, as it needs 1.85 times more concrete and gives only a trifling increase in strength, so that dirt filling, which is more economical, is preferable. Referring to the fact that both concrete arches, with or

TABLE XIX.—TESTS OF ARCHES WITH SPANS OF 8 FEET 10.2 INCHES.

No.	Load per Square Foot, Pounds.	CONCRETE ARCH. Thickness, 3⅜ ins. Rise, 9.5 ins. Dead Load, 3,140 lbs.		MONIER ARCH. Thickness, 1 15/16 ins. Rise, 10¼ ins. Dead Load, 2,460 lbs.		MONIER ARCH, CONCRETE FLOOR. Thickness, 2⅛ ins. Rise, 10¼ ins. Dead Load, 5,420 lbs.	
		Movement of Arch.		Movement of Arch.		Movement of Arch.	
		Vertical.	Lateral.	Vertical.	Lateral.	Vertical.	Lateral.
1	102	.04	.008	.04	.01	.03	.004
2	204	.09	.015	.08	.02	.07	.008
3	307	.17	.04	.16	.04	.13	.02
4	409	.24	.07	.24	.06	.19	.03
5	512	.33	.11	.34	.08	.28	.05
6	614	.42	.16	.43	.11	.37	.07
7	716	.53	.20	.55	.13	.43	.10
8	819	.64	.25	.67	.17	.56	.14
9	921	.79	.31	.83	.22	.71	.18
10	1,024	.95	.38	1.01	.28	.86	.23
11	1,126					1.03	.29
12	1,331					1.21	.35

Remarks.—At No. 2, for the concrete arch, a horizontal fissure in the unloaded half occurred. On No. 6, for the same arch, a crack at the abutment occurred. This arch totally failed at 1,109 lbs. per square foot, one-half of the arch being loaded. At No. 10, for the 1 15/16 ins. Monier arch, failure occurred with 1,217 lbs. per square foot, one-half of the arch being loaded. At No. 10 of the 2⅛ ins. Monier arch, a longitudinal crack appeared. This arch failed with eccentric load of 1,320 lbs. per square foot.

without wire netting, have shown almost the same behavior, the report further says:

Compared with a plain concrete arch, a concrete arch with one wire netting has shown but little more carrying capacity, so that the difference cannot have any importance to the ordinary building practice.

It is also concluded that a little thicker concrete arch may replace it.

To understand this condemnation of the Monier arch, it must be considered that, besides the wire netting, it is composed of a more expensive concrete, to wit, mortar 1:3. Engineers have for a long time recommended Monier arches with two wire nettings for large spans, as it is evident that wire is needed on both sides of the arch and not in the center, but it is nearly impossible to use two nettings in a small span. This point has been fully justified by the test of the Monier arches with two nettings of a span of 75 ft. and 15 ft. rise, which may be well to insert here, so as not to make a faulty impression. The failure occurred under a one-sided load.

Material.	Thickness.		Breaking Load per Square Foot.
	Center.	Haunches.	
Brick.	2 ft.	3 ft. 6 ins.	590 lbs.
Stone.	2 "	3 " 6 "	655 "
Concrete.	2 " 3 ins.	2 " 3 "	735 "
Monier, 2 nets.	1 " 1 "	2 "	1,300 "

The conclusion of the greatest importance deduced from this series of experiments is that the elastic theory was entirely sustained, the actual deflections agreeing with the computations.

TABLE XX.—TESTS OF ARCHES WITH SPANS OF 13 FEET 3.6 INCHES.

No.	Load, Lbs. per Square Foot.	MONIER ARCH. Thickness, $1\frac{15}{16}$ ins. Rise, $15\frac{1}{2}$ ins. Dead load of arch, 2,215 lbs. Dead load of filling, 4,160 lbs.		CONCRETE ARCH. Thickness, $3\frac{15}{16}$ ins. Rise, $15\frac{7}{8}$ ins. Dead load of arch 4,420 lbs. Dead load of filling, 4,560 lbs.		MELAN ARCH, SPAN 13′ 11″. Thickness, $3\frac{1}{8}$ ins. Rise, 11.4 ins. Dead load of arch, 4,860 lbs. Dead load of filling, 1,549 lbs.	
		Movement of Arch.		Movement of Arch.		Movement of Arch.	
		Vertical.	Lateral.	Vertical.	Lateral.	Vertical.	Lateral.
1	102	.02	.01	.01	.0	.05	.0
2	205	.05	.02	.03	.004	.09	.004
3	307	.10	.04	.07	.02	.14	.02
4	409	.15	.07	.11	.04	.19	.04
5	512	.24	.11	.17	.07	.21	.07
6	614	.33	.16	.21	.14	.25	.08
7	717	.40	.23	.23	.18	.42	.12
8	819	.77	.31			.54	.15
		1.08	.39				
9	1,024					.58	.11
10	1,028					.67	.09
11	1,412					.80	.05
12						.30	.02

Remarks.—The $1\frac{15}{16}$-in. Monier arch at No. 4 developed a crack at the abutment, and at No. 5 a second crack in the unloaded half. At No. 8 the movement indicated took place under 822 lbs., and at No. 9 the arch failed after $2\frac{1}{2}$ hours, under 895 lbs. per square foot, including dead load. The concrete arch developed a crack at No. 4, throughout the entire length of the arch, at the abutment. The movement indicated at No. 7 for the concrete arch took place under 662 lbs. This arch failed under 812 lbs. per square foot, including dead load. The Melan arch of $3\frac{1}{8}$ in. thickness developed at test No. 10 a hair crack on the under side of the arch about 11 ft. 8 ins. long, visible near the crown; this crack did not widen later after a pause of 18 hours.

The value of E_c, as determined by tests made in connection with these experiments, is very much less than has been customarily assumed, and was first stated by Prof. Boek, who had charge of this division of the work, at only 750,000. This value was used by Herr

Von Emperger, but since the publication of the full report, and after a more careful study of the results therein given, Mr. Thacher concludes that the value of $E_c = 1,400,000$ is warranted for concrete in large masses and that small specimens give a higher value.

As E_c has such a wide range for varying conditions, it should be fixed by special experiments for very important works, so that the conditions of the experiments will coincide as far as possible with those of the actual work. A high value of E_c assumed in calculations gives proportionately high stresses in the concrete and low stresses in the steel, and since the steel in an arch ring of reinforced concrete cannot be stressed up to an amount anywhere near its safe working unit stress without exceeding the permissible stress in the concrete, for safety in calculating the stresses E_c should be assumed at its maximum probable value. On the other hand, if E_c is assumed at its minimum value there will be some saving in the amount of steel required, but this will not be very great compared with the entire cost of the structure. In the large masses of the rings of arch bridges the value of E_c will certainly be much less than in ordinary beams, slabs, and test specimens, and the available data on the subject fully warrants the assumption that $e = E_s \div E_c = 20$.

Table XXI gives the results of the experiments made with a Wünsch arch by the Hungarian government in 1891. The arch had a clear span of 16 ft. 6 ins. and a rise of 20 inches. It was designed for a load of 80 lbs. per square foot in addition to its own weight. Fig. 22 shows this test arch in plan, elevation, and sections, and gives the sizes and arrangement of the reinforcing metal.

The points at which the deflections given in the table were measured are indicated on the drawing. There was no measurable horizontal movement of the abutment, but this may have been due to the straight horizontal extrados.

TABLE XXI.—RESULTS OF TESTS OF WÜNSCH ARCH.

Total Load, Pounds.	Load per Square Foot, Pounds.	Deflection at Point		
		II. Inches.	III. Inches.	IV. Inches.
2,844	88	....	....	0.04
5,049	105	....	0.04	0.04
8,356	258	....	0.16	0.16*
11,673	360	0.04	0.27	0.27†
16,082	470	....	0.59	0.55‡
18,287	565	0.75	2.64	1.97§
33,101	1,010	....	15.75	

* Between Sections I, II, and IV, V fine cracks.
† Crack unchanged.
‡ Crack open to 0.08.
§ Cracks increase slowly until the breakdown.

Table XXII gives the results of some experiments made in New York by Mr George Hill in 1894 and 1895. Numbers 76 and 77 were arched steel T and I beams, respectively, without any concrete, and 78 to 87, inclusive, were Melan arches. The latter were about ninety days old when tested.

TABLE XXII.—RESULTS OF TEST OF MELAN ARCHES MADE IN 1894 AND 1895.

No.	Span.	Rise.	Width.	Areas.		Loads.			Deflection.
				Total.	Loaded.	Total.	Per Sq. Ft. Area.	Per Sq. Ft. Loaded.	
	Ft.	Ins.	Ft.	Sq. Ft.	Sq. Ft.	Lbs.	Lbs.	Lbs.	Ins.
76	6	7.00	0.25	......	......	14,000	....	....	1.62
77	6	6.66	0.20	......	......	14,000	....	....	0.87
78	6	6.97	6.00	36	3.16	4,150	115	131	0
79	6	6.97	3.00	18	3.24	5,320	296	1,642	0
80	6	6.97	6.00	36	9.90	12,945	360	1,310	0
81	6	8.53	4.00	24	1.00	40,000	1,667	40,000	0.43
82	6	7.28	3.00	18	1.25	55,000	3,055	44,000	0.47
83	6	7.28	3.00	18	1.25	58,750	3,260	47,000	0.59
84	7	7.52	6.00	36	1.00	55,000	1,528	55,000	0.58
85	6	7.28	3.00	18	1.50	31,250	1,735	20,850	0.49
86	.....	.....	......	......	......	32,500	....	....	0.52
87	6	7.28	6.00	36	1.00	70,000	1,945	70,000	0.10

No.	Thickness.	Material.	Areas in Compression.		Tie-rods, Tension.	Ultimate Shear.	Horizontal Thrust.		
			Steel.	Concrete.			Total.	Per Square Inch of Material. Steel.	Per Square Inch of Material. Concrete.
	Ins.		Sq. Ins.	Sq. Ins.	Lbs.	Lbs.	Lbs.	Lbs.	
76	3	Steel T	1.95	......	84,250	53,000	36,000	18,450	
77	4	Steel I	2.05	......	84,250	53,000	37,800	18,450	
78	4	Reinforced concrete	6.15	212	168,500	106,000	5,360		
79	4	"	4.10	106	84,250	53,000	13,740		
80	4	"	6.15	212	168,500	106,000	33,400		
81	3	"	1.95	70.5	168,500	106,000	80,450	18,450	880
82	4	"	4.10	106	168,500	53,000	135,800	36,900	933
83	4	"	4.10	106	168,500	53,000	145,000	36,900	1,019
84	4	"	2.05	106	168,500	106,000	153,800	18,450	1,277
85	4	"	......	106	84,250	53,000	87,900	......	839
86	....	"	......	......	....	....	....	......	....
87	4	"	2.05	106	168,500	106,000	180,800	18,450	1,530

Note.—In the calculations here made, the ultimate tensile strength of the tie-rods is assumed to be 50,000 lbs. per square inch. The ultimate shearing strength of the rivets in single shear is taken to be 30,000 lbs. per square inch.

Approximate Method for Computing the Stresses in an Arch Ring of Reinforced Concrete.—The following short method of calculating the stresses in reinforced-concrete arches was given by Herr Von Emperger, who stated that it gives results that compare well with the exact method of the elastic theory:

Let the dead load in pounds per square foot $=w'$;
" live " " " " " " $=w$;
" span in feet $=l$;
" rise in inches $=r$;
" thickness of concrete in inches $=d$, this generally equal to height of beams;
" distance between beams in inches $=a$; and further:

	Concrete.		Steel.	
Modulus of elasticity	E_c	750,000	E_s	30,000,000
Section	A_c	ad	A_s	
Moment of inertia	I_c	$\frac{ad^3}{12}$	I_s	
Moment of resistance	S_c	$\frac{ad^2}{6}$	S_s	
Coefficient of distribution	α		β	

$$\alpha+\beta=1.$$

The stresses in both materials, or, what is practically the same, the total loads W, can be assumed distributed, $W=W_c+W_s$, if it answer the following proportion:

$$\frac{W_c}{W_s}=\frac{E_cI_c}{E_sI_s}=\frac{d^3a}{480I_s}=\gamma. \quad . \quad . \quad . \quad . \quad . \quad . \quad . \quad . \quad (1)$$

This put in,

$$W=W_c+W_s=\alpha W+\beta W=\frac{\gamma}{1+\gamma}W+\frac{1}{1+\gamma}W,$$

or

$$\alpha=\frac{\gamma}{1+\gamma} \text{ and } \beta=\frac{1}{1+\gamma}. \quad . \quad . \quad . \quad . \quad . \quad . \quad (2)$$

The span and rise being given, we have to assume the size of beam d and its distance a. The proportion $\frac{a}{d}$ is limited by the necessity that the concrete body has to act as a main and also as a cross beam. It carries $W_c=\alpha(w'+w)$ as a main girder and transmits $W_s=\beta(w'+w)$ to the iron beam.

With reference to the elastic theory, in cases where spans in two vertical plans occur, by reducing the allowed stress in concrete to 42 lbs. per square inch we get

$$\frac{a}{d}\leqq\sqrt{\frac{\frac{4}{3}\times 42\times 144}{\beta(w'+w)}}.$$

$$\leqq\sqrt{\frac{8064}{\beta(w'+w)}}. \quad . \quad . \quad . \quad . \quad . \quad . \quad . \quad . \quad (3)$$

If we find the a of our choice is not too large, we proceed to get the stresses in the main girders. A force, H, which is not centrally applied to a section, A, produces consequently a moment, M, with the following stresses in the outer fibers:

$$f=\frac{H}{A}\pm\frac{M}{S}.$$

We have only to use our signs tabulated above and we find the stress per square inch in the concrete,

$$f_c=\alpha\left(\frac{H}{d}\pm 6\frac{M}{d^2}\right), \quad . \quad . \quad . \quad . \quad . \quad . \quad . \quad . \quad (4)$$

and in the steel beam,

$$f_s=\beta\left(\frac{H}{A_s}\pm\frac{M}{S_s}\right)a. \quad . \quad . \quad . \quad . \quad . \quad . \quad . \quad . \quad (5)$$

It can be seen at a glance that by proper choice of α and β, which mainly depend on a and I_s, the tension in the concrete or the pressure in the steel can be reduced to its proper limits.

To get the highest tension and pressure, we must consider, as previously stated, a half-loaded span.

The horizontal thrust is equal to

$$H=\frac{1}{8}\left(w'+\frac{w}{2}\right)\frac{l^2}{r_1}. \quad . \quad . \quad . \quad . \quad . \quad . \quad . \quad . \quad (6)$$

r_1 we call the rectified rise, and is found by equation (7).

$$r_1=r+\frac{15}{16}\frac{d^3a+12I_s}{(da+A_s)r}. \quad . \quad . \quad . \quad . \quad . \quad . \quad . \quad . \quad (7)$$

The moment in a section $\frac{3}{16}l$ from the center is

$$-M=\frac{9}{1024}wl^2+\frac{1}{3}H(r_1-r). \quad . \quad . \quad . \quad . \quad . \quad . \quad . \quad (8)$$

For a full-loaded span

$$H=\tfrac{1}{8}(w'+w)\frac{l^2}{r_1}. \qquad (9)$$

The moment at the crown is

$$M=\tfrac{1}{3}H(r_1-r). \qquad (10)$$

The moment at the abutment is

$$-M=\tfrac{2}{3}H(r_1-r). \qquad (11)$$

The last three equations are of use so far as the spandrels be considered, while for ordinary use the equations (1) to (8) will be sufficient.

For the abutment itself we have to compute H from equation (9), and its weight, and to see that the resultant keeps within the inner third of the base; and further, that the foundation has a proper area for the total weight.

Compared with other arch calculations this is extremely simple and correct enough for practice as a comparison with an exact method, since we have introduced the moment of inertia and the section in equation (1).

While a number of the earlier Melan arches built in America were designed according to the above approximate method, the Concrete-Steel Engineering Company, successors to the Melan Arch Construction Company, now use the more exact elastic method exclusively. The approximate method is safe enough for all ordinary spans, say up to 100 ft., but it is not economical. Its chief value now is in the preparation of preliminary estimates and in getting an approximate thickness of ring to assume for use in the calculation by the elastic method, so that in correcting the thickness of the ring by the latter method the dead load will not be materially altered. It cannot be recommended for exact or important work, and when a sufficient number of computations by the elastic method have been accumulated, approximations can be interpolated from them more satisfactorily than they can be calculated by the approximate method.

Style or Shape of the Arch Ring.—Herr Von Emperger assumed the critical point of the ring to be at a distance of three-sixteenths of the span from the center and proportioned the ring at that point by his equation (8). He then gave the ring a uniform depth for a length of three-eighths of the span at the center by making that part of the extrados parallel to and concentric with the intrados. From the points three-sixteenths of the span each side of the center to the skewbacks the lines of the extrados were tangent to the curve of the central part, the depth of the ring thus gradually increasing from those points to the skew-backs (see Fig. 24). This rule only holds good for rings in which the intrados is segmental, which was the case in nearly all of Herr Von Emperger's designs. With a ring of this shape, when the span is short or the ratio of the rise to the span relatively large, all the stresses can be properly provided for without great difficulty or much waste of material, but in long-span flat arches the moment in the ring near the springing becomes so great, the pressure line sometimes passing entirely outside of the ring, that the proper adjustment of the sections is by no means easy and the amount of material required is excessive.

The first reinforced-concrete arches in America to be designed strictly according to the elastic method were those in the Topeka

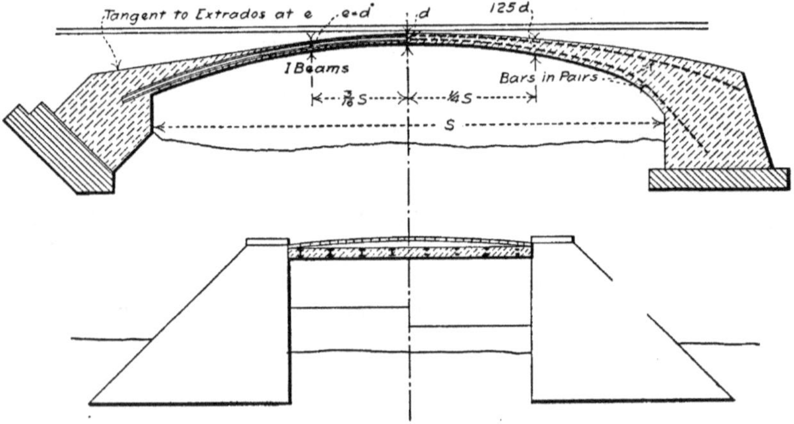

Melan Arch with Von Emperger Style of Ring. Thacher's Arch and Style of Ring.

FIG. 24.

bridge. After determining the lines of pressure, Mr. Thacher introduced sharp curves in the intrados at the ends and used a considerably larger radius for the extrados in order to fit the shape of the ring to the pressure line and avoid moments producing excessive tension in the concrete.

In his succeeding designs Mr. Thacher has generally used three and rarely five centered curves for the intrados and a length of radius for the extrados that gives a depth of ring at the two quarter points, distant one-quarter of the span from the center, of about one and a quarter times the depth at the crown (see Fig. 24). This saves some material at the crown over Herr Von Emperger's form of ring.

In an article in the *Engineering News* of September 21, 1899, Mr. Thacher says:

An economical concrete-steel arch is usually neither a segment nor a parabola, and the moment of inertia at the end is frequently thirty times as great at the center, which is very important to consider. In late designs for concrete-steel arches the writer (Mr. Thacher) has endeavored to have the line of pressure lie within the middle third of the arch ring, and *have no tension on the concrete from dead or live load.* By a due amount of patience this can usually be done, not only as regards dead and live loads, but temperature stresses as well. The writer, however, has never succeeded in doing this economically with a segmental arch. He prefers and generally uses with solid spandrel filling a three-centered arch, as it is less troublesome to construct, and, he thinks, is more pleasing to the eye than a five-centered or elliptical arch, THOUGH THE LATTER (ELLIPTICAL) WILL FREQUENTLY TAKE LESS MATERIAL. Arches with solid spandrel filling should be flat at the center and sharper at the ends, approaching an ELLIPSE; while arches with open spandrel spaces should be sharp at the center and flatter at the ends, approaching a parabola, or, which is better, sharp at the ends and center and flat at the haunches.

The author's experience, which includes the design of a number of elliptical arches, several of which have been built, agrees with Mr. Thacher's on the point that elliptical arches are economical, but differs from his in that they have not proved to be more difficult to construct than three-centered arches.

It seems almost unreasonable to say that a three-centered curve, with its abrupt change of radius, is more pleasing to the eye than an elliptical curve, when we consider that the fairest curve of all, the circle, presents the appearance of an ellipse from every point of view except those on a line perpendicular to its plane at the center.

In very many cases, if not in all, with a semiellipse for the intrados and a segmental extrados, the neutral axis can be made to coincide with the pressure line as nearly as is possible with any other curves. The arches in Prof. Burr's design No. 3 for the "Memorial Bridge" were designed on these lines, and the pressure line, with the live load on one-half of the span only, lay so close to the neutral surface that the moments at all intermediate points were almost negligible.

Whether or not the elliptical, parabolic or three-centered curve is the best for a solid reinforced-concrete arch, it is certain that a segmental curve does not come anywhere near what is required by the true line of pressure.

Before leaving the subject of the shape of arch rings it may be well to give a brief description of the principal styles and the uses to which they are best adapted.

The original arch, at least in Europe, is the semicircular, or Roman, illustrated in Fig. 25. Where the available height permits of its use it is economical, because the horizontal thrust is less than in any flatter arch, and it can often be employed with good effect.

While the pointed arch is said to have been used much earlier in Oriental countries, it did not appear in Europe until about the fifth or sixth century A.D. There are four principal forms of the Gothic arch, namely, the equilateral, lancet, depressed, and four-centered or Tudor arches, which are shown in Figs. 26, 27, 28, and 29, respectively. The pointed arches are rarely used in bridges, but are particularly

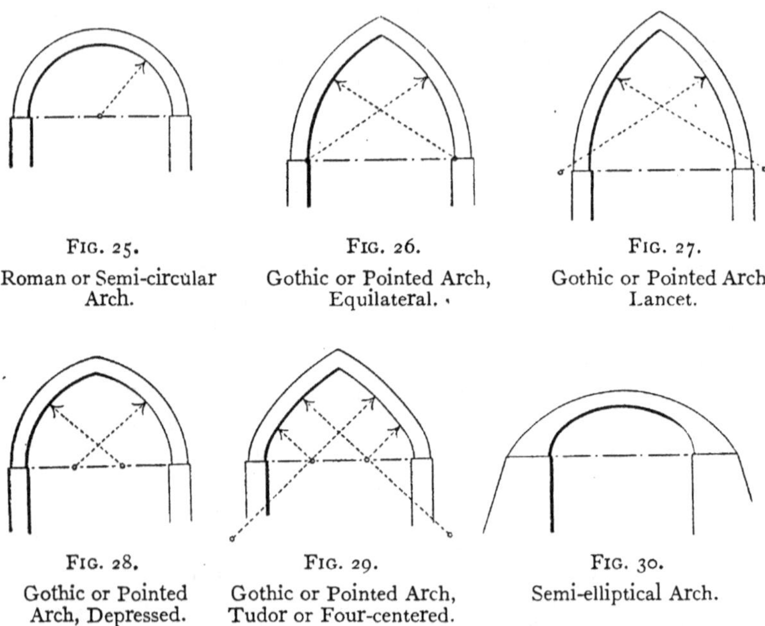

FIG. 25. Roman or Semi-circular Arch.

FIG. 26. Gothic or Pointed Arch, Equilateral.

FIG. 27. Gothic or Pointed Arch, Lancet.

FIG. 28. Gothic or Pointed Arch, Depressed.

FIG. 29. Gothic or Pointed Arch, Tudor or Four-centered.

FIG. 30. Semi-elliptical Arch.

adapted to cases where there is a heavy load directly over the center of the span.

Fig. 30 shows a flat arch in which the intrados is a semiellipse and the extrados segmental.

Two-hinged and three-hinged arches are illustrated in their theoretical form in Figs. 31 and 32. These are segmental arches, and properly so, and, in the opinion of the author, whenever it is necessary to build segmental-arch bridges of any considerable span they should have either two or three hinges. There can be no moment at the hinges except that due to friction in the hinges, and therefore the depth at the springing, instead of being much greater than other parts of the ring, need only be sufficient to transmit the thrust to the hinge.

Hinged arches have been quite extensively used in Europe with reinforced concrete as well as with stone voussoir and plain concrete rings, with the object of preventing cracks due to the movement of abutments or piers. The American practice, however, is to expend a little more on the foundations to make them secure, which can generally be done,

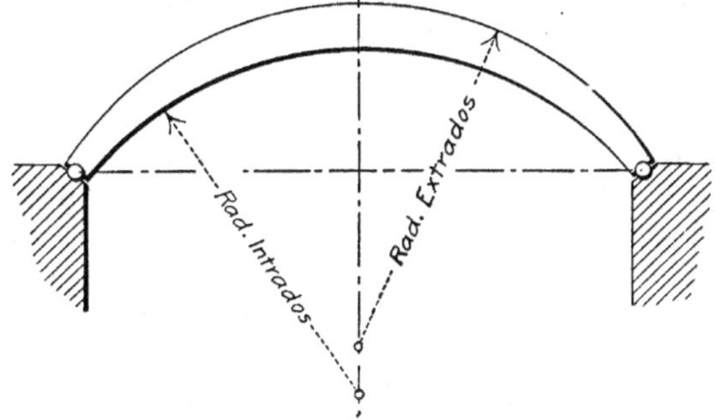

FIG. 31.—Two-hinged Segmental Arch.

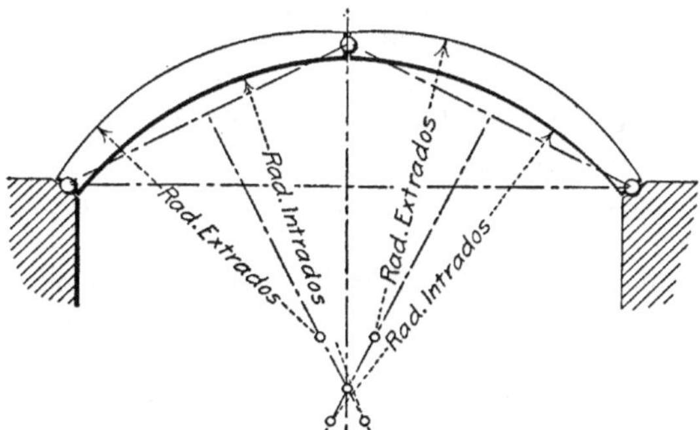

FIG. 32.—Three-hinged Segmental Arch.

and avoid the complication of hinges. Hinges reduce rigidity and increase deflections.

In both the Roman and Gothic arches the intrados is tangent to the verticals at the springing lines. This feature is preserved even in the flat Tudor arch. In a flat arch ring without hinges the intrados must be tangent to the vertical as in the semielliptical ring, or nearly so, as in the three-centered or multi-centered curves or heavy moments, and generally tension will exist in the ends of the ring near and at the skew-backs. The semielliptical ring not only economically fulfills the statical requirements, but has more points in common with the classic Roman semicircular ring than any other flat arch. It is claimed however, by many architects that the elliptical curve if too flat is not pleasing. An arch of 55 ft. span and 11 ft. rise, built from a design by the author, had a semielliptical intrados in which the minor axis was $\frac{4}{10}$ the major axis. The elliptical arches in the Zanesville *Y* bridge are much flatter than this, but they are not full semiellipses. The elliptical arches in the Memorial Bridge competition had a rise of one-fourth the span; that is to say, the minor axis was one-half the major axis.

The general practice of European designers of voussoir arches has been to increase the radial depth of the ring from the crown to the skew-back. This is commonly done in segmental rings by making the radial depth of the joints divided by the secant of the angle between the joint radius and the vertical constant throughout the ring. This agrees fairly well with the theoretical shape of ring required by the moments and thrusts; that is to say, the theoretical ring of a solid arch always increases in depth from the crown to the springing.

An approximate depth of the ring at the crown for reinforced-concrete arches may be found by the formula $K=.0075(S+10V)$, where K= the depth of the crown, S= the span, and V= the rise of the intrados, all in feet. This should only be used to find the trial ring and weights for use in the elastic method, and only then when more reliable data are not available. It will, however, generally give dimensions so close that the assumed dead load will be near enough to the final weights of the adjusted ring to make a recalculation unnecessary.

The formula given in Trautwine's Pocket-book is not correct and cannot be relied upon. It does not conform to the principle governing the pressure line in arch rings, that the greater the rise (or the less the radius, R) the greater the moments, and the flatter the arch (or the greater the radius) the less will be the moments, other things being equal. But Trautwine's formula, $K=0.2+\frac{1}{4}\sqrt{R+0.5S}$, makes the depth of ring vary with the radius. Except in rare cases of very long spans or very flat arches the question of direct crushing due to thrust does not come up, or at least is not the determining factor in fixing the depth of the ring. The amount of the bending moment is generally the chief factor to be considered in designing an arch ring.

The Elastic Theory.—The elastic theory, which, since its verification by the Austrian experiments, may be termed the law of flexure in solid arches, affords an exact method for computing the stresses in an arch of any form or style and of constant or variable cross-sections that is fixed at the skew-backs and continuous at the crown. It is also applicable to arches hinged at the ends only, for which case the solution is very simple if the friction in the hinges is neglected. The theory holds good for any material or combination of materials that is homogeneous throughout the length of the ring from skew-back to skew-back. Even for stone voussoir arches the method of the elastic theory gives more accurate results than any other that has been published, provided the line of pressure does not pass outside of the middle third and that the joints are thin, laid in good Portland-cement mortar, and allowed to harden before the centers are struck.

Briefly, the method consists in finding the true equilibrium polygon for any given arch and system of loading and its true position in the arch ring. The conditions which when fulfilled give these results are (1) that the tangents to the neutral line of the ring at the springing are fixed in direction; (2) that the span is invariable; and (3) that the vertical deflection of one springing line with respect to the other is zero. Expressed mathematically these conditions are:

$$\Sigma\frac{Ms}{EI}=0;$$

$$\Sigma\frac{Mys}{EI}=0;$$

and

$$\Sigma\frac{Mxs}{EI}=0;$$

in which s is the length of a small division or segment of the ring measured on the neutral line and y and x are respectively the ordinate and abscissa of the center of s with the origen at one springing. For a reinforced-concrete ring these equations of condition take the form

$$\Sigma\frac{Ms}{E_c(I_c+eI_s)}=0;$$

$$\Sigma\frac{Mys}{E_c(I_c+eI_s)}=0;$$

and

$$\Sigma\frac{Mxs}{E_c(I_c+eI_s)}=0.$$

Let H = the horizontal thrust, which is constant throughout the ring for any given loading;

t = the intercept of the ordinate between the neutral axis and the line of pressure;

$M=Ht$;

and $e=E_s\div E_c$, practically constant throughout.

Then if we make $s \div (I_c + eI_s)$ constant, by construction, the term $\frac{Hs}{E_c(I_c + eI_s)}$ may be placed outside of the sign of summation and we have

$$\Sigma t = 0, \quad \Sigma ty = 0, \quad \text{and} \quad \Sigma tx = 0.$$

In most cases it will be sufficiently exact to assume that $I_c \div I_s$ is constant throughout the ring, and therefore that making $s \div I_c$ constant fulfills the condition.

Since $I_c = b'h^3 \div 12$, where b' is the thickness of the slice of the ring considered, usually 12 inches (1 foot), and h is the radial depth of the ring, we may construct $s \div h^3$ constant.

While the elastic theory is exact, the approximations introduced in its practical application are not mathematically so. But it is a well-known fact that approximations or assumptions are introduced in every analysis of structural work, either in the methods of computation or in fixing the data. Considering the entire problem, beginning with the assumption of the loading and resulting in the final determination of the intensity of the stresses, the degree of accuracy secured by this method is as high if not higher than those in general use for computing the stresses and sections of the most simple elements of engineering and architectural structures.

Theoretically the ring should be divided into an infinite number of segments, making s infinitely small; therefore the larger the number of segments into which the ring is divided the more accurate will be the result.

When the variations of the values of $I_c \div I_s$ for different parts of the ring are greater in percentage than the probable error in the assumption of the loads or than the maximum permissible error in the final stresses, $s \div (I_c + eI_s)$ must be made constant.

In important cases of sufficient magnitude the value of E_c may be predetermined by experiments carefully conducted to represent the local conditions, which should then be closely followed in the execution of the work.

The conjugate pressure of the earth-fill always acts to increase stability, and may be neglected or it may be avoided by constructing the arch with hollow spandrels. With very short spans and with very flat arches, unless very long, hollow spandrels are of course not practicable. But great refinement in computing the stresses or rigid economy in the design of the ring are not so important in very short spans, and in very flat arches the conjugate pressure is so small that its effect may be neglected, as it is always on the side of safety. The flatter the arch the less will be the amount of the conjugate pressure.

Open Spandrels.—With hollow or open spandrels the roadway is supported by floor arches, beams, or slabs carried by transverse walls or columns. When columns are used they may be built of masonry, steel, or reinforced concrete, and so disposed as to favorably distribute the loads on the ring.

The architectural treatment of open spandrel construction is usually in the arcade style, and properly so, although the colonnade style may sometimes give satisfactory results. In many cases the effect is far better than can be secured by any treatment of the large blank spaces of spandrel walls, and greater diversity in design is permitted.

A few hollow spandrel arches have been built with longitudinal walls, giving the appearance of solid spandrels. When the longitudinal spaces are spanned by floor arches it is customary to use tie-rods to resist the outward thrust on the outside walls. The drainage of the enclosed spaces requires as much attention as in solid-filled spandrels, and the conjugate pressure of the earth fill is replaced by that of the walls, both of which are avoided when transverse walls or columns are used. In solid spandrel arches of reinforced concrete, the ring has a tendency to deflect more than the spandrel walls on account of the greater depth of the latter, and in a number of cases, notably in the Juanadias Bridge in Porto Rico, wide cracks have opened up between the ring and the spandrel walls. The spandrel wall then becomes an independent arch unsupported by the ring. This may be prevented by constructing the walls with slip joints or by tying the ring to the walls with vertical rods. Neither remedy is entirely satisfactory, but the second is particularly objectionable because it changes the stresses throughout the structure and makes them difficult, if not impossible, to analyze. This objection applies, but in a lesser degree, to the longitudinal walls of a hollow spandrel.

Open spandrels with transverse walls or columns offer the following additional advantages. The dead load and consequently the stresses are reduced, thus requiring less material both in the ring and the abutments, and reducing the cost of the structure by a considerable amount. The ring is loaded at definite points, which may be made to correspond with the points of loading assumed in the calculations. The lateral stresses due to the thrust of the earth-fill are entirely eliminated. Water cannot collect in the spandrel spaces. The extrados is open for inspection, and the ring can be examined on all sides at any time. Any kind of pavement can be laid as soon at the ring and spandrel supports are completed without waiting for the fill to settle. With open spandrels the ring may be built in two or more ribs, with far greater facility and economy than with solid-filled spandrels.

The Ribbed Ring.—By constructing the arch ring in several parallel ribs from 25 to 35 per cent. of the cost of the ring can be saved. The consequent reduction in weight will also permit a further saving to be made in the material and cost of the abutments. By giving the ribs a radial depth slightly greater than would be used for a ring of the full width of the bridge, a more economical distribution of material results, and the moment of inertia of a section of the ring is increased. In bridges with very wide roadways carrying trolley tracks, the ribs supporting the heavier loads of the tracks, etc., may be made deeper or thicker than the others, and by thus proportioning each rib to its particular duty the maximum of economy is attained.

The modified Wünsch arch illustrated in Fig. 23 is particularly well adapted to the construction of arch rings in parallel ribs. The transverse latticed struts, which may be built of very light material, should extend over the full width of the ring and be enclosed in concrete between the ribs to form reinforced-concrete transverse struts or beams. The author has worked out two complete designs on this line, using open spandrels with reinforced-concrete columns and ribs of variable depth proportioned to the loads. While neither of these has been built, very satisfactory details were obtained without any unusual difficulties and the greatest economy resulted without the sacrifice of any essentials.

Abutments.—While the elastic theory is based on the assumption that the abutments are rigidly immovable, the effect of a slight horizontal movement on the stresses may be computed by precisely the same method as will be given for thermal stresses, the application of which will readily be seen.

It can be shown that the elastic yielding of the material of the abutments does not produce an appreciable effect on the stresses on account of their relatively large section and moment of inertia as compared with the section of the ring.

Care should be taken to keep the angle between the resultant thrust and the perpendicular to the bed of each layer or course of concrete inside of the angle of repose, or say 30 degrees. If founded in soft material on piles, about half of the latter should be driven at an angle with the vertical of not less than 20 degrees, and if practicable 30 degrees.

Thermal Stresses.—If D is the deflection of the crown due to the change of length of the ring with changes of temperature, and H the corresponding horizontal thrust, then

$$H=\frac{DE_c(I_c+eI_s)}{s\Sigma xy}. \quad . \quad . \quad . \quad . \quad . \quad . \quad . \quad . \quad (12)$$

By tabulating the values of xy for all the segments, s, of the ring from one springing to the other the solution is extremely simple, since $\frac{E_c(I_c+eI_s)}{s}$ is constant.

The following is another method, the demonstration of which is given by Prof. Cain:

Let l = the span of the neutral axis;

c = change of temperature of the ring in degrees Fahrenheit;
r = rate of expansion per degree Fahrenheit;
$m=(\Sigma_0^n y)\div n$;
n = number of divisions or segments of the ring, s;

then

$$H=\frac{E_c lrc}{\Sigma y^2-m\Sigma y}\frac{I_c-eI_s}{s}. \quad . \quad . \quad . \quad . \quad . \quad . \quad . \quad (13)$$

The summations of y^2 and y may be taken for half the span and multiplied by two.

The thrust on any section will be the component of H perpendicular to that section, and the bending moment will be

$$M=H(y-m). \quad . \quad . \quad . \quad . \quad . \quad . \quad . \quad . \quad (14)$$

A full demonstration of the elastic theory and its practical application is given in "Theory of Steel-Concrete Arches," by Prof. William Cain, and also in "Theory of Solid and Braced Elastic Arches," by the same author (Van Nostrand's Science Series Nos. 42 and 48). The method used in the examples that will be given is essentially that of Prof. Cain, with the construction of the deflection polygons from Prof. Burr's "Stresses in Bridge and Roof Trusses." The method of finding the summations by tabulation is due to Mr. Thacher. The modifications introduced in the second and third examples by the author reduce the number of scaling operations on the diagram and give greater accuracy. Further reference may be had to "Researches in Graphical Statics," by Prof. H. T. Eddy.

Having found the thrusts and bending moments, the intensities of the stresses in all parts of the concrete and of the steel reinforcement are easily computed by the following formulas, published in the *Engineering News* of Sept. 21, 1899, by Mr. Edwin Thacher. The notation is mostly the same as that used in Chapter II, but to facilitate reference is given here in full.

E_c = coefficient or modulus of elasticity of concrete.
E_s = " " " " " " steel.
e = $E_s \div E_c$.
A_c = area of section of concrete one inch wide, square inches.
A_s = " " " " steel in width b, " "
a = " " steel per inch width = $A_s \div b$, " "
I_c = moment of inertia of concrete, A_c, about the neutral axis of the combination.
I_s = moment of inertia of steel, a, about the neutral axis of the combination.
f_c = intensity of stress in the concrete.
f_s = " " " " " steel.
h = depth of concrete in inches.
h' = " " steel " "
u = distance from neutral axis of the combination to outer fiber of the concrete, inches.
v = distance from neutral axis of the combination to outer fiber of steel, inches.
T = thrust on section one inch wide, pounds.
M = bending moment on section one inch wide, foot-pounds.
P = pressure on line of pressure on section one inch wide, pounds.
t' = distance from neutral axis of combination to line of pressure in inches, taken normal to line of pressure.
k = distance from center of gravity of steel to bottom of concrete, inches.
b = distance from center to center of steel members in the direction of the width of the arch, inches.

For sections in which the steel reinforcement is symmetrically disposed about the center of gravity of the concrete,

$$f_c = \frac{T}{A_c + ea} \pm \frac{6hM}{I_c + eI_s}. \qquad (15)$$

$$f_s = \frac{eT}{A_c + ea} \pm \frac{6eh'M}{I_c + eI_s}. \qquad (16)$$

For sections in which the steel is not symmetrically disposed about the center of gravity of the concrete,

$$f_c = \frac{T}{A_c + ea} \pm \frac{Pt'u}{I_c + eI_s} \qquad (17)$$

or .

$$f_c=\frac{T}{A_c+ea}\pm\frac{12uM}{I_c+eI_s}. \qquad (18)$$

$$f_s=\frac{eT}{A_c+ea}\pm\frac{ePt'v}{I_c+eI_s}. \qquad (19)$$

or

$$f_s=\frac{eT}{A_c+ea}\pm\frac{12evM}{I_c+eI_s}. \qquad (20)$$

Ht may be written for Pt' in (17) and (19).

The height of the neutral axis of the combination above the soffit of the arch or bottom of the concrete is equal to

$$\left(\frac{A_ch}{2}+eak\right)\div(A_c+ea). \qquad (21)$$

The section of the steel reinforcement should not be less than one square inch for each square foot of the cross-section of the concrete at the crown, or make A_s equal to or greater than $hb\div 144$, h being the depth at the crown.

Some engineers proportion the steel to take the entire bending moment in the ring without assistance from the concrete and not exceeding the limit of elasticity of the steel. Such a distribution of stresses cannot occur unless the concrete fails, and in a properly designed ring in which the line of pressure is everywhere within the middle third the precaution is not necessary. To comply with this condition, the limit of elasticity of the steel being 36,000 pounds per square inch,

$$I_s\geqq Mv\div 3{,}000, \qquad (22)$$

and for sections in which the steel is symmetrically disposed about the center of gravity of the concrete

$$a\geqq M\div 1{,}500h' \qquad (23)$$

or

$$A_s\geqq Mb\div 1{,}500h'. \qquad (24)$$

EXAMPLE 1.—This example shows the determination of the stresses and sections in detail for an arch for a highway or city bridge, to consist of two duplicate spans with multi-centered intrados, approximating the semiellipse.

Fig. 33 is the graphic diagram showing the line of pressure and thrusts. The clear span of the arches, $L=42'.5$, and the elevation of the street grade is just 4 feet above high water. The elevations of the profile and the limiting points on the arches are as follows:

Elevation of grade of roadway +275.40 ft.
" " top of arch ring at crown +274.15 "
" " intrados " " +273.40 "
" " high water +271.40 "
" " foundation +255.00 "
" " center line or neutral axis at crown +273.775 ft.
" " " " " " " " springing +268.525 "

Rise $= r$ (or vers sine) of neutral axis ($=\frac{1}{8}$ of the span, nearly) . . . . 5.25 ft.

$$\text{Radius of center line} = R = \frac{r^2 + (L^2 \div 4)}{2r} = \frac{5.25^2 + (42.5^2 \div 4)}{2 \times 5.25} = 45.63 \text{ ft.}$$

The ring was proportioned and subdivided into 10 divisions or segments, "s," so as to make $s \div h^3$ constant, as given in Table XXIII, the correct spacing being obtained on the third trial.

TABLE XXIII.

Point.	First Trial.			Second Trial.			Third Trial.*		
	h	h^3	s	h	h^3	s	h	h^3	s
1	1.60	4.10	7.20	1.56	3.80	9.00	1.40	2.74	12.00
2	1.20	1.73	3.04	1.10	1.33	3.15	.94	.83	3.64
3	1.06	1.19	2.10	.97	.91	2.16	.82	.55	2.41
4	.97	.91	1.60	.89	.70	1.66	.78	.47	2.06
5	.90	.73	1.30	.83	.57	1.35	.77	.46	2.01

* Correct.

The loads are then computed by Table XXIV for each division "s," and for greater accuracy in scaling them on the load line the summation of the weights from the center to each springing are given. The live load is placed over half the span, to give maximum or nearly maximum moments.

TABLE XXIV.—LOADS.

Points.	1	2	3	4	5	6	7	8	9	10
Ring.	2,520	520	300	250	230	230	250	300	520	2,520
Spandrel filling.	3,810	630	340	270	250	250	270	360	680	4,150
Pavement.	1,100	360	240	205	200	200	205	240	360	1,100
Total dead load.	7,430	1,510	880	725	680					
Live load.	.....	....	...	....	....	200	205	240	360	1,120
Total, loaded side.	......	....	.....	......	.....	880	930	1,140	1,920	8,890
Summation loads *	11,225	3,795	2,285	1,405	680	880	1,810	2,950	4,870	13,760

* For convenience in plotting load line.

The following operations have reference to Table XXV and the diagram, Fig. 33:

Assume a pole and construct the trial equilibrium polygon. Scale ordinates to center line of arch and to equilibrium polygon at points of loading or centers of divisions "*s*," and set down results in columns 2 and 3 of Table XXV. Scale horizontal distance of points from V' and set down results in column 4. Column 5 = (column 3) × (column 4) and are moments of ordinates $= Mx$. The true closing line for trial polygon must make $\Sigma M = 0$ and $\Sigma Mx = 0$. (ΣMy must also be equal to 0.) Trial closing line for equilibrium polygon may be taken anywhere, but it is convenient to take it so as to make $\Sigma M = 0$, which is done by making $vn = v'n' = (\Sigma$ column 3) ÷ (number of points) = 83.99 ÷ 10 = 8.399. In like manner the closing line for arch $K_1 - K_2$ is found by dividing (Σ column 2) by the number of points, which gives the ordinates $K_1V = K_2V' = 44.36 \div 10 = 4.436$.

To make $Mx = 0$: Draw diagonal nv' in parallelogram $nn'v'v$, making the triangles nvv' and $nn'v'$. Find the ordinates (or more properly the intercepts of the triangles). This can be most easily done by setting a slide rule to $vn \div$ span × (column 4) = 8.399 ÷ 42.5 × (column 4), and set down the results in column 6. Only the figures for triangles nvv' are shown, the other triangle, $nn'v'$, is just the same except that it is reversed and the figures opposite point 1 would be opposite point 10, etc. Only one triangle is required when the trial closing line is drawn parallel to $v - v'$, as in the present case, since they are equal.

Multiply column 6 by column 4 and set down the products (or moments) in column 7. Lay off the distance from the center to T and T' on the diagram horizontally, making the distance from the

$$\text{center} = \frac{\Sigma \text{column } 7}{\Sigma \text{column } 6} - \frac{\text{span}}{2} = \frac{1034.30}{42.02} - \frac{42.5}{2} = 3.36.$$

The vertical through these points will pass through the centers of gravity of the intercepts of the two triangles.

Next take the distance $\dfrac{\Sigma \text{column } 5}{\Sigma \text{column } 3} - \dfrac{\text{span}}{2} = \dfrac{1803.93}{83.89} - 21.25 = 0.23$ ft. and lay it off horizontally left from the center, erect a perpendicular through the point and mark it R. This is the center of gravity of the ordinates of the equilibrium polygon. On this line lay off $r - r' = vn = v'n' = 8.40$ (or to be exact equals 8.399). Assume pole "o" anywhere and draw $o - r'$ from pole to point r' and note its intersection t' with the perpendicular through the center of gravity of the triangle; draw $o - r$ and then draw $r' - t$ parallel to or. Connect t and t' and draw $o - s$ parallel to $t - t'$; scale off distances $r - s$ and $s - r'$.

Now locate *true* closing line by the end ordinates found by these equations:

$nvv' : rs :: vn : vm$ and $nn'v' : sr' :: v'n' : v'm'$;
42.0 : 44.5 :: 8.4 : 8.9 " 42.0 : 39.5 :: 8.4 : 7.9.

Lay off $vm=8.9$ and $v'm'=7.9$; connect $m-m'$, which is the true closing line, with $\Sigma Mx=0$. Take off the intercepts mpm' between $m-m'$ and polygon and set down in column 8, noting proper sign plus and minus. Do the same with the intercepts between arch "a" and K_1-K_2, K_1aK_2 and set down with proper sign in column 9.

Take the sum of intercepts of arch, column 9, for points 1 and 10 and set down in column 10 opposite point 1. Do the same for points 2 and 9, 3 and 8, 4 and 7, and 5 and 6. In the same way construct column 11 from column 8. Lay off values of column 10 on horizontal line through crown of arch to the right of center as a load line and construct polygon, getting the closing line $l=2.03$. By the same process construct the deflection polygon for equilibrium polygon on the left side of center line, getting $l'=2.75$.

Correct column 8 by multiplying values in column 8 by $\frac{l}{l'}=\frac{2.03}{2.75}$ and set down results in column 12. ΣMy is now$=0$.

Then take the difference of columns 9 and 12, changing the sign of column 9, and set results in column 13.

The horizontal thrust or true pole distance is the assumed pole distance multiplied by $\frac{l'}{l}=12{,}000 \text{ lbs.} \times \frac{2.75}{2.03}=16{,}300 \text{ lbs.}$

Multiply column 13 by this horizontal thrust (16,300 lbs.) and the results are the bending moments in arch ring in foot-pounds, which are set down in column 14.

The thrusts and shears are obtained graphically by drawing the rays from the true pole parallel to the tangents to the center line of arch and to the line of pressure as shown.

The line of pressure may be plotted by two methods, the one checking the other. First lay off on the ordinates from the center line or neutral surface the values found in column 13. Connect these points and we have the line of pressure. Secondly, the rays from the true pole to the load line are parallel to the line of pressure at the respective points. This is omitted on the diagram because the scale is too small to show it distinctly.

The true pole is found by drawing a line from the assumed pole to the load-line parallel to $m-m'$, the true closing line, and by laying off horizontally from this point on the load-line the true pole distance or horizontal thrust$=16{,}300$ lbs. found above.

TABLE XXV.

1	2	3	4	5	6	7	8	9	10	11	12	13	14	15
Point.	Ordinates, Arch.	Equilibrium, Polygon.			Triangles.		Ordinates for Closing Lines.		Ordinates for Deflection Polygons, Both Sides.		True Ordinates, Col. 8 × $\frac{l}{l'} = \times \frac{2.03}{2.75}$.	Measure of Moments, Difference of Cols. 9 and 12.	Moments, Foot-pounds, Col. 13 × 16,300.	Thrusts.
		Ordinates.	Lever Arms.	Moments.	Ordinates.	Moments.	*mpm'*	*K'aK''*	Arch.	Equilibrium Polygon.				
V'	0.00	0.00					(−7.9)	(−4.44)			(−5.82)	−1.38	−22,500	
1	2.43	5.41	5.55	30.03	1.09	6.05	−2.62	−1.97	−3.94	−5.47	−1.94	+0.03	+ 490	
2	4.48	8.17	13.09	106.95	2.59	33.91	0.00	+0.04	−3.86	−5.40	0.00	−0.04	− 650	
3	4.92	8.88	16.08	142.80	3.17	50.98	+0.61	+0.48	−2.90	−4.06	+0.45	−0.03	− 490	
4	5.13	9.22	18.28	168.53	3.60	65.81	+0.89	+0.68	−1.54	−2.18	+0.66	−0.02	− 330	
5	5.22	9.45	20.31	191.93	4.00	81.24	+1.07	+0.77	−0.00	−0.00	+0.79	+0.02	+ 330	
6	5.22	9.54	22.26	212.36	4.39	97.73	+1.11	+0.77			+0.82	+0.05	+ 820	16,400
7	5.13	9.47	24.32	230.31	4.80	116.74	+0.99	+0.68			+0.73	+0.05	+ 820	
8	4.92	9.26	26.54	245.77	5.24	139.07	+0.73	+0.48			+0.54	+0.06	+ 980	16,450
9	4.48	8.66	29.53	255.73	5.83	172.16	+0.07	+0.04			+0.05	+0.01	+ 160	
10	2.43	5.93	37.02	219.52	7.31	270.61	−2.85	−1.97			−2.10	−0.13	− 2,120	18,150
V	0.00	0.00					(−8.9)	(−4.44)			(−6.56)	−2.12	−34,600	20,750
Summation *	44.36	83.99		1803.93	42.02	1034.30	+ = − ±5.47	+ = − ±3.94			+ = −			

* Summations from 1 to 10, figures in parentheses excluded.

To proportion the ring to the stresses apply one of the formulas (15) to (20), using the one applicable to the sections and most convenient. An inspection of columns 14 and 15 of Table XXV shows that the critical points are 6, 8, 10, and V. The moments and thrusts of Table XXV are for a section or slice 1 foot thick. These must first be reduced to a section 1 inch thick, as in Table XXVI.

TABLE XXVI.

Governing Points.	Thrust T for Sec. 1″ Wide.	Moment M for Sec. 1″ Wide.	Depth Assumed for Ring.	Length on Arch "S."	Depth of Ring Used.
V	1,730	2,880	....	....	42″
10	1,510	176	13″	12.0	13″
8	1,370	82	$7\frac{3}{4}$″	2.41	
6	1,365	68	$7\frac{1}{8}$″	2.01	
Center	....	....	7″	....	7″

When it is found that the depth of ring "h" at any point has to be changed, the depth of all other points, except at the springing lines, V' and V, must also be changed in order to keep $S \div h^3$ constant, and therefore h^3 must be changed in the same ratio at all points. Since any change whatever at V' and V does not affect the conditions, the depth at these two points may be varied without any reference to the other points of the ring, and by increasing or diminishing the section on either one or both sides of the neutral axis. The reason for this is that the length of the intercepts mpm' and K_1aK_2 are not altered by any change at V or V'.

It will often require two or three and sometimes more trials to find the right section for the critical points. Those sections given below are only the final ones, the trial sections which gave too high stresses or which were excessive in strength having been discarded. To illustrate possible variations in methods the sections are worked out by the formulas numbers (1), (2), (4), and (5), given in the approximate method, which, however, for this part of the problem are exact and reliable.

Point 10.—5″ I beams 36″ c. to c., *h at crown*=7″, h=13″, h'=5″, I_s=0.336, and a=0.079.

$$\frac{L_c}{L_s}=\frac{E_cI_c}{E_sI_s}=\frac{h^3}{480I_s}=y=\frac{2,197}{162}=13.6.$$

$$\alpha=\frac{y}{1+y}=\frac{13.6}{14.6}=0.93.$$

$$\beta=\frac{1}{1+y}=\frac{1}{14.6}=0.07.$$

$$f_c = \left[\frac{1{,}510}{13} \mp \frac{72 \times 176}{169}\right] 0.93 = \begin{cases} +178 \text{ lbs. per sq. in.} \\ +\ 38 \text{ `` `` `` ``} \end{cases}$$

$$f_s = \left[\frac{1{,}510}{.079} \mp \frac{6 \times 5 \times 176}{.336}\right] 0.07 = \begin{cases} +2{,}440 \text{ lbs. per sq. in.} \\ +1{,}710 \text{ `` `` `` ``} \end{cases}$$

Point 6. $y=\frac{362}{162}=2.24,\ \alpha=\frac{2.24}{3.24}=0.69,\ \beta=\frac{1}{3.24}=0.31,$

$$f_c = \left[\frac{1{,}365}{7} \mp \frac{72 \times 68}{49}\right] 0.69 = \begin{cases} +204 \text{ lbs. per sq. in.} \\ +\ 65 \text{ `` `` `` ``} \end{cases}$$

$$f_s = \left[\frac{1{,}365}{.079} \mp \frac{6 \times 5 \times 68}{0.336}\right] 0.31 = \begin{cases} +7{,}230 \text{ lbs. per sq. in.} \\ +3{,}500 \text{ `` `` `` ``} \end{cases}$$

Point V. $h=42'',\ I_c=6{,}174.$

From equation (15),

$$f_c = \left[\frac{1{,}730}{45.16} \mp \frac{17{,}280 \times 42}{6{,}187.45}\right] = \begin{cases} +154 \text{ lbs. per sq. in.} \\ -\ 78 \text{ `` `` `` ``} \end{cases}$$

and from equation (16),

$$f_s = [38 \pm 14]40 = \begin{cases} +2{,}080 \text{ lbs. per sq. in.} \\ +\ 960 \text{ `` `` `` ``} \end{cases}$$

EXAMPLE 2.—This is a 72-foot span for a single-track railroad bridge. There should generally be at least 3 feet of ballast and filling between the top of the arch and the bottom of the ties, and 5 or 6 feet would be better. It is usual to consider the wheel concentrations as distributed over three ties, and they may be taken with perfect safety as uniformly distributed longitudinally on the arch over all the space between adjacent wheels.

The graphic diagram of Fig. 34 and the Tables XXVII to XXX inclusive are similar to those of Example 1 and require no additional explanation except on the one point in which they differ.

Instead of the graphic construction for finding the ratios *vn* to *vm* and *v'n'* to *v'm'*, the following analytical method is used.

The closing line of the arch is located as before by the equation $K_1V=K_2V'=\frac{266.66}{18}=14.813$ and $vn=v'n'=\frac{234.89}{18}=13.05;\ vn+v'n'=26.10;\ CR=(8{,}518.22 \div 234.89)-36=0.266$ and $CT=CT'=(5177.77 \div 117.45)-36=8.082;\ TR=8.082-0.266=7.816.$

$$RT'=8.082+0.266=8.348,$$

$T-T'=16.164$, from which we get

$$vm=26.10\frac{8.348}{16.164}=13.48,$$

and

$$v'm'=26.10\frac{7.816}{16.164}=12.62.$$

TABLE XXVII.—LOADS PER FOOT OF WIDTH.

Live load 5,000 lbs. per lineal foot of track plus 50,000 lbs. excess at head of train, assumed as distributed on 16⅔ ft. width of arch.

Points.	L'gth.	Ring, 150 lbs. per Cu. Ft.		Fill, 120 lbs. per Cu. Ft.		Track 25 lbs. per Sq. Ft.	Total Dead Load, lbs.	Live Load, Pounds + 3,000 Excess.	Totals, lbs.	Summations, lbs.
		Depth.	Lbs.	Depth.	Lbs.					
1	8.4	4.8	6,050	16.5	16,630	210	22,890	....	22,890	22,890
2	5.6	3.3	2,780	11.1	7,450	140	10,370	....	10,370	33,260
3	4.2	2.7	1,720	8.5	4,290	110	6,120	....	6,120	39,380
4	3.65	2.4	1,330	6.9	3,020	90	4,440	....	4,440	43,820
5	3.3	2.25	1,130	5.8	2,300	80	3,510	....	3,510	47,330
6	3.0	2.12	950	5.0	1,800	75	2,825	....	2,825	50,155
7	2.7	2.00	810	4.6	1,490	70	2,370	1,000	3,370	53,525
8	2.6	1.97	780	4.3	1,340	65	2,185	1,000	3,185	56,710
9	2.55	1.95	750	4.1	1.250	60	2,060	1,000	3,060	59,770
Total	......	......	16,300		39,570	900	56,770	(3,000)	59,770	
10	2.55	......	......	......	......	....	2,060	760	2,820	2,820
11	2.6	......	......	......	......	....	2,185	780	2,965	5,785
12	2.7	......	......	......	......	....	2,370	810	3,180	8,965
13	3.0	......	......	......	......	....	2,825	900	3,725	12,690
14	3.3	......	......	......	......	....	3.510	990	4,500	17,190
15	3.65	......	......	......	......	....	4,440	1,090	5,530	22,720
16	4.2	......	......	......	......	....	6,120	1,250	7,370	30,090
17	5.6	......	......	......	......	....	10,370	1,670	12,040	42,130
18	8.4	......	......	......	......	....	22,890	2,520	25,410	67,540
Total..	......	......	......	......	......	....	56,770	10,770	67,540	
Grand total.		......	......	......	......	....	113,540	13,770	127,310	

Assume pole distance at 56,000 lbs.

Then in place of scaling the intercepts mpm' they are computed by Table XXIX as follows:

The values in column 2 multiplied by $\frac{13.48}{72}$ give the values in column 3, and the values in column 2 *inverted* multiplied by $\frac{12.62}{72}$ give the values in column 4.

Column 5 is the sum of columns 3 and 4, and column 7 gives the intercepts mpm', by subtracting the values in column 5 from those in column 6, which latter are taken from column 3 of Table XXVIII.

The values in column 7 of Table XXIX are set down in column 8 of Table XXVIII.

Column 8 of Table XXIX gives the values for the ordinates or load line of the deflection polygon for the equilibrium polygon and are set down in column 11, Table XXVIII.

The thrusts and moments for a slice one inch thick are given in Table XXX, which also illustrates the method of adjusting the ring from the assumed depth at the crown of 1.94 feet to the required depth of 1.48 feet, keeping $s \div h^3$ constant.

TABLE XXVIII.

1	2	3	4	5	6	7	8	9	10	11	12	13	14	15
Points.	Arch Ordinates.	Equilibrium Polygon.			Triangles.		Ordinates for Closing Lines.		Ordinates for Deflection Polygons; Both Sides.		True Ordinates Col. 8 $\times\frac{l}{l'}=$ Col. 8 $\times\frac{17.81}{15.25}$.	Measure of Moments Difference of Cols. 9 and 12.	M in Foot-pounds, Col. 13 × H = Col. 13 × 47,950.	Thrusts.
		Ordinates.	Lever-arms.	Moments.	Ordinates.	Moments.	mpm'	k_1ak_2	Arch.	Equilibrium Polygon.				
V'	0.0	0.0	0.0		0.0		−12.62	−14.81			−14.74	+0.07	+ 3,400	77,000
1	4.71	4.31	3.85	16.59	.70	2.69	−8.35	−10.12	−20.24	−17.23	−9.75	+0.37	+17,700	66,400
2	10.80	9.23	11.08	102.27	2.01	22.27	−3.52	−4.01	−28.26	−24.27	−4.11	−0.10	− 4,800	
3	13.70	11.70	16.02	187.43	2.90	46.45	−1.11	−1.11	−30.48	−26.29	−1.30	−0.19	− 9,100	54,000
4	15.44	13.21	19.94	263.41	3.61	71.98	+0.36	+0.63	−29.22	−25.28	+0.42	−0.21	−10,100	51,600
5	16. 6	14.28	23.40	334.15	4.24	99.21	+1.38	+1.79	−25.64	−22.22	+1.61	−0.18	− 8,700	
6	17.42	15.03	26.57	399.33	4.82	128.04	+2.10	+2.61	−20.42	−17.76	+2.45	−0.16	− 7,700	
7	17.97	15.60	29.40	458.63	5.33	156.69	+2.63	+3.16	−14.10	−12.33	+3.07	−0.09	− 4,300	
8	18.26	15.97	32.08	512.32	5.81	186.37	+2.96	+3.44	− 7.22	− 6.31	+3.46	+0.02	+ 1,000	
9	18.43	16.19	34.63	560.64	6.28	217.46	+3.16	+3.61	− 0.00	− 0.00	+3.69	+0.08	+ 3,800	48,000
10		16.22	37.37	66.12	6.77	252.98	+3.15	+3.61			+3.68	+0.07	+ 3,400	
11		16.15	39.92	644.66	7.24	289.01	+3.06	+3.44			+3.57	+0.14	+ 6,700	48,000
12		15.93	42.60	678.60	7.72	328.84	+2.80	+3.16			+3.27	+0.11	+ 5,300	
13		15.53	45.43	705.47	8.23	373.86	+2.36	+2.61			+2.76	+0.15	+ 7,200	
14		14.88	48.60	723.14	8.81	428.14	+1.68	+1.79			+1.96	+0.17	+ 8,200	50,000
15		13.90	52.06	723.60	9.44	491.44	+0.65	+0.63			+0.76	+0.13	+ 6,200	
16		12.38	55.98	693.00	10.15	568.21	−0.91	−1.11			−1.06	+0.05	+ 2,400	
17		9.83	60.92	598.83	11.04	672.53	−3.52	−4.01			−4.11	−0.10	− 4,800	
18		4.55	68.15	310.03	12.35	841.60	−8.88	−10.12			−10.37	−0.25	−12,000	69,400
V	0.0	00.00	72.00		13.05		−13.48	−14.81			−15.74	−0.93	−44,600	82,400
Summation 1 to 18	266.66	234.89		8518.22	177.45	5177.77								

$$H=56{,}000\frac{15.25}{17.81}=47{,}950.$$

TABLE XXIX.

1	2	3	4	5	6	7	8
Point.	Lever-arm.	Triangles.		Summa-tion.	Table 28, Col. 3.	Table 28, Col. 8.	Table 28 Col. 11.
		mvv'	$mm'v'$				
V'	0.0	0.0	12.62	12.62	0.0	−12.62	
1	3.85	.72	11.94	12.66	4.31	−8.35	−17.23
2	11.08	2.07	10.68	12.75	9.23	−3.52	−24.27
3	16.02	3.00	9.81	12.81	11.70	−1.11	−26.29
4	19.94	3.73	9.12	12.85	13.21	+0.36	−25.28
5	23.40	4.38	8.52	12.90	14.28	+1.38	−22.22
6	26.57	4.97	7.96	12.93	15.03	+2.10	−17.76
7	29.40	5.50	7.47	12.97	15.60	+2.63	−12.33
8	32.08	6.01	7.00	13.01	15.97	+2.96	− 6.31
9	34.63	6.48	6.55	13.03	16.19	+3.16	− 0.00
10	37.37	7.00	6.07	13.07	16.22	+3.15	
11	39.92	7.47	5.62	13.09	16.15	+3.06	
12	42.60	7.98	5.15	13.13	15.93	+2.80	
13	45.43	8.51	4.66	13.17	15.53	+2.36	
14	48.60	9.10	4.10	13.20	14.88	+1.68	
15	52.06	9.75	3.50	13.25	13.90	+0.65	
16	55.98	10.48	2.81	13.29	12.38	−0.91	
17	60.92	11.41	1.94	13.35	9.83	−3.52	
18	68.15	12.76	.67	13.43	4.55	−8.88	
V	72.00	13.48	0.00	13.48	0.00	−13.48	

Point 9.—Steel, 12″ I beam at $31\frac{1}{4}$ pounds; $b=36''$, $a=9.3\div36$, $I_s=222\div36$. From equations (15) and (16),

$$f_c=\left[\frac{4,000}{28.1}\pm\frac{34,100}{713}\right] = +191 \text{ lbs. per sq. in.}$$

$$f_s=\left[143\pm\frac{23,000}{713}\right]\times40=+7,000 \text{ " " " "}$$

Point 14:

$$f_c=\left[\frac{4,170}{29.85}\pm\frac{79,500}{865}\right] = +232 \text{ lbs. per sq. in.}$$

$$f_s=\left[140\pm\frac{49,000}{865}\right]\times40=+7,870 \text{ " " " "}$$

Point 18:

$$f_c=\left[\frac{5,780}{40.35}\pm\frac{180,000}{2,497}\right] = +215 \text{ lbs. per sq. in.}$$

$$f_s=\left[143\pm\frac{72,000}{2,497}\right]\times40=+6,850 \text{ " " " "}$$

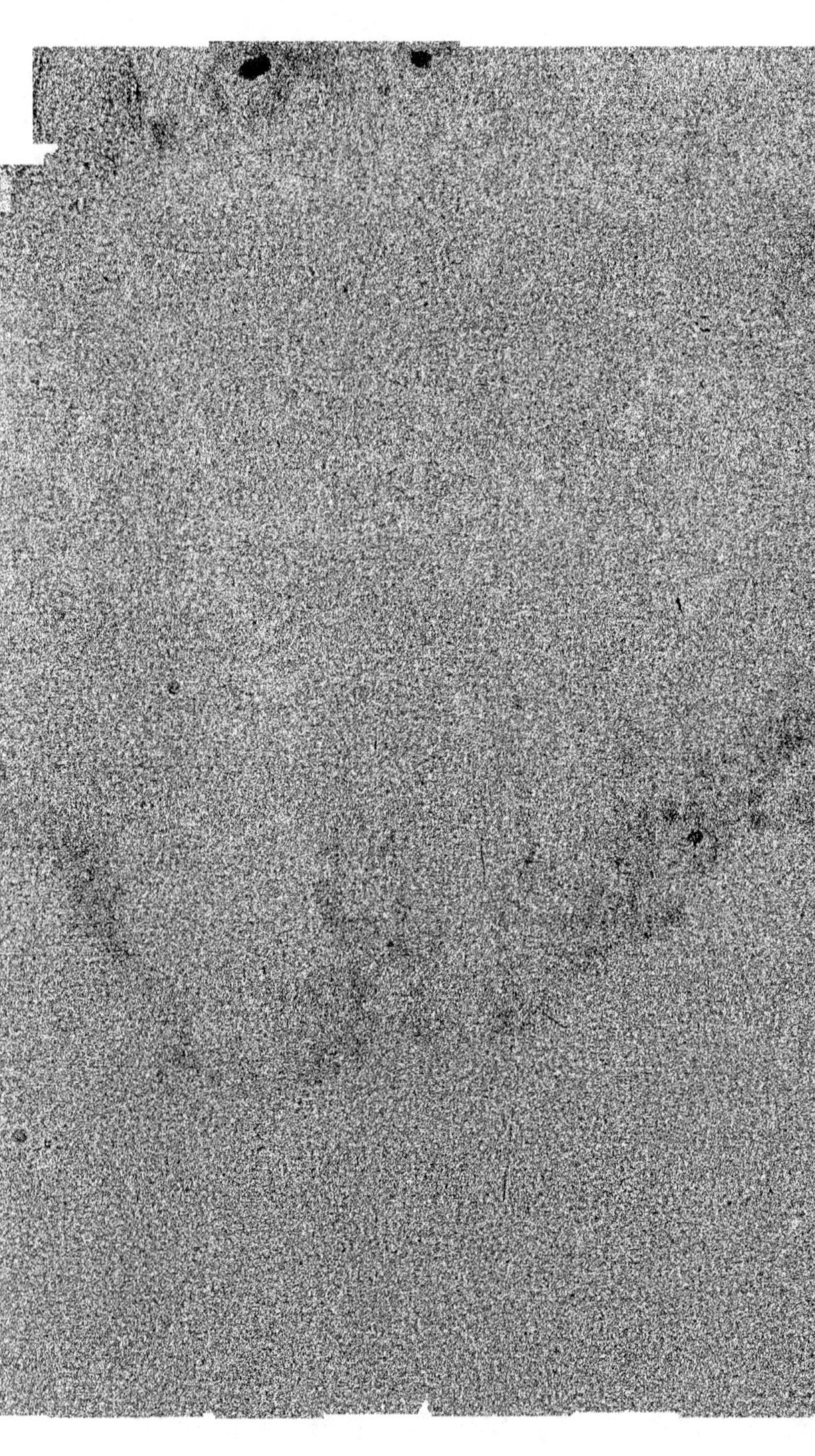

Point 1:

$$f_c=\left[\frac{5,530}{40.35}\pm\frac{264,000}{2,497}\right] \quad = \quad +244 \text{ lbs. per sq. in.}$$

$$f_s=\left[138\pm\frac{106,000}{2,497}\right]\times 40=+7.230 \text{ `` `` `` ``}$$

Point V:

$$f_c=\left[\frac{6,860}{55.35}\pm\frac{1,000,000}{7,840}\right]=\begin{cases}+252 \text{ lbs. per sq. in.}\\ -\ \ 4 \text{ `` `` `` ``}\end{cases}$$

$$f_s=\left[124\pm\frac{268,000}{7,840}\right]\times 40=+6,300 \text{ lbs per sq. in.}$$

In both of the above examples, 1 and 2, the value of e was taken at 40. This is considerably higher than is generally used at the present time. The best available data indicate that in arches values ranging from $e=20$ to $e=15$ are near to the truth.

TABLE XXX.

Governing Points.	T for 1″.	M for 1″.	h assumed.	s.	Trial.			h^3 in inches.	I_c.
					h.	h^3.	h in inches.		
V	6,860	3,720					45	91,125	7,593
1 18	5,530 5,780	1,470 1,000	3.3′	12.0	2.5	15.625	30	27,000	2,250
14	4,170	680	2.15	3.32	1.63	4.33	19½	7.415	618
9	4,000	320	1.95	2.48	1.48	3.24	17¾	5,592	466
Crown			1.94				17¾		

EXAMPLE 3.—This is for a semicircular arch culvert of 15 ft. clear span, for a public highway. The diagram, Fig. 35, gives the loads and the graphical construction for finding the thrusts and moments. The latter is exactly similar to that of Example 2, Fig. 34, and does not require any further explanation.

The analytical part of the analysis, given in Table XXXI, is similar to that of Example 2, except that in the present case the work is somewhat simplified by combining the operations in one table. In the previous example Table XXIX was introduced for finding the ordinates *mpm′*, while in this case the same operations are performed by inserting the columns 8, 9, and 10 of Table XXXI, which correspond to columns 3, 4, and 5 of Table XXIX. The labor of copying the figures of five columns from one table to another is thus saved, and the chance of errors in copying reduced.

$$K_1V=K_2V'=\frac{\Sigma \text{ column } 2}{\text{number of points}}=\frac{25.28}{8}=3.16;$$

$$vn=v'n'=\frac{\Sigma \text{ column } 3}{\text{number of points}}=\frac{28.99}{8}=3.624;$$

$$vn+v'n'=7.248;$$

$$CR=\text{distance from center}=\frac{\Sigma \text{ column } 5}{\Sigma \text{ column } 3}-\frac{\text{span}}{2}$$

$$=\frac{219.57}{28.99}-7.5=7.56-7.50=0.06;$$

$$CT=CT'=\frac{\Sigma \text{ column } 7}{\Sigma \text{ column } 6}-\frac{\text{span}}{2}=\frac{135.01}{14.51}-7.5$$

$$=9.3-7.5=\pm 1.80;$$

$$TR=CT-CR=1.80-0.06=1.74;$$

$$RT'=CT'+CR=1.80+0.06=1.86.$$

Therefore

$$TT'=TR+RT'=1.74+1.86=3.60;$$

$$vm=(vn+v'm')\frac{RT'}{TT'}=7.248\times\frac{1.86}{3.60}=3.75;$$

$$v'm'=(vn+v'm')\frac{TR}{TT'}=7.248\times\frac{1.74}{3.60}=3.50.$$

Therefore

$$vm+v'm'=7.25.$$

The computation of columns 8, 9, 10, and 11 of Table XXXI follow in the same way as that of Table XXIX.

The assumed pole distance is 4,150 lbs., and

$$H=4,150\frac{l'}{l}=4,150\frac{1.66}{1.63}=4,225.$$

The moments and thrusts, columns 17 and 18, are found precisely as before. All other operations are identical with those of examples 1 and 2.

TABLE XXXI.

1	2	3	4	5	6	7	8	9	10
		Equilibrium Polygon.			Triangles.		Intercepts of the		
Point.	Arch Ordinates.	Ordinates.	Lever-arm.	Moments Col. 3 × Col. 4.	Ordin'tes vm × span Col. 4 $\frac{3.624}{15}$ × Col. 4.	Moments Col. 6 × Col. 4.	Triangle mvv' $\frac{mv}{15}$ × Col. 4 $\frac{3.75}{15}$ × Col. 4.	Triangle $mm'v'$ $\frac{m'v'}{15}$ × Col. 4 $\frac{3.50}{15}$ × Col. 4.	Trapezoid $mm'v'v$ Col. 8 + Col. 9.
V'	0.00	0.00	0.00	0.00	0.00	0.00	0.00	(3.50)	(3.50)
1	1.66	2.02	1.43	2.89	0.35	0.50	0.36	3.17	3.53
2	3.27	3.58	3.95	14.13	0.95	3.75	0.99	2.58	3.57
3	3.75	4.17	5.45	22.70	1.32	7.20	1.36	2.23	3.59
4	3.96	4.46	6.85	30.60	1.66	11.39	1.72	1.90	3.62
5	3.96	4.51	8.15	36.80	1.97	16.05	2.04	1.60	3.64
6	3.75	4.30	9.55	41.10	2.31	22.07	2.39	1.27	3.66
7	3.27	3.76	11.05	41.60	2.67	29.55	2.77	.92	3.69
8	1.66	2.19	13.57	29.75	3.28	44.50	3.39	.33	3.72
V	0.00	0.00	15.00	....	(3.62)	....	(3.75)	.00	(3.75)
Σ to 18	25.28	28.99	60.00	219.57	14.51	135.01			

1	11	12	13	14	15	16	17	18
	Ordinates for Closing Lines.		Moments for Deflection Polygon.					
Point.	mpm'. Col. 3 − Col. 10.	$K'aK''$. Col. 2 − $K'V$.	Arch Summation. Col. 12	Equilibrium Polygon Summation. Col. 11.	True Ordinates. Col. 11 × $\frac{l}{l'}$ Col. 11 × $\frac{1.63}{1.66}$.	Measure of Moments = Difference of Columns 12 and 15.	Moments. in Ft.-lbs. Col. 16 × H = Col. 16 × 4225.	Thrusts.
V'	(−3.50)	(−3.16)	....	....	(−3.44)	−0.28	−1180	7200
1	−1.51	−1.50	−3.00	−3.04	−1.48	+0.02	+84	5600
2	+0.01	+0.11	−2.78	−2.96	−0.01	−0.12	−506	4700
3	+0.58	+0.59	−1.60	−1.74	+0.57	−0.02	−84	4400
4	+0.84	+0.80	−0.00	−0.03	+0.83	+0.03	−127	4250
5	+0.87	+0.80	....	....	+0.86	+0.06	+253	4250
6	+0.64	+0.59	....	....	+0.63	+0.04	+170	4400
7	+0.07	+0.11	....	....	+0.07	−0.04	−170	4800
8	−1.53	−1.50	....	....	−1.50	−0.00	....	5800
V	(−3.75)	(−3.16)	....	....	(−3.68)	−0.52	−2200	7600

Centers and False Works.

Success in arch construction depends nearly or quite as much on the centers and their supports as it does on the design and construction of the abutments, piers, and arch ring.

The supports for the centers should be as unyielding as it is possible to make them. If piles are used they should be as well driven as permanent foundation piles, and they should not be more heavily loaded. Sills should only be used when they may be laid on bedrock or on an equally good foundation.

The centers themselves may be divided into two classes, those which are supported at every intersection of the *braces* or web members with the lower chord or *tie-beam*, and those which act as a truss in spanning an opening to give under clearance. The latter are sometimes required to provide water-way or to keep traffic open under the arch during construction and are called "retroussée" or "cocket" centers. The width of the span may be only a narrow opening or series of openings, or it may be equal to the span of the arch, with the centers supported by the abutments or piers.

When retroussée or cocket centers have to be used, extraordinary care must be exercised to avoid excessive and injurious deflections, under the charges in the loading that will occur during construction. Deflection can be reduced to a minimum by properly designing the centers, and deformations during the progress of construction of the ring can be largely prevented by temporarily loading the centers at the crown with cement or other material, and gradually removing the temporary load as the weight of the ring comes on the centers.

The *back-pieces* of the ribs are usually formed by spiking together several planks so as to break joints, the top edges of which are cut to a curve parallel with the intrados and the thickness of the lagging below or inside of it, so that the top of the lagging will coincide with the intrados. The *back-pieces*, *tie-beams*, and *braces* are bolted together to form trussed ribs. This construction is shown in Fig. 287 illustrating the centers of the Topeka bridge. The centers should be given an amount of camber about equal to the deflection of the arch ring under full loading, so that the ring will never have a length, measured on the neutral surface, shorter than that assumed in the computations of the stresses and sections.

To facilitate striking the centers, they should be supported on folding wedges or sand-boxes. If the latter are used, the sand should be fine, clean, and perfectly dry, and they should be sealed around the plunger with cement mortar. Such sand-boxes were used under the centers of both the Topeka and Jacaguas bridges. In the latter case they were not entirely successful, which may be accounted for by the poor quality of sand used and the failure to seal them around the plungers.

In most cases the most convenient form of lagging for concrete arches has been found to be 2-in. by 4-in. pieces set on edge. They are narrow enough to conform to the curve of the ring, and are deep enough, 4 ins., to carry the weight of the ring between the ribs when the latter are 5 ft. to 6 ft. between centers, without deflecting sufficiently to give an uneven surface to the soffit of the arch. If lightly toe-nailed at not more than two or three places, the salvage will more than pay for the extra material, and a lesser number of costly ribs are required, as the latter can be spaced further apart than when thin lagging is used.

In important cases the deflection of the lagging and other beam supports should be investigated by the formula

$$D = 5Wl^3 \div 384EI,$$

in which W is the weight in pounds uniformly distributed, l the length of the span from center to center of supports in inches, E the coefficient or modolus of elasticity of the timber, and I the moment of inertia of the cross-section in inches. E may be taken at 1,200,000 for white pine and spruce and at 1,600,000 for yellow pine or oak.

Where deflection is not in question, the lagging and any other beams used in the centers and forms may be proportioned by Table XXXII for white pine or spruce, and by Table XXXIII for yellow pine or oak. The braces and struts may be proportioned by Table XXXIV.

It is the usual practice with concrete arches not to strike the centers until thirty days after the ring is closed, but in a number of cases they have been struck within one week. Care should be used in striking the centers to let them down or relieve them of load gradually and uniformly and without shock.

In some cases the supports are carried up through the *tie-beams* and support the wedges directly under the *back-pieces*, as illustrated in Figs. 281 and 282. This arrangement is effective and economical and has been successfully employed.

It is important that the supports and centers be well braced in all directions, but Figs. 280 to 286, inclusive, sufficiently illustrate this point.

TABLE XXXII.—SHOWING CAPACITY OF TIMBER BEAMS IN BENDNG MOMENTS (FOOTS-POUNDS) FOR 1,000 LBS. PER SQ. IN. FIBER STRAIN.

For White Pine or Spruce.

Width, Inches.	Depth of Beam in Inches.								
	6	7	8	9	10	12	14	15	16
2	1,000	1,361	1,778	2,250	2,778	4,000	5,444	6,250	
2½	1,250	1,701	2,222	2,813	3,472	5,000	6,806	7,813	8,889
3	1,500	2,042	2,667	3,375	4,167	6,000	8,167	9,375	10,667
3½	1,750	2,382	3,111	3,938	4,861	7,000	9,528	10,938	12,444
4	2,000	2,722	3,556	4,500	5,556	8,000	10,889	12,500	14,222
4½	2,250	3,062	4,000	5,063	6,250	9,000	12,250	14,063	16,000
5	2,500	3,403	4,444	5,625	6,944	10,000	13,611	15,625	17,778
5½	2,750	3,743	4,889	6,188	7,639	11,000	14,972	17,188	19,556
6	3,000	4,083	5,333	6,750	8,333	12,000	16,333	18,750	21,333
7		4,764	6,222	7,875	9,722	14,000	19,056	21,875	24,889
8			7,111	9,000	11,111	16,000	21,778	25,000	28,444
9				10,125	12,500	18,000	24,500	28,125	32,000
10					13,889	20,000	27,222	31,250	35,556
12						24,000	32,667	37,500	42,667

TABLE XXXIII.—SHOWING CAPACITY OF TIMBER BEAMS IN BENDING MOMENTS (FOOT-POUNDS) FOR 1,500 LBS. PER SQ. IN. FIBER STRAIN.

For Yellow Pine or Oak.

Width, Inches.	Depth of Beam in Inches.								
	6	7	8	9	10	12	14	15	16
2	1,500	2,042	2,667	3,375	4,167	6,000	8,167	9,375	
2½	1,875	2,552	3,333	4,219	5,208	7,500	10,208	11,719	13,333
3	2,250	3,063	4,000	5,063	6,250	9,000	12,250	14,063	16,000
3½	2,625	3,573	4,667	5,906	7,292	10,500	14,292	16,406	18,667
4	3,000	4,083	5,333	6,750	8,333	12,000	16,333	18,750	21,333
4½	3,375	4,594	6,000	7,594	9,375	13,500	18,375	21,094	24,000
5	3,750	5,104	6,667	8,438	10,417	15,000	20,417	23,438	26,667
5½	4,125	5,615	7,333	9,281	11,458	16,500	22,458	25,781	29,333
6	4,500	6,125	8,000	10,125	12,500	18,000	24,500	28,125	32,000
7		7,146	9,333	11,813	14,583	21,000	28,583	32,813	37,333
8			10,667	13,500	16,667	24,000	32,667	37,500	42,667
9				15,188	18,750	27,000	36,750	42,188	48,000
10					20,833	30,000	40,833	46,875	53,333
12						36,000	49,000	56,250	64,000

TABLE XXXIV.—SHOWING UNIT STRESSES FOR TIMBER POSTS.

$\frac{L}{d}$.	Yellow Pine. $f=1{,}075-138\frac{L}{d}$.	White Pine. $f=675-107\frac{L}{d}$.
1.0	937	568
1.1	924	557
1.2	910	546
1.3	896	535
1.4	882	524
1.5	868	513
1.6	855	502
1.7	841	491
1.8	827	480
1.9	813	470
2.0	799	460
2.1	786	450
2.2	772	439
2.3	758	428
2.4	744	417
2.5	730	406
2.6	717	395
2.7	703	384
2.8	689	374
2.9	675	364
3.0	661	354
3.1	647	343
3.2	633	332
3.3	619	321
3.4	605	311

L = length of member in feet.
d = least dimension of member in inches.
f = safe stress in pounds per square inch.

Some engineers have proposed to make the steel reinforcement of the ring heavy and stiff enough to carry the lagging and concrete during construction, without the use of centers or temporary supports. Such a scheme may be successfully carried out on a small scale, and it is very generally used in floor construction, but for large or important arch bridges the author regards it as experimental and even hazardous. It is certain that deformations would be taking place continually during construction, and they could not but have an injurious effect on the concrete of the ring, the amount of which could not be estimated.

The general subject of centers is treated in considerable detail in a work by John B. McMaster, entitled "Bridge and Tunnel Centers," Van Nostrand's Science Series.

PART II.

REPRESENTATIVE EXAMPLES.

CHAPTER VI.—FOUNDATION CONSTRUCTION IN REINFORCED CONCRETE.

REINFORCED concrete is employed in a variety of forms in foundation work. Only two forms have, however, met with general acceptance. The first of these is the spread foundation so extensively employed in tall-building construction, and the second is the plate capping for piles. Comparatively recently concrete-steel piles have been made and driven successfully, and they promise to have a greater development as experience with them grows and begets confidence. There are a few examples also of concrete-steel caisson-work, but these are exceptional uses. A minor use of the material which has met with considerable favor is a sheathing or armoring for timber piles in teredo-infested waters.

Spread Foundations.—The spread foundation of reinforced concrete has attained its highest development in tall-building construction. It is employed in three forms: the simple column footing, combined footings for two or more columns, and continuous footings covering the whole foundation area. In principle these forms are similar; the purpose of each is to provide a slab or plate of concrete and embedded steel, which is rigid under concentrated column load and so will distribute this load over an area of ground capable of carrying it without overloading the soil. Various systems of concrete-steel are employed in constructing each of these forms of footing, but they may be arbitrarily divided into heavy grillages of steel beams embedded in concrete and into footings of concrete with skeletons composed of small elements, such as rods, bars, and metal mesh.

Beam Grillages.—In the earlier forms of beam grillages steel railway rails superimposed in layers at right angles to each other and embedded in concrete were generally employed. A succeeding step in the development was to replace the top layer of rails with rolled I beams. A typical single column footing of this construction is shown by Fig. 36. This foundation was designed to carry a column load of 1,166,000 lbs. The allowable pressure on the soil was 3,000 lbs. per square foot, which gave a footing 22 ft. 8 ins.×17 ft. 3 ins. in plan. For the top course of the grillage 15-in. I beams were employed, and for the other courses 75-lb. railway rails. The final step in the develop-

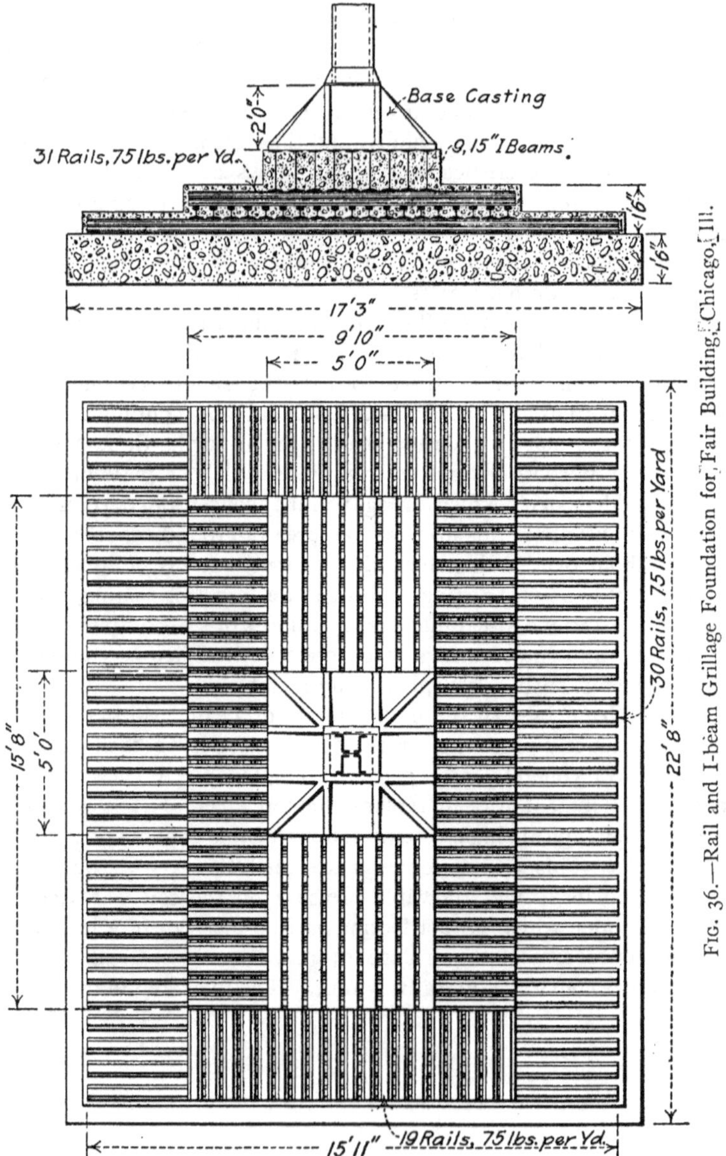

FIG. 36.—Rail and I-beam Grillage Foundation for Fair Building, Chicago, Ill.

ment of beam grillages was the employment of I beams for the entire grillage. Fig. 37 shows the continuous grillage foundation employed for a 19-story building in San Francisco, Cal. The foundation area of this building was 98×102 ft. This footing consists of a 2-ft. bottom layer of plain concrete. On this was set a layer of 15-in. I beams,

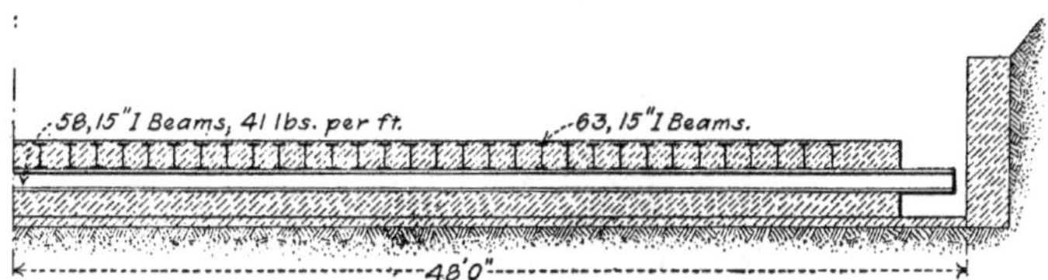

FIG. 37.—Continuous I-beam Grillage Foundation for Spreckles Building, San Francisco, Cal.

spliced end to end to a length of 96 ft. Concrete was filled in flush with the tops of these beams, and on it a second layer of 15-in. beams spliced to a length of 91.5 ft. was placed at right angles to the first layer. These beams were then filled with concrete like the first layer. This operation completed a footing 54 ins. thick of concrete reinforced by two layers of 15-in. I beams. The columns were carried on this footing by means of rafts of distributing girders.

Footings with Mesh and Bar Reinforcements.—A number of constructions employed for spread foundations differ only in the form of the reinforcing bars or netting used, and these may be considered together. Monier construction, as employed for wall and column footings and for continuous footings, consists of the regular Monier network embedded in the tension side of the plate. For wall footings the netting is placed so that the carrying-rods are perpendicular to the wall and the spacing-rods parallel to it. For column footings both sets of bars are made carrying-bars. Footings for heavy loads are reinforced by two or more layers of netting placed one above the other with layers of concrete between them. Expanded metal, as employed in spread foundations, is used substantially in the same manner as Monier netting. This is in fact the almost universal form of all foundation footings employing a reinforcement in the shape of a metal mesh or netting. The same general uniformity of location and arrangement exists in constructions with bar reinforcement. The reinforcing rods or bars are arranged in layers, the bars of which are at right angles to each other and all of which are embedded in the tension side of the footing-slab for ordinary loads and partly on the tension side and partly on the compression side for heavy loads. Fig. 38 shows the regular arrangement of tension-bars for one of the column footings for the concrete-steel Ingalls Building at Cincinnati, O. The footing is square in plan with pyramidal sides and is reinforced at the bottom

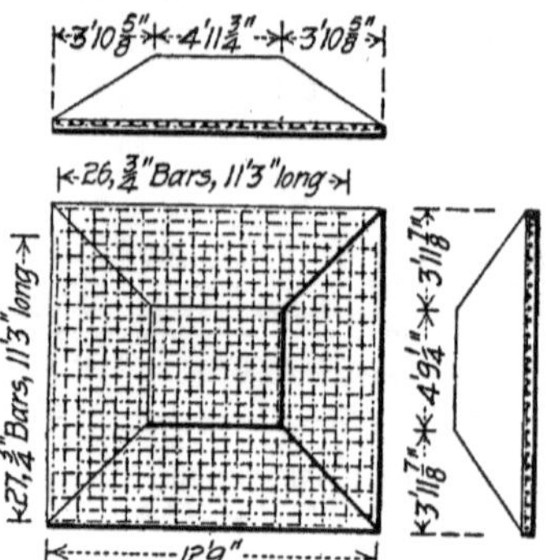

FIG. 38.—Single Column Footing with Bar Reinforcement, Ingalls Building, Cincinnati, Ohio.

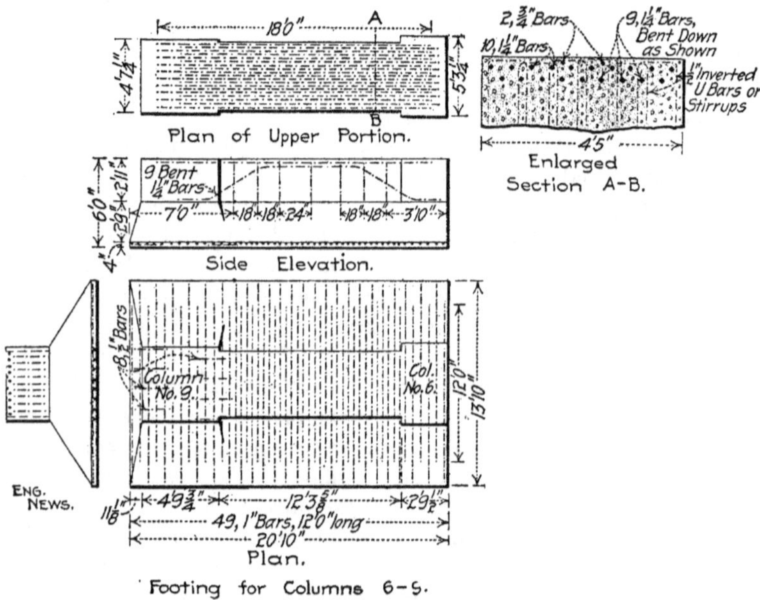

FIG. 39.—Combined Footing for Two Columns, Ingalls Building, Cincinnati, Ohio.

by two layers of ¾-in. twisted square bars, the bars in each layer being spaced 6 ins. apart. The combined footings for two columns employed in the same building, Fig. 39, are rectangular in plan with a wide batter on the sides and a narrow one on the ends, and are reinforced at both top and bottom by transverse and longitudinal rods. Fig. 40 shows a combined footing for unequally loaded columns which was

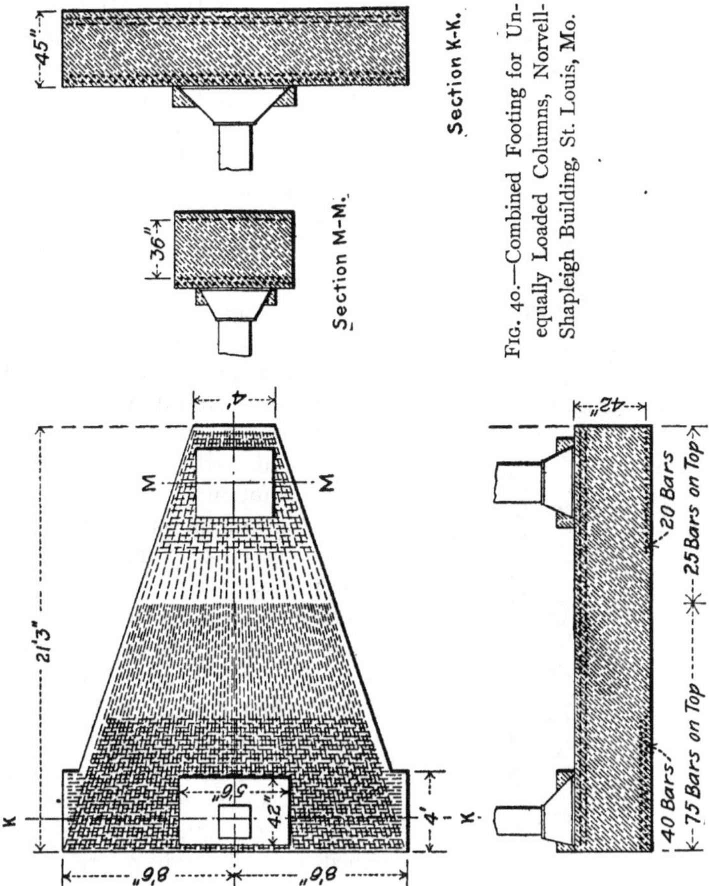

Fig. 40.—Combined Footing for Unequally Loaded Columns, Norvell-Shapleigh Building, St. Louis, Mo.

employed in the Norvell-Shapleigh Hardware Company's building at St. Louis, Mo. The reinforcing-bars were of the corrugated form described on p. 334. The total load carried by this footing is 580 tons distributed over an area of 232 sq. ft.

Footings with Bar and Stirrup Reinforcement.—A number of European constructions for spread foundations employ rod and stirrup systems of reinforcement. The most notable of these is the Hennebique, whose application to a single column footing is shown by Fig. 41. This particular footing was employed in a building erected at Lisle, France, and it carries a column load of 130 metric tons. It is square in plan and is reinforced by two courses of round iron bars laid at right angles to each other, the lower bars having stirrups extending upward into the concrete. When a stiffer footing is required Mr. Hennebique adopts a bar reinforcement in which straps are laid flatwise and parallel in horizontal planes. The straps in each plane are at right angles to those in the adjacent planes, and a layer of concrete separates the different layers of straps. For continuous footings Mr. Hennebique employs a modified form of his ribbed flat-plate floor construction in which the ribs are heavily reinforced on the compression edge in addition to the usual tension-bars. Fig. 42 shows the details of a continuous spread foundation of this construction.

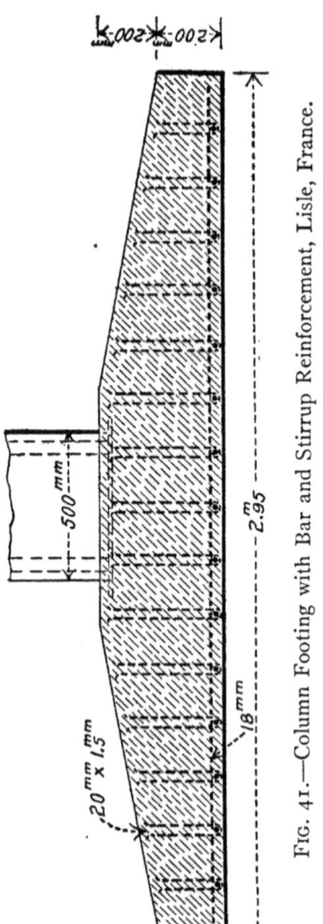

Fig. 41.—Column Footing with Bar and Stirrup Reinforcement, Lisle, France.

Pile Foundations with Reinforced Concrete Caps.—The use of reinforced concrete to cap and embed the tops of pile foundations is simply an adaptation of concrete-steel to another of the purposes for which plain concrete is employed. The most important examples of it are furnished by tall-building practice, but bridge piers have been founded in the same way. The same reason influences the adoption of concrete-steel caps for pile foundations that influences its adoption so largely for spread foundations for tall buildings; it gives a footing of the same stiffness with a smaller area and much less thickness than masonry or unreinforced concrete. The usual construction of pile foundations with reinforced concrete caps is to embed and cover the tops of the piles with a rather thick bed of concrete. On this platform are placed two or more layers of I beams which are filled between and covered with concrete exactly as in spread foundations of beam grillages. A bold example of pile foundation with reinforced concrete cap is shown by Fig. 43, which is a portion of the foundation

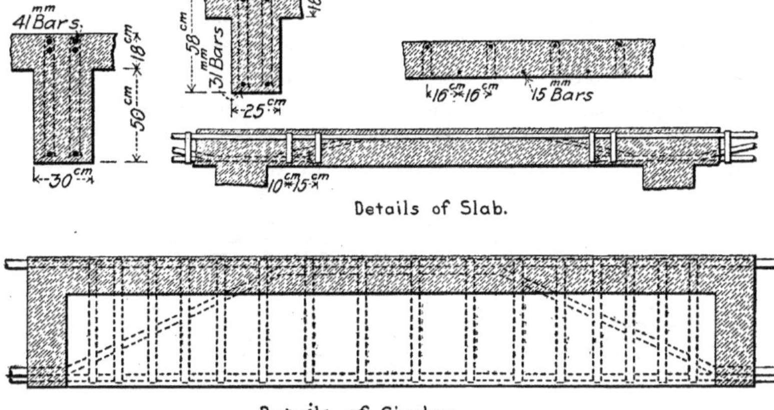

FIG. 42.—Continuous Footing, Slab, and Girder Construction.

for an 11-story Chicago office building. The foundation area was 40 ft. wide and 165 ft. long, and owing to the conditions it was undesirable to approach close to the walls of the buildings adjoining the long sides of the lot. Each transverse row of columns is carried on a

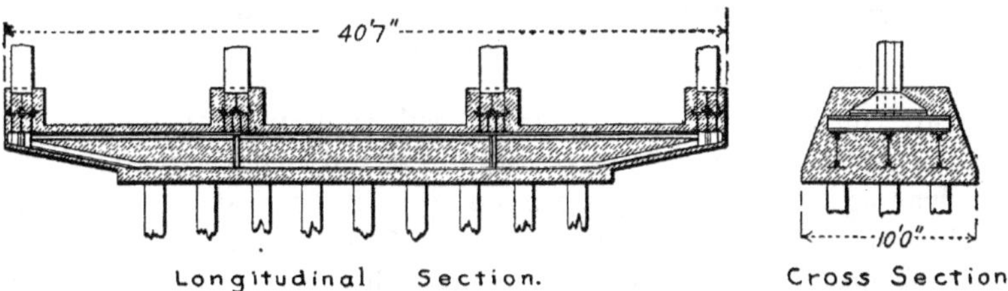

FIG. 43.—Concrete-steel Capping for Timber Piles, 11-story Building, Chicago, Ill.

separate footing, and these footings each consist of three parallel plate girders embedded in a rectangular slab of concrete which rests on three transverse rows of piles, with its ends bracketing out beyond the outermost piles. Fig. 44 shows the reinforced concrete capping employed for the pile foundations for the masonry piers of the Penrose ferry bridge over the Schuylkill River at Philadelphia, Pa. The reinforcement consists of a layer of expanded metal of 6-in. mesh and No. 4 gage plate. The metal was secured to each pile top by a staple.

Concrete-steel Pile Foundations.—The construction of foundations with concrete-steel piles does not differ materially from the familiar construction of timber-pile foundations; the piles are driven in rows or in clusters and capped with a platform or grillage of concrete or concrete-steel on which the masonry or steel superstructure is built in the usual manner. In construction concrete-steel piles are not essentially different from columns of that material; they may be employed for the same purposes and in much the same manner as timber piles. Compared with timber piles they have several advantages and some disadvantages. Their most important advantage, perhaps, is that

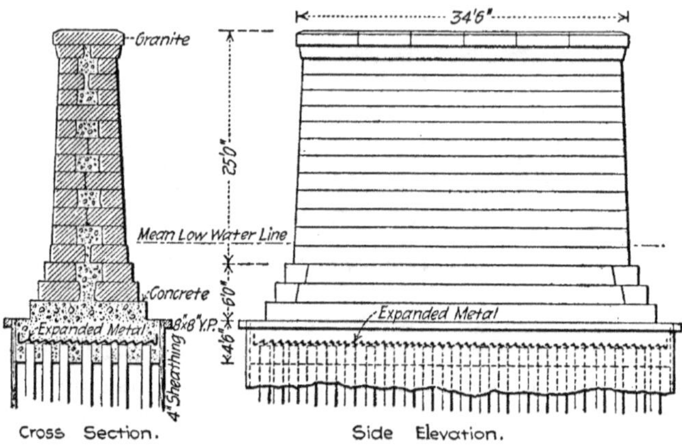

FIG. 44.—Concrete-steel Capping for Pile Foundation, Penrose Ferry Bridge, Philadelphia, Pa.

they are equally durable in dry or wet soil, while piles of timber must be constantly submerged to remain preserved. Their most serious disadvantages are that they are more expensive in first cost and more troublesome to drive than timber piles. In recent practice, however, the difficulty of driving concrete-steel piles has been largely overcome, and with care it is probable that such piles can be driven in almost any soil that timber piles will penetrate. Pile for pile concrete will usually cost more than timber. Often, however, its use will effect a saving in the total cost of the foundation, since, not having to cut the piles off below ground-water level, the depth of excavation and volume of masonry may be reduced. Another advantage of concrete-steel piles is the readiness and efficiency with which they may be bonded monolithically with the grillage or capping of concrete or reinforced concrete which embeds their tops. In a word, concrete-steel piles promise to combine the advantages of timber piles with some of the advantages of masonry piers, and in particular their independence of moisture conditions and their freedom from the ravages of the teredo make them particularly suitable for marine work.

Cast-pile Construction.—The illustrations of Figs. 45 to 50 show the principal forms of concrete-steel piles which have been employed in construction. The writers know of no instance of a cast reinforced concrete pile of circular section being used. Such a form of pile would possess certain obvious advantages over rectilinear sections, but it would be more expensive because of the greater complexity of the molds and of the reinforcement required in its construction. All the piles illustrated are rectilinear in cross-section and all but one are quadrilateral. Fig. 45 shows the form of bearing pile designed

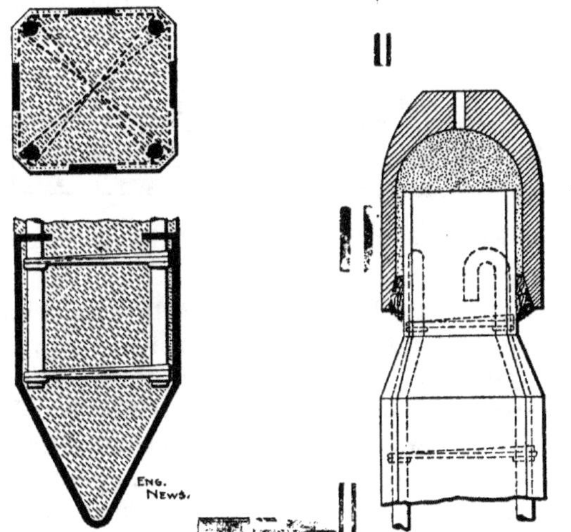

FIG. 45.—Reinforced Concrete Pile of Hennebique Construction.

and used in a number of important structures by the French engineer Hennebique. It is square in section, the point being pyramidal and the top being shouldered in with a round section above the shoulder to permit the application of a driving-cap. The point is shod with a cast-iron shoe consisting of a point and four side straps which bend in at their tops to lock into the concrete. The reinforcement consists of four round soft-iron rods placed at the four corners and wired together at intervals as shown by the drawing. In some instances this reinforcement is increased by four additional rods placed midway of the sides. The piles are also sometimes constructed with a circular hole running lengthwise of them at the center. This hole bifurcates into two, reaching to the beveled surfaces at the point, and serves for a water-jet during driving and as a means of forcing grout into the soil surrounding the point when the pile has penetrated to its final position. It is filled previous to capping the pile. In the construction of sheet piles Mr. Hennebique changes the cross-section from a square to a rectangular slab and employs six instead of four reinforcing-bars, the two additional bars being placed one midway of each broad face of the pile.' The point is formed by beveling one of the narrow faces so as to form a wedge edge with the opposite face, and both narrow faces are slotted by a semicircular groove running from point to shoulder. These semicircular grooves form, when the piles are driven as sheeting, a circular hole from top to bottom between each pair of piles. This hole serves for a water-jet during driving and afterwards for injecting grout into the soil around the points. The final filling of the circular holes forms a monolithic plate of the sheeting. The drawings Fig. 46

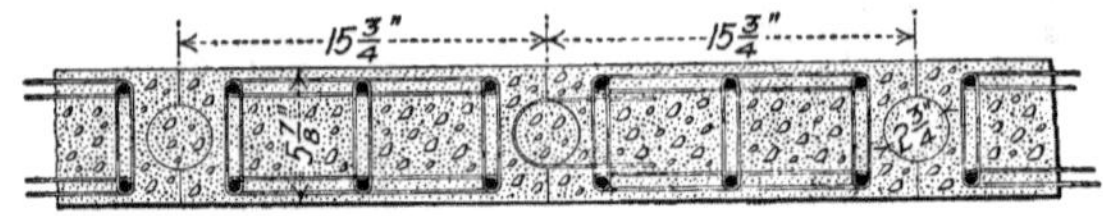

FIG. 46.—Reinforced Concrete Sheet Piling of Hennebique Construction.

show the form of sheet pile whose construction has been described. Another concrete-steel sheet pile of European design is shown by Fig. 47. The reinforcement consists of two parallel I beams tied together at intervals of about 7.7 ft. by means of plate and angle cross-ties. One of the narrow faces is grooved and the other has a tongue to provide for the interlocking of the adjacent piles when they are driven as sheeting. The particular pile illustrated was designed for a quay wall, and the anchor-rod shown at the top was carried back into the filling of the quay and fastened to a massive concrete anchor. Fig. 48 shows a form of concrete-steel pile which was employed in the foundation construction for a court-house built near Berlin, Germany. As will be seen, the pile is triangular in section, with the apices of the triangle cut off, and is reinforced by three 1-in. iron rods placed at the corners and wired together at intervals of 8 ins. by ties of ¼-in. wire. At the bottom of the pile the ends of the three rods converge to a point and are welded together. The concrete mixture was 1 part Portland cement and 3 parts river gravel. A reinforced concrete pile of American design is shown by Fig. 49. The pile is square in cross-section with a pyramidal point and flat top, and is made of 1–2–4 Portland-cement concrete. The reinforcement consists of four $1\frac{7}{16}$-in. rods placed at the corners and extending down into the pyramidal point, where they take seat on the cast-iron shoe-point. This point is anchored to the pile by wrought-iron straps bent in at their tops. Throughout the shaft of the pile the reinforcing-rods are bound together at intervals by wire ties. At the top of the pile the concrete is carried 2 ins. above the tops of the reinforcing-rods, but after the piles are driven this top concrete may be broken off, leaving the rods projecting

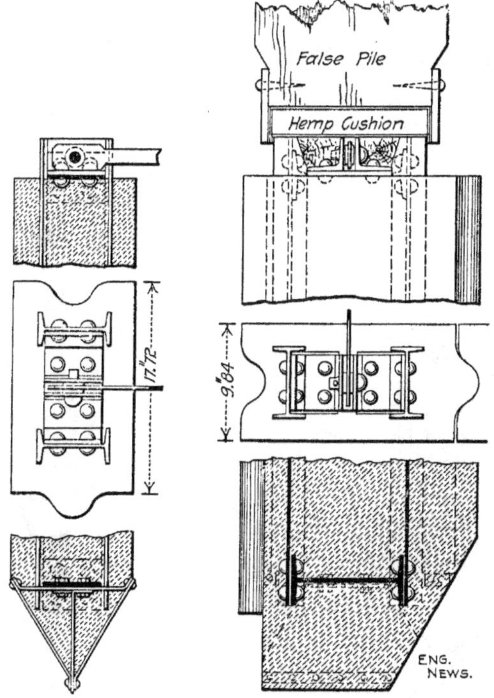

FIG. 47.—Concrete Sheet Pile with I-beam Reinforcement.

to bond the pile with its capping. Fig. 50 shows the form of concrete-steel pile employed in constructing a wharf at Novorossisk, Russia. Altogether twenty-four of these piles were employed. They were each 15.75 ins. square and were reinforced as shown by the drawings. The reinforcement weighed about 45 lbs. per lineal foot of pile, and the total weight of the piles per lineal foot was about 262 lbs. Each pile had to support a section of wharf floor 269 sq. ft. in area and to carry a load of 85,584 lbs.

Concrete-steel Piles Built in Place.—The forms of reinforced concrete piles so far described are molded complete, and after hardening

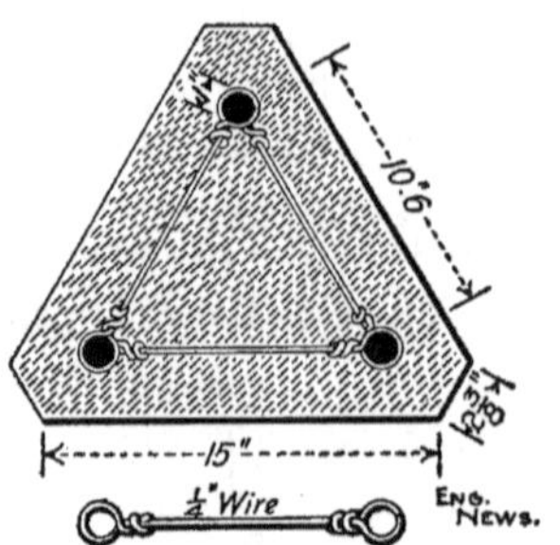

Fig. 48.—Reinforced Concrete Pile for Court-house Foundation, Berlin, Germany.

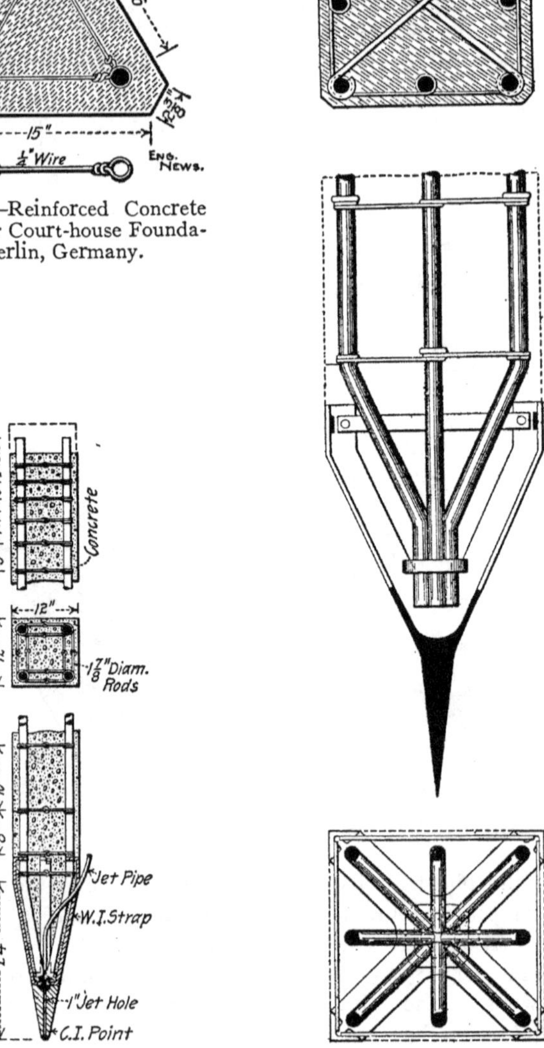

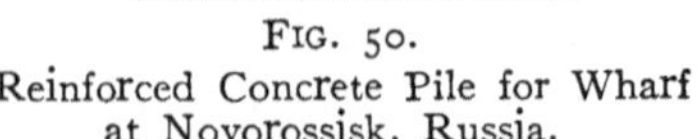

Fig. 49.
Reinforced Concrete Pile for Hallenbeck Building, New York City.

Fig. 50.
Reinforced Concrete Pile for Wharf at Novorossisk, Russia.

are driven like timber piles. A number of inventors have recently devised methods of constructing the piles in place, and these are fully described in a succeeding chapter. Two forms of reinforced concrete piles which are designed to be constructed in place are described in the following paragraphs. So far as the author knows these forms of concrete-steel piles have never been used in actual construction, but plain concrete piles of the same form and constructed in the same manner, excepting that the reinforcement was not inserted, have been employed in a number of important structures. Fig. 51 shows trans-

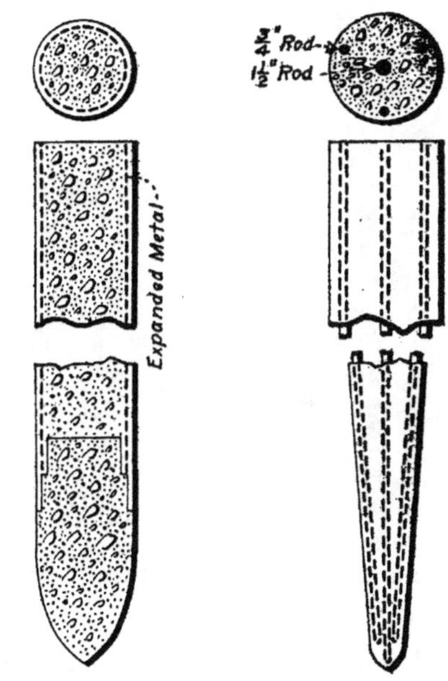

FIG. 51.—Simplex Pile Built in Place.

FIG. 52.—Raymond Pile Built in Place.

verse horizontal and vertical longitudinal sections of the reinforced concrete pile designed by the enigneers of the Simplex Concrete Piling Company of Philadelphia, Pa. This pile is reinforced by a circumferential cylinder of expanded metal. The reinforcing metal has a thickness of $\frac{5}{16}$ in. and a 3-in. mesh. Fig. 52 shows the concrete-steel pile made by the Raymond Concrete Pile Company of Chicago, Ill. The reinforcement consists of a round bar about $1\frac{1}{2}$ ins. in diameter extending lengthwise of the pile along its axis, and of three smaller bars spaced equidistant around the circumference. These outside bars are generally $\frac{3}{4}$ in. in diameter. The pile is usually made with a uniform taper from butt to point.

Examples of Foundations.—Fig. 53 shows the reinforced concrete pile foundation and pier of the Brumath bridge in Alsace-Lorraine. The piles, pier, and spans are all of Hennebique construction. As will be seen, the pier rests on eleven piles which are in one piece with it, this monolithic construction being obtained by practically continuing the reinforcing-rods up through the pier. The piles used

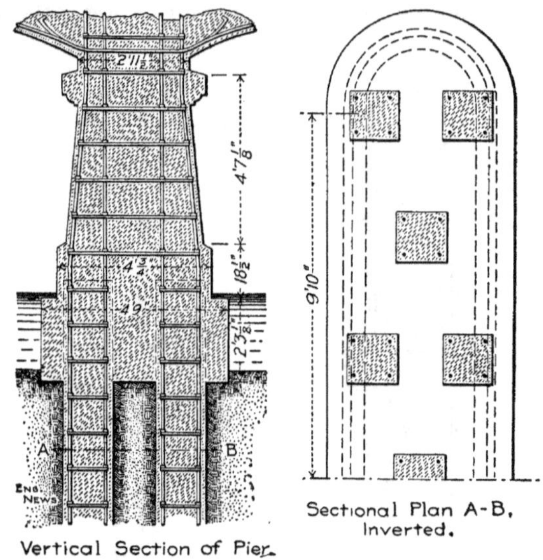

FIG. 53.—Reinforced Concrete Pile Foundation, Brumath Bridge, Alsace-Lorraine.

were 14 and 16 ft. long and 16 ins. square, and the foundation-bed consisted of about $6\frac{1}{2}$ ft. of swamp mud overlying coarse gravel. With an 8,800-lb. hammer and a drop of 20 ins. the piles penetrated the soft material from 8 to 12 ins. at each blow; in the gravel with a drop of hammer of $2\frac{1}{3}$ to $4\frac{1}{4}$ ft. the penetration was 2 ins. at first and gradually decreased to 0.4 in. The ultimate bearing power of these piles was figured by Ritter's formula,

$$W=\frac{h}{e}\,\frac{Q^2}{Q+q}+Q+q,$$

to be 546,000 lbs. each, the values taken being

Q = weight of hammer in pounds = 8,800;
q = weight of piles in pounds = 8,800;
h = height of fall in inches = 48;
e = final penetration in inches = 0.4.

This foundation, though small, shows the unity of structure which is readily obtained when reinforced concrete piles are combined with a capping or superstructure of the same material. A concrete-steel

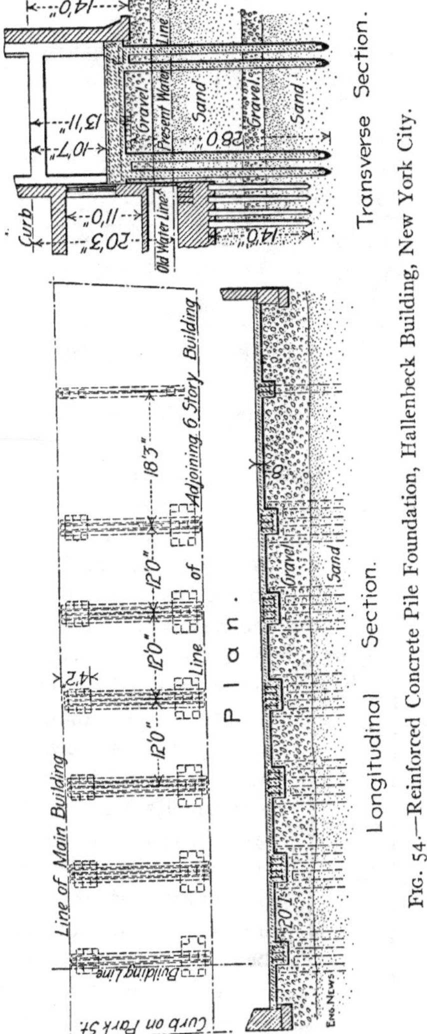

FIG. 54.—Reinforced Concrete Pile Foundation, Hallenbeck Building, New York City.

foundation of more pretentious character is shown by the drawings of Figs. 54 and 55. This foundation was designed for a ten-story addition to the Hallenbeck Building in New York City. Fig. 54 is a general plan and sections of the foundation and illustrates its arrangement and character quite plainly. The piles were of the construction

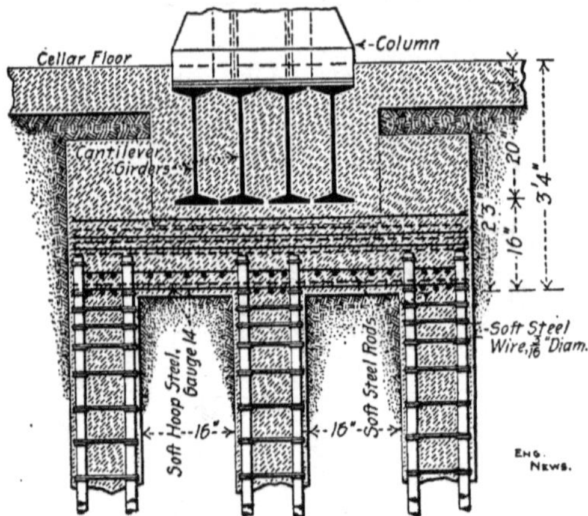

Fig. 55.—Reinforced Concrete Pile Capping, Hallenbeck Building.

shown by Fig. 49. The piles of each group of two, four, or six piles were capped with a reinforced concrete slab constructed as shown by Fig. 55. The piles were calculated to sustain 80,000 lbs. each, made up of 36,000 lbs. per square foot allowed by the New York Building Laws for concrete piers and an additional load of 44,000 lbs. borne by the four reinforcing-rods.

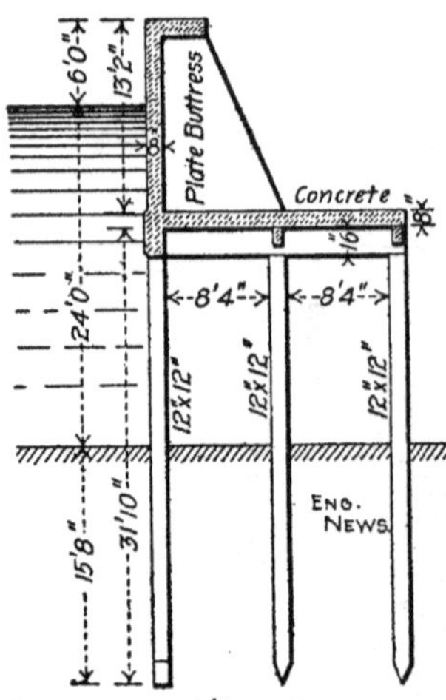

Fig. 56.—Reinforced Concrete Quay Wall Carried on Reinforced Concrete Piles, Southampton, England.

Fig. 56 is a diagram cross-section of a quay wall somewhat over 400 ft. long which was built for the London and Southwestern Railway at Southampton, England, in 1898. Longitudinally of the wall the supporting piles were spaced 3 ft. 7 ins. apart, and those of the front row under the face of the wall were filled between with concrete-steel sheet piles. Both bearing and sheet piles were of Hennebique construction. The bearing piles were 12×12 ins. in section and were reinforced by eight $1\frac{1}{8}$-in. bars connected by $\frac{3}{16}$-in. ties. A wharf carried by concrete-steel piles and constructed at Southampton in 1898 is shown in cross-section and part longitudinal section by Fig. 57. This wharf consists of two parts at right angles to each other; one part is 90 ft. long and 30 ft. wide and the other part is 100 ft. long and 46 ft. wide. The foundation piles are 12 ins. square and are each reinforced by four rods; they carry a superstructure composed of a braced frame-

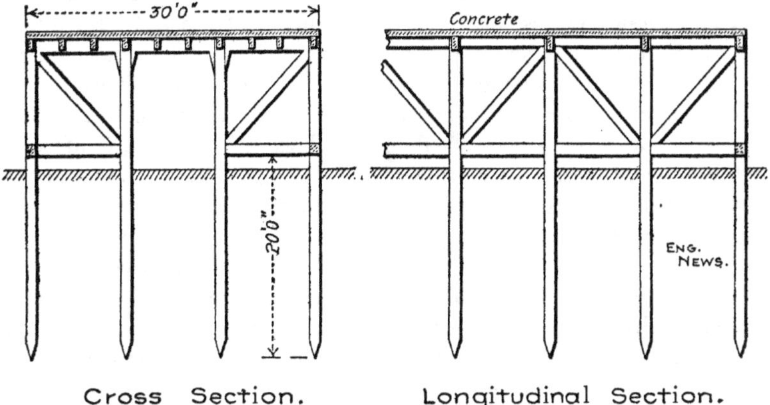

FIG. 57.—Reinforced Concrete Wharf Carried on Reinforced Concrete Piles, Southampton, England

work and a platform of reinforced concrete. The pier was designed for a load of about 575 lbs. per square foot.

Open-caisson Foundations.—A use of concrete-steel in foundation work, which has been practiced in Europe, is its substitution for steel or brick curbing in sinking pits for foundations. Two modes of procedure are followed in this work; one is to build the reinforced concrete curb wall in place and the other is to mold cylinders of concrete-steel and sink them one on top of the other as metal cylinders are sunk in similar cases. An example of the first mode of procedure is furnished by the foundations for the Municipal Theatre at Berne, Switzerland, constructed by Mr. Paul Simons, civil engineer, of the same city. This building was founded on piers which penetrated a layer of made ground into the solid soil beneath. These piers were rectangular in horizontal section with a flared bottom to gain a large bearing area. To construct them a pit was sunk and curbed as the excavation progressed with concrete reinforced by horizontal rods spaced 10 cm. (3.9 ins.) apart vertically. The mode of procedure was to excavate a depth of pit varying with the nature of the soil and then set up inside the excavation a frame or form the annular space outside of which was filled with concrete to form the curbing. As soon as a section of curbing had hardened the frame was knocked down and another section of pit dug and curbed in a manner similar to the first. Each section of curbing was bonded to the sections above and below by short hooked rods or dowels embedded in the concrete. After completion to their full depth the pits were filled with concrete.

In constructing the new road bridge over Cockle Creek, near Sydney, New South Wales, Mr. E. M. De Burgh, M. Inst. C. E., made use of Monier cylinders for constructing two double-cylinder piers on a foundation consisting of about 5 ft. of sand and silt overlying hard blue clay. The cylinders were made 3½ ft. internal diameter and 3 ft. 7 ins. long with a shell 2⅛ ins. thick. They were reinforced with a layer of steel-wire netting of No. 16 gage wire woven to 1½-in. mesh, and with two spirals of No. 8 gage wire wound completely round the cylinder, the turns being 1 in. apart. To couple the cylinders together end to end six pairs of bars 1¾×¼ in. were inserted between the two spirals lengthwise of the cylinder and spaced 60° apart circumferentially. The bars forming each pair were blocked apart by spacers so as to leave at six points at each end of the cylinder a sort of mortise. In placing the cylinders together the mortises were made to coincide and a fish-plate was inserted so as to extend one-half downward into the lower cylinder and one-half upward into the upper cylinder. Slots were cut in the mortise-bars and fish-plate into which steel keys were inserted locking the two cylinders together. The coupling construction is shown by Fig. 58, and Fig. 59 shows the cylinder construction proper. In sinking, the bottom cylinder was provided with a cast-iron cutting edge, and they were sunk by excavating inside and adding successive lengths exactly as cast-iron cylinder piers are commonly sunk. After sinking the cylinders were filled with concrete. These cylinders were sunk to a depth of 36 ft. below water level. The cylinders delivered cost 24 s. per lineal foot.

Armored Timber Piles.—In waters infested by the teredo cylinders of reinforced concrete have been successfully employed for armoring timber piles. A notable example of this construction is furnished by the road bridge built over Cockle Creek, near Sydney, New South Wales. Five three-pile bents were armored with Monier cylinders. The formation penetrated by the piles was sand mixed with vegetable matter overlying a stiff blue clay to a depth of about 5 ft. The piles were of ironbark about 40 ft. long, 14 ins. in diameter at the point and 18 ins. at the butt, and they were driven about 15 ft. into the clay stratum. From the level of the clay to high-water level four hard-wood battens were spiked longitudinally of the pile and 90° apart circumferentially. The pile was coated with Stockholm tar previous to driving, and after being driven a platform was attached above high-water level, and on this was erected, by threading over the head of the pile, a sufficient number of Monier pipes 21 ins. in diameter to reach the length of the battened portion of the pile. The joints between the adjacent lengths of pipe were connected by means of a collar of Monier netting covered with cement plaster as indicated by Fig. 60. The several

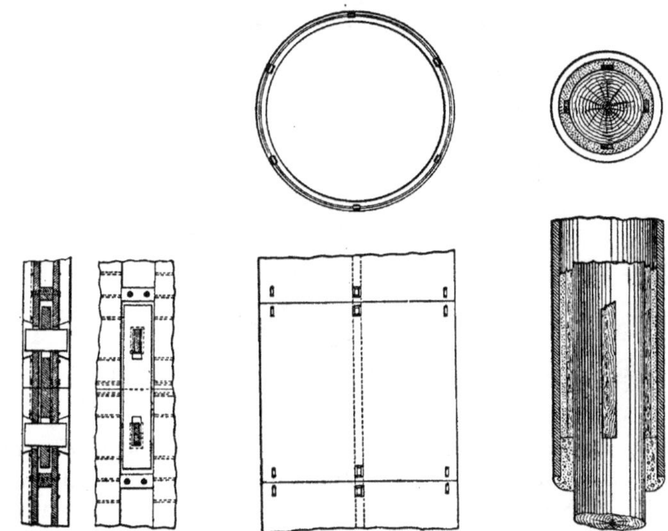

FIG. 58.—Coupling for Sections of Reinforced Concrete Cylinder Foundation.

FIG. 59.—Reinforced Concrete Cylinder for Cylinder Foundation.

FIG. 60.—Reinforced Concrete Armoring for Timber Piles, Cockle Creek Bridge, New South Wales.

lengths of pipe were joined up to form a continuous cylinder. The platform was removed and the cylinder lowered by means of hooks under the bottom edge until it rested on the bottom of the stream. A jet of water from a 1½-in. pipe was then worked around the bottom of the casing to loosen the underlying material, and pressure being applied by means of screw-jacks at the top the casing was forced down to the clay stratum. The batten-pieces on the pile kept the casing and pile concentric, and after sinking the annular space was scoured out by means of the jet and filled with clean sand with 9 ins. of concrete mortar at the top to form a cap. The construction is clearly shown in Fig. 60. The Monier pipes were 1⅜ ins. thick, of cement mortar with a reinforcement consisting of a 1¼-in. square-mesh netting of No. 16 gage wire. A form of timber-pile armoring adopted by the State Harbor Commission, San Francisco, Cal., for pier construction in teredo-infested waters is shown by Fig. 61. The piles are designed to take all the load and are driven in clusters of three and sawed off, one 2 ft., one 6 ft., and one 8 ft. below the wharf level. After the piles are driven, wooden-stave cylinders of 3-in. planks are

FIG. 61.—Reinforced Concrete Armoring for Timber Piles, Wharf Construction, San Francisco, Cal.

placed over them and driven 10 ft. or 12 ft. into the mud. These cylinders are then sealed at the bottom and pumped out, after which a cylinder of No. 16 gage expanded metal 6 ins. less in diameter is set inside each and the remaining space is filled with concrete. It is expected that the wooden-stave sheathing will ultimately be destroyed by the teredo, leaving the concrete armored piles as the permanent structure.

CHAPTER VII.—REINFORCED CONCRETE IN BUILDING CONSTRUCTION.

THE principal structural items which combine to form buildings are foundations, columns, floors, partitions, walls, and roof. Reinforced concrete is employed for all these and also for stairways, corbels, projecting galleries, vaults, cupolas, and ornamental shapes. In Continental Europe buildings of concrete-steel have been constructed in greater number and variety than in the United States. In this country its use as a material for foundations and for fire-proof floors has been highly developed for a number of years, but it is only recently and following European precedents that whole buildings of concrete-steel have been erected in any considerable number. The pioneer in this development was Mr. Ernest L. Ransome, Assoc. Am. Soc. C. E., and some of the early buildings of concrete-steel erected by this engineer on the Pacific Coast are notable examples of such construction. Most of Mr. Ransome's work has, however, been done within the last few years. All the other systems of concrete-steel building construction employed in the United States are foreign systems imported direct or the adaptations from foreign practice. The Hennebique and De Vallière systems of construction belong to the latter class. In the United States buildings constructed wholly of concrete-steel vary in character and purpose from factories and shops to tall office buildings, and from small residences to churches and court-houses. European practice covers quite as wide a range of structure.

Probably the most ambitious example of concrete-steel building construction in the United States is the 16-story office building known as the Ingalls Building at Cincinnati, Ohio. This building is 100×50½ ft. in plan and 213 ft. high from sidewalk to cornice. It is constructed entirely of reinforced concrete except the marble and brick veneering of the walls, the metal window-frames, and the inside trimming. Many of the separate details of this building are described in the succeeding pages of this chapter, and a description of the building as a whole will be found in *Engineering News* of July 30, 1903.

The following sections take up in order floors, columns, walls, stairways, roofs, and other items of building construction and give representative examples of each.

Floors.

Reinforced concrete floor constructions may be divided into two classes. The first class comprises those constructions which serve simply as a filling between the girders and beams of a floor framework of steel; the second class includes those constructions in which the girders and beams are of reinforced concrete and monolithic with the floor plate proper, all steel floor framing being done away with. Between these two separate and distinct classes of constructions there is a third form of construction, or a sub-class, which partakes somewhat of the nature of each of the two main classes. This construction is designed to eliminate the small beams or joists of the ordinary floor framework and to depend for support on the main beams and girders alone; in effect it is a long-span construction of the first class. Floor constructions belonging strictly to the second class, or all-concrete constructions as they are commonly designated, have been more extensively developed in Europe than in America. Indeed until quite recently the Ransome construction was the only true representation of this class that was strictly an American development. Within a few years, however, other systems have been worked out by American engineers and have been employed to a considerable extent. The backwardness of American engineers in comparison with their European brethren in developing all-concrete constructions has, however, been more than counterbalanced by their activity in devising constructions of the first class in both short-span and long-span forms. The number of these constructions is legion, but only a few of them have gained wide recognition and need be specifically considered.

Considering first the various forms of floors of the first class, these may be enumerated as follows: (1) Flat-plate floors in which the floor-plate is carried on the top flanges of the floor-beams; (2) flat-plate floors in which the floor-plate is carried on the bottom flanges of the floor-beams; (3) flat-plate floors in which the floor-plate is flush with the tops of the floor-beams or embeds the upper flanges; (4) arch-ring floors sprung between the lower flanges of the floor-beams, and, (5) flat-topped arches with springing lines on the bottom flanges and top flush with or just above the top flanges. In floors of the first and third forms some form of haunching or boxing is required to protect the floor-beams below the floor-plate; these two forms of plate floors and both forms of arch floors also require some form of suspended ceiling construction if a flat ceiling is desired. The floors of the second and fourth forms require a filling above the floor-plates to protect the portions of the floor-beams above them. In floors of the first form the floor-plate is commonly continuous over a large part or the whole of the floor framework.

Floors of the second class, or of all-concrete construction, all have the form of ribbed plates or flat-topped arches. In the plate forms the stiffening-ribs take the place of the steel beams and joists of constructions of the first class, and they usually consist of a series of parallel heavy ribs, which form girders resting on the building columns and walls, and of similar series of smaller ribs at right angles to and supported by the main ribs or girders. These ribs are monolithic with the floor-plate, which is continuous between walls except that expansion joints are sometimes introduced at intermediate points. In some constructions the main ribs or girders are arched, this form of floor being designed and employed for long and heavy spans.

Columbian Construction.—The standard Columbian floor consists of a flat plate of concrete reinforced by straight parallel bars extending from beam to beam. It is employed for spans up to 24 ft., but for spans longer than 16 ft. the saving in metal is small and it rapidly decreases as the span lengthens. The thickness of the concrete plate and the depth, sectional profile, spacing, and method of support of the reinforcing-bars differ for different spans and loads. Generally the depth of the plate is made 1½ ins. greater than the depth of the bars, but this limit is frequently varied from, especially in long-span construction. The reinforcing-bars are of structural steel and are rolled to the special sections described on p. 336. In short-span floors they are suspended from the floor-beams by stirrups, which are hung over the floor-beam with a leg on each side which is perforated to receive the end of the bar. In long-span floors the floor-beams are omitted and the reinforcing-bars are fastened directly to the webs of the main floor-girders by riveted angles. Stone or slag concrete is employed to embed the bars; no cinder concrete is employed except as a veneering to receive the direct action of flame in case of fire. The proportions of the mixture employed are generally 1–2–5, but with slag aggregate the proportion of sand is usually increased to 2½ or 3 parts. The mixture is made very wet and if a slag aggregate is employed it is thoroughly drenched with water before mixing. The constructions described are illustrated by Figs. 62 and 63. Fig. 62 shows the con-

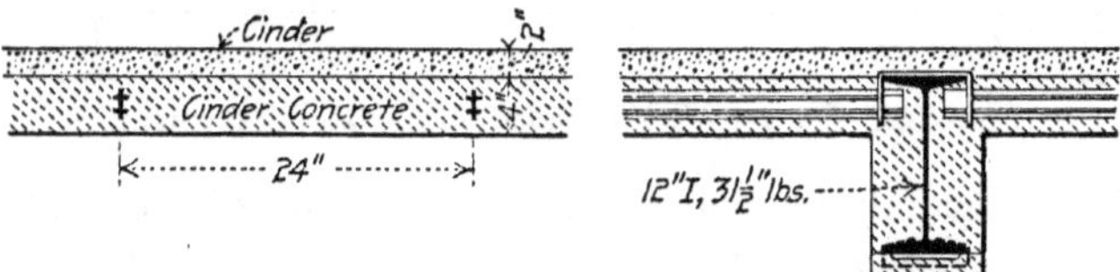

FIG. 62.—Short-span "Columbian" Floor Construction.

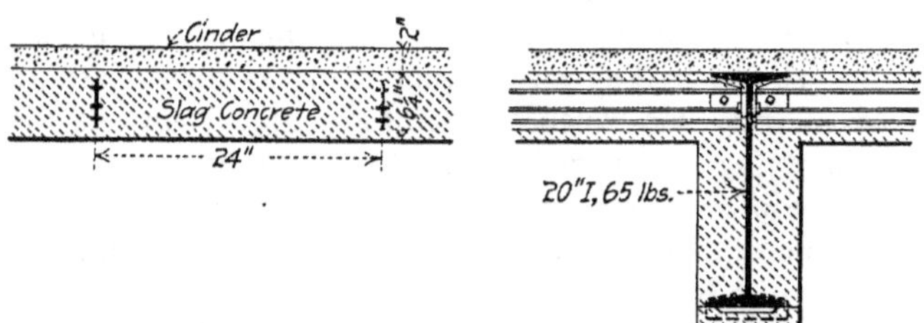

FIG. 63.—Long-span "Columbian" Floor Construction.

struction of a 6-ft. span floor using 2-in. bars spaced 24 ins. apart and embedded in a 4-in. plate. The bars are hung from the floor-beams by stirrups. Fig. 63 shows a 16-foot space floor using 4½-in. bars spaced 24 ins. apart and embedded in a 6¼-in. plate. The reinforcing bars are connected to the beams by riveted angles.

Expanded Metal Constructions.—Floors with expanded metal reinforcement are built in the form of flat plates, flat-topped arches, and flat plates with stiffening-ribs. Approximately speaking the flat-plate construction is employed for all spans up to 8 ft., for ordinary loads, and ribbed plates are employed for all greater spans; the arch construction is a heavy floor form for spans up to 8 ft. or 9 ft. This division must be regarded as a very elastic one. There is no hard and fast line dividing the span-lengths for which the different forms of floors are employed in actual practice; ribbed plates have been frequently used for spans considerably less than 8 ft., and flat plates have been employed for much greater spans and in some instances for heavy floors for the construction of which the division calls for flat-topped arches. The division is approximately exact, however; beyond spans of 8 ft., plate construction is of doubtful economy as compared with ribbed plates and for ordinary loads the ribbed plate has no advantages over the plain flat plate for spans less than 8 ft.

Flat-plate Floors.—The standard construction of flat-plate floors with expanded metal reinforcement is shown by Fig. 64, but this con-

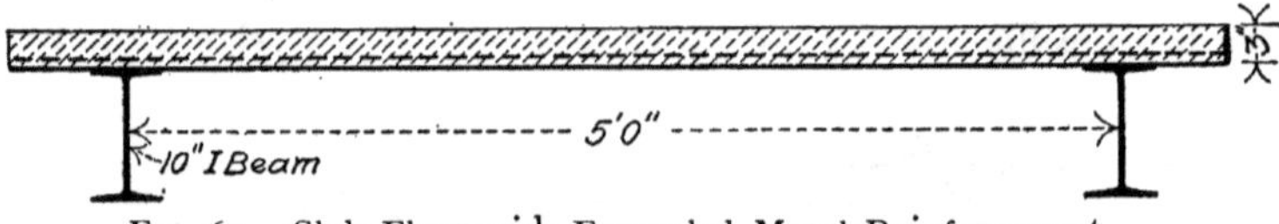

FIG. 64.—Slab Floor with Expanded Metal Reinforcement.

struction is frequently modified by locating the top of the reinforced plate flush with the tops of the I beams or by carrying it on the lower flanges of these beams and filling above with a meagre concrete made with cinders or some other aggregate of light weight. In the modified constructions the reinforcing metal, which is made continuous over the I beams as shown by Fig. 64, is a separate sheet for each span. In many instances, however, when the reinforced plate embeds the top flange the reinforcing sheets are continued over the beams, and given a downward curve or sag between beams, thus securing continuity over supports and the reinforcement of the tension side of the plate. For a standard floor of 6 ft. span, the plate is made 3 ins. thick and is composed of a mixture of 1 part cement, 2 parts sand, and 6 parts anthracite-coal cinders, with the reinforcement embedded in the lower inch of the plate. Cinder concrete of the composition stated weighs about 84 lbs. per cubic foot.

In the special cases where flat-plate floors have been designed for spans of considerably more than 8 ft., a special arrangement and disposition of the reinforcement has usually been adopted. As an illustration reference may be made to a design recently developed for a warehouse floor to carry 400 lbs. per square foot. This floor had a span of 11 ft. 1 in. and consisted of a reinforced slab of stone concrete 6½ ins. thick, with 1½ ins. of cinder concrete on its under side. The reinforcement was embedded in the stone-concrete slab at a depth of 5½ ins. from its top. This reinforcement was expanded metal of No. 4 gage and 8-in. mesh, and the two 8-ft. lengths were lapped at the middle of the span, the metal being double for a distance of 5 ft. 5 ins. There was also a sheet of expanded metal bent over and behind each I beam and then downward toward the center of the span. These beam sheets were placed near the top of the slab. The composition of the slab concrete was as follows: For the top 1 in., 1 part cement, 1 part sand, and 1 part ¼-in. broken stone; for the next 5½ ins., 1 part cement, 2½ parts gravel, and 6 parts anthracite cinders. While the floor described is essentially an individual example of long-span flat-plate floor design, it is fairly typical in illustrating the nature of the reinforcement which is employed for such spans.

The occasional use of the flat-plate form for heavy floors in place of flat-topped arches is fairly illustrated by the floor built to carry the boilers and generating machinery in the power plant of the Queens Borough Electric Light and Power Company, at Far Rockaway, N. Y. This power-house was located on ground offering a very poor base for machinery foundations and it was necessary to excavate to a depth of 10 ft. or more to hard-pan. Upon this hard material, concrete bases were laid 4 ft. wide and 18 ins. deep and running the entire length of the building. These bases were spaced 6 ft. to 7 ft. from center to center and upon them brick piers 20 ins. square were built spaced about 6 ft. apart and rising to within 8 ins. of the level of the boiler-house floor. A false floor was then built flush with the tops of the piers and over the whole was spread a 2-in. layer of concrete. While still soft this was covered with a layer of expanded metal, then with another 2-in. layer of concrete upon which was placed a second layer of expanded metal. The floor was finished by a 4-in. layer of concrete, making its total thickness 8 ins. The engine foundations and boiler housings were placed directly on this floor.

Flat-topped Arch Floor.—The segmental flat-topped arch floor for heavy loads calls only for brief description. As ordinarily employed the arch is sprung between I beams. Fig. 65 shows the design of the St. Louis Expanded Metal Company. Other designs differ in detail from this. In some practice the sheets of expanded metal are bent to curve and set on the lower flange of the I beams. When this is done no center is built, the concrete being deposited directly upon the expanded metal sheet, which is stiff enough to carry the temporary load, and the soffit being plastered to embed the metal. This floor belongs to the composite type discussed in Chapter I. A good example of the construction is furnished by the boiler-room floors of the Waterside

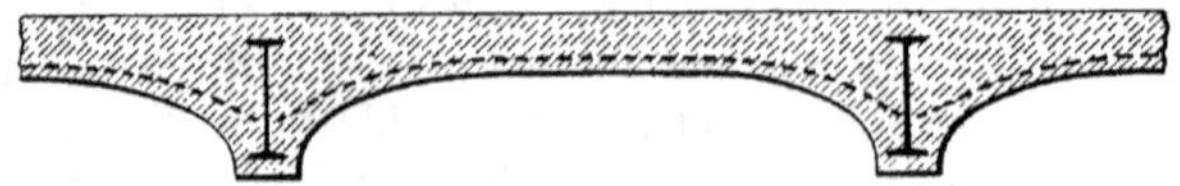

FIG. 65.—Segmental Arch Floor with Expanded Metal Reinforcement.

station of the New York Edison Company. Here the arches are of 5 ft. span and 10 ins. rise and are built between 15-in. I beams, with a thickness of 5 ins. at the crown.

Ribbed-plate Floors.—The earliest form of ribbed floor of expanded metal construction was the Golding floor with arch ribs. Fig. 66

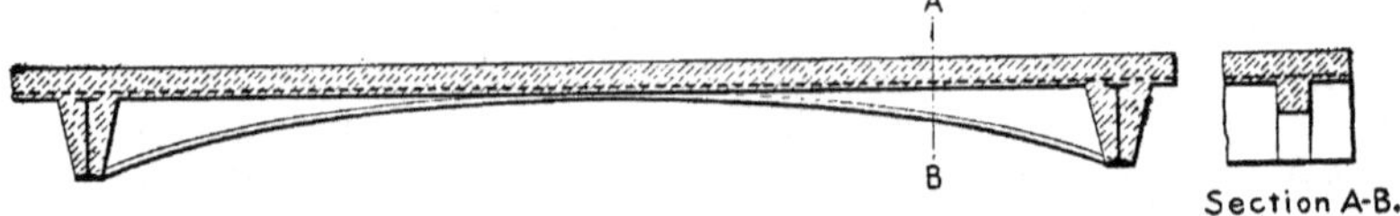

FIG. 66.—"Golding" Floor with Expanded Metal and Channel Reinforcement.

shows this form of construction, which is adapted to spans from 8 ft. to 18 ft. in length. The arch ribs are carried by soffit channels, which are sometimes exposed, as shown by the illustration, and sometimes covered, usually by a wrapping of expanded metal and plaster. The floor-plate reinforcement is substantially the same as in plain flat-plate construction. The ribs are usually placed about 4 ft. apart and the distance between ribs is assumed to be the span in figuring the thickness of plates employed. Fig. 67 shows a ribbed-plate construction designed for spans up to 30 ft. In this floor the plate is reinforced by expanded metal in the usual way and the stiffening-ribs have a corrugated-bar reinforcement.

Hennebique Constructions.—Floor construction with Hennebique reinforcement has been elaborately developed in France. Four forms of construction are in general use: plain flat plates, ribbed flat plates, plain flat-topped arches, and arch rib construction. All of these forms are monolithic and take their support directly on the main building walls and columns without the intermediary of a steel framework of beams and girders.

Plain Flat Plates.—Plain flat-plate floors of Hennebique construction are generally limited to spans of 5 m. (16.4 ft.), but in exceptional cases where the space covered is approximately square and support is had for all four edges of the plate much greater spans are considered permissible. As a rule, however, ribbed or arch constructions are employed for spans greater than 5 m. (16.4 ft.). In designing flat-

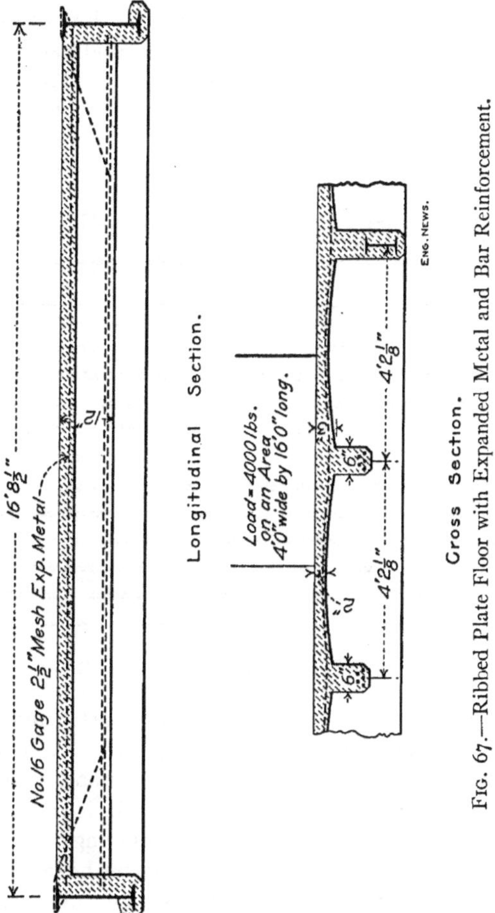

FIG. 67.—Ribbed Plate Floor with Expanded Metal and Bar Reinforcement.

plate floors, the thickness of the floor-plate is arbitrarily chosen and then the reinforcement is determined by calculation. Three forms of reinforcement are employed. Of these the form shown by Fig. 68 is the most characteristic. It consists of alternate straight and curved bars placed parallel to each other and in the direction of the span. The straight bars are located close to the bottom of the plate and are tied to the concrete above by flat iron stirrups. The ends of these bars are turned up to anchor them to the concrete. The curved bars start near the top of the plate at each end and bend downward so that in the middle of the span they are in the same horizontal plane as the straight bars. These curved bars are without stirrups. The distance apart of the bars is determined by calculation; generally there are from three to six per meter width of the plate and their diameter varies from 8 to 20 mm. (0.32 in. to 0.8 in.). The stirrups are of 20×1.5 mm. (0.8×0.06 in.) flat iron. At the ends of its spans the plate is encased in the wall masonry. The particular floor illustrated

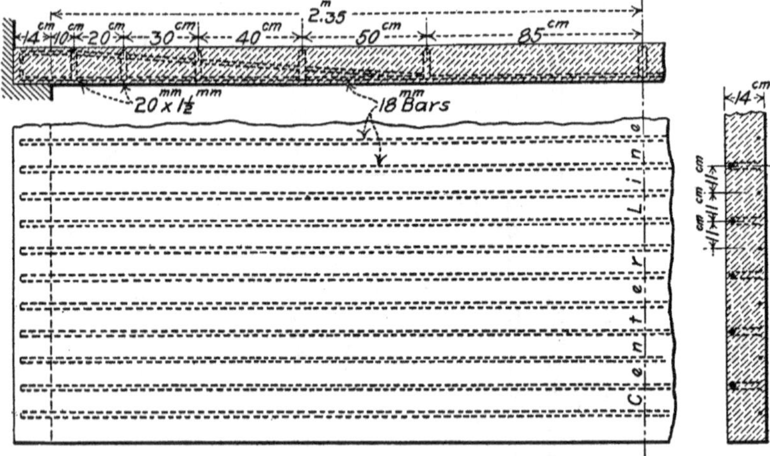

FIG. 68.—Hennebique Flat Plate with Single Reinforcement.

by Fig. 68 is one of 4.7 m. (15.42 ft.) span constructed in 1897 for a bank building at Basel, Switzerland. The second and third forms of flat-plate reinforcement are both designed for floors covering approximately square openings and supported on all four edges, and both are similar in having two sets of parallel reinforcing-bars at right angles to each other. Fig. 69 shows one of these forms of reinforcement. As will be seen the bars in one direction are straight and are placed alternately at the top and the bottom of the plate. The bottom bars are provided with stirrups. The bars at right angles to these staggered bars are alternately straight and bent as in the floor shown by Fig. 68, but the straight bars are without stirrups. The second form of double reinforcement is exactly like that of Fig. 69, except that the curved bars are replaced by straight bars at the top of the plate.

Ribbed-plate Floors.—The form of Hennebique floor most commonly employed is the ribbed-plate floor. Generally two sets of stiffening-ribs are employed, a set of main ribs running parallel with the span and a set of secondary ribs running transversely of the span between main ribs. The main ribs are spaced generally from 3 m. to 5 m. (9.84 ft. to 16.4 ft.) apart and the secondary ribs not exceeding 3.5 m. (11.5 ft.) apart. The span of the main ribs is usually kept

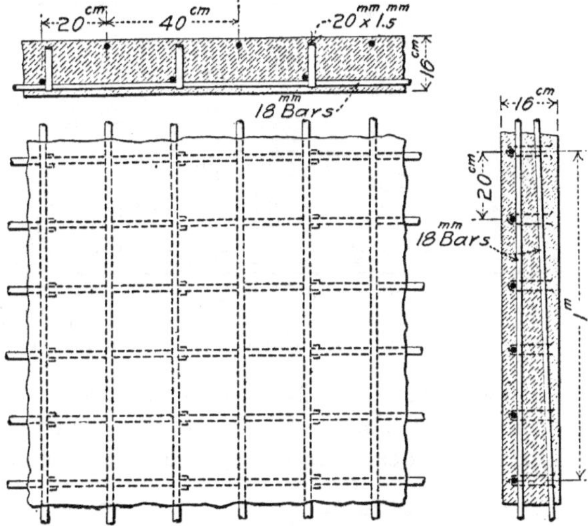

FIG. 69.—Hennebique Flat Plate Floor with Double Reinforcement.

down to 5 m. to 7 m. (16.4 ft. to 23 ft.) if possible, but has been made as much as 14 m. (45.9 ft.). In designing, the thickness of the plate and the depth of the ribs are determined arbitrarily and the reinforcement is figured to correspond. The form and arrangement of the reinforcement for the ribs are indicated by Fig. 70, which shows sections of the floors for the Palais de Justice at Viviers, France. As shown, each main rib is reinforced by rods arranged in pairs, each of which is composed of a straight bottom rod and a bent top rod. The diameter of the rods never exceeds 50 mm. (1.97 ins.) and the number of pairs employed depends upon the total section of reinforcement required. The lower bar of each pair is provided with stirrups reaching vertically upward into the concrete. The bars are so spaced that there is from 3 cm. to 6 cm. (1.2 ins. to 2.4 ins.) of concrete between them at all points, and the bottom bars are kept at least 25 mm. (1 in.) from the bottom of the concrete. The main ribs are firmly anchored into the supporting-walls by the reinforcing-rods. The reinforcement of the secondary ribs is practically the same as that of the main ribs, but seldom consists of more than one pair of bars. These are, however, made continuous through the main ribs from side to side of the span. The reinforcement of the plate proper is always one or the

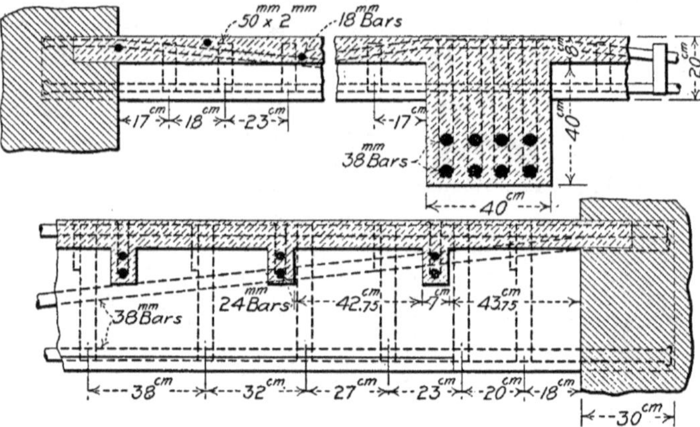

FIG. 70.—Hennebique Floor with Double System of Beams.

other of the forms already described. In some cases the secondary ribs are omitted. Fig. 71 shows such a construction, which was adopted at the Petit Palais des Beaux Arts at the Paris Exposition of 1900.

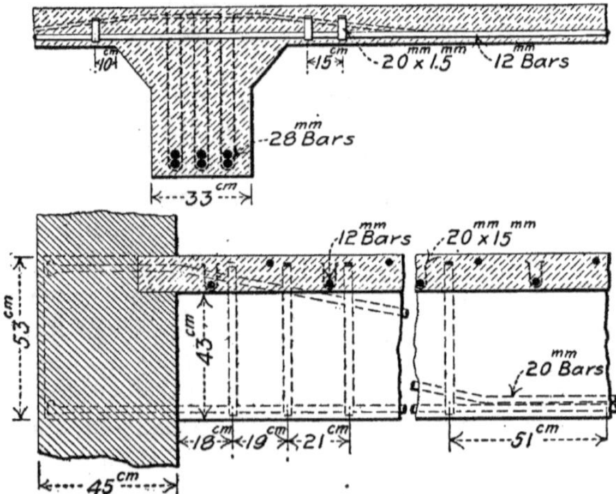

FIG. 71.—Hennebique Floor with Single System of Beams.

This floor has a span of 7.35 m. (24.11 ft.) for the ribs, which are spaced 2.316 m. (7.58 ft.) apart, and was designed for a load of 800 kgs. per square meter.

Plain-arch Floors.—A good example of plain-arch floor construction with Hennebique reinforcement is shown by Fig. 72. Floors

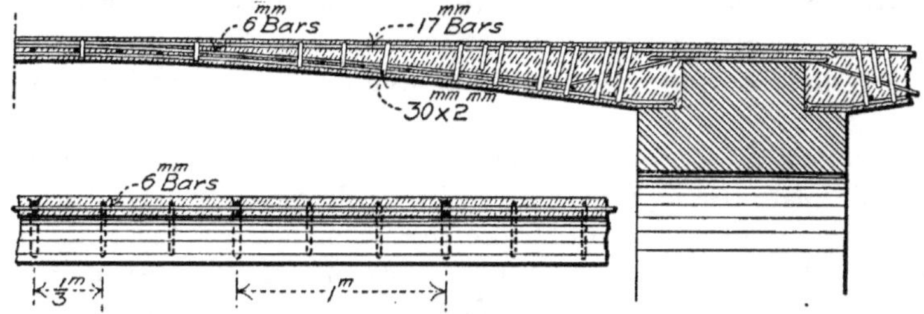

FIG. 72.—Hennebique Plain Arch Floor.

of this type are employed for long spans or for very heavy loads. The particular floor illustrated was employed in the Petit Palais des Beaux Arts at the Paris Exposition of 1900 to span two adjacent galleries of 6.1 m. (20 ft.) span each, and was designed for a load of 1,100 kgs. per square meter. The arrangement of the reinforcement and the dimensions of the arch are clearly shown by the drawings.

Arch-ribbed Floors.—The plain arch is less commonly used in

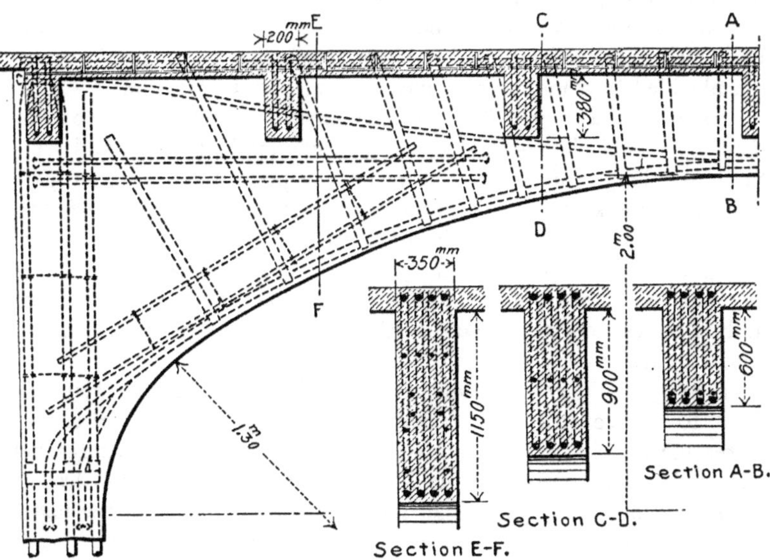

FIG 73.—Hennebique Arch Ribbed Floor.

Hennebique construction than are arch ribs carrying a flat plate. Fig. 73 shows a floor of 7.91 m. (25.95 ft.) span of this type which was built for a machine-shop at Nantes, France. Another construction consists of springing plain arches between the main arch ribs.

Monier Constructions.—There are several forms of floor construction in which Monier netting is used as a reinforcement. They may be divided into flat-plate floors, arch-plate floors, and long-span constructions.

Flat-plate Floors.—The simplest form of flat-plate floor consists of a thin plate resting on the tops of steel floor-beams; this construction makes it necessary to enclose the beam beneath the plate in a boxing of Monier slabs or in a haunching of concrete. Another construction consists in placing the floor-plate on the tops of the lower flanges of the beams and filling above with cinder or coke concrete. In some cases both top and bottom plates are used and filled between with meagre concrete or the space is left empty. Floors of the forms just described are suited only to short spans. The beam spacing employed in Europe is generally from 2 m. (6.56 ft.) to 2.5 m. (8.2 ft.). The thickness of the plates is generally from 4 cm. (1.57 ins.) to 10 cm. (3.94 ins.). The carrying-rods are from 5 mm. to 10 mm. (0.18 in. to 0.39 in.) in diameter and are spaced 5 cm. (1.87 ins.) apart, and the distributing-rods are from 3 mm. to 6 mm. (0.12 in. to 0.24 in.) in diameter and are wired to the main rods with 1 mm. (0.03 in.) wire. Both wrought-iron and steel rods are employed. The reinforcement is embedded in a mortar of 1 part cement and 3 parts sand.

Arch-plate Floors.—The ordinary form of arch-plate floor consists of a single plate arched to a rise of one-tenth the span and set on the lower flanges of the floor-beams. The plate is made about 5 cm. (1.98 ins.) thick and is filled above with a 1 part cement and 8 parts aggregate concrete. This type of construction is considered suitable for spans of 5 m. (16.4 ft.) and loads of 1,200 kgs. per square meter (250 lbs. per square foot). When a stronger floor is desired, two arched plates, set one above the other and filled between with meagre concrete, are employed.

Long-span Floors (*Koenen Floor*).—To adapt Monier reinforcement to the construction of floors of long span without materially increasing the depth and dead weight of the construction, Mr. Koenen, one of the firm controlling the Monier patents in Germany, designed the floor shown by Fig. 74. In this floor the main rods of the netting are anchored to the I beams and given a slight sag between beams, while the concrete is molded to a section of uniform strength longitudinally and completely encases the beams. The same construction is adapted to floor-spans between walls by encasing the ends in the wall masonry and providing anchors for the reinforcement. In Germany this construction is employed particularly for floors for warehouses, factories, and similar buildings, and with it spans of 6.5 m. (21.32 ft.) are built with a thickness at the center of only 2 dcm. (7.86

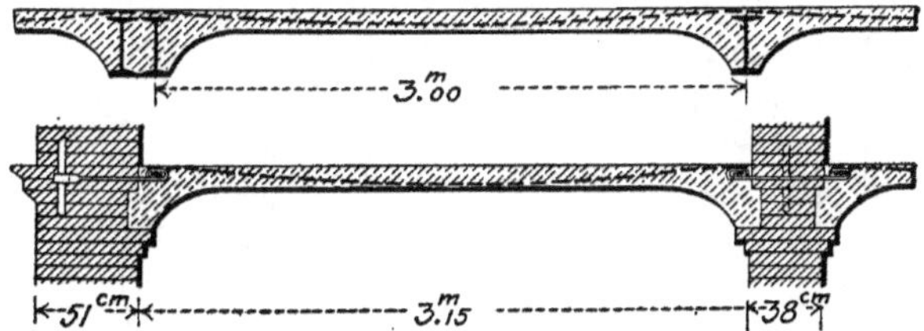

FIG. 74.—Long-span Floors with Monier Reinforcement (Koenen Floor).

ins.). The reinforcement is embedded in a 1 part cement and 4 parts sand mortar.

Ransome Construction.—The floor construction developed by Mr. Ernest L. Ransome and the firms employing his system of reinforcement is a ribbed flat-plate construction. It is best described by specific examples. The drawings of Fig. 75 illustrate a Ransome floor built for the foundry of the Eastwood Company, at Paterson, N. J. This floor is carried by steel columns spaced 11 ft. apart in rows 22 ft. apart. It consists of girders along the rows of columns, of floor-beams at right angles to the girders and 5½ ft. apart, and of a flat plate 4 ins. thick carried by the girders and beams and in one piece with them. This construction divided the floor into panels 5½×22 ft. in dimension. In several places the beams on opposite sides of the girders were staggered, as shown by the part plan. The reinforcement of the floor-plate consists of ¼-in. square twisted rods spaced 6 ins. apart and running parallel to the main girders; that of the main girders consists of two 1¼-in. twisted square rods, and that of the beams consists of two 1¼-in. twisted square rods. The concrete was composed of 2 parts cement, 3 parts sand, and 4 parts ½-in. screened stone. The floor-plate had a ¾-in. finishing-coat of 1 part cement and 2 parts sand mixed with 50 per cent. beach gravel up to ½ in. in size.

The floor described covered an 88×100 ft. main building and a 22×100 ft. annex, and was divided into 22×11 ft. blocks by expansion-joints over each girder and every other beam. The beams were made in single solid sections, those under the expansion-joints having one upper corner rebated to receive the edge of the adjacent slab, as is shown by the drawings of Fig. 75. The joint between beam and slab was not separated; the fresh concrete of the slab was placed directly on the hardened concrete of the beam, so that the joint was in reality

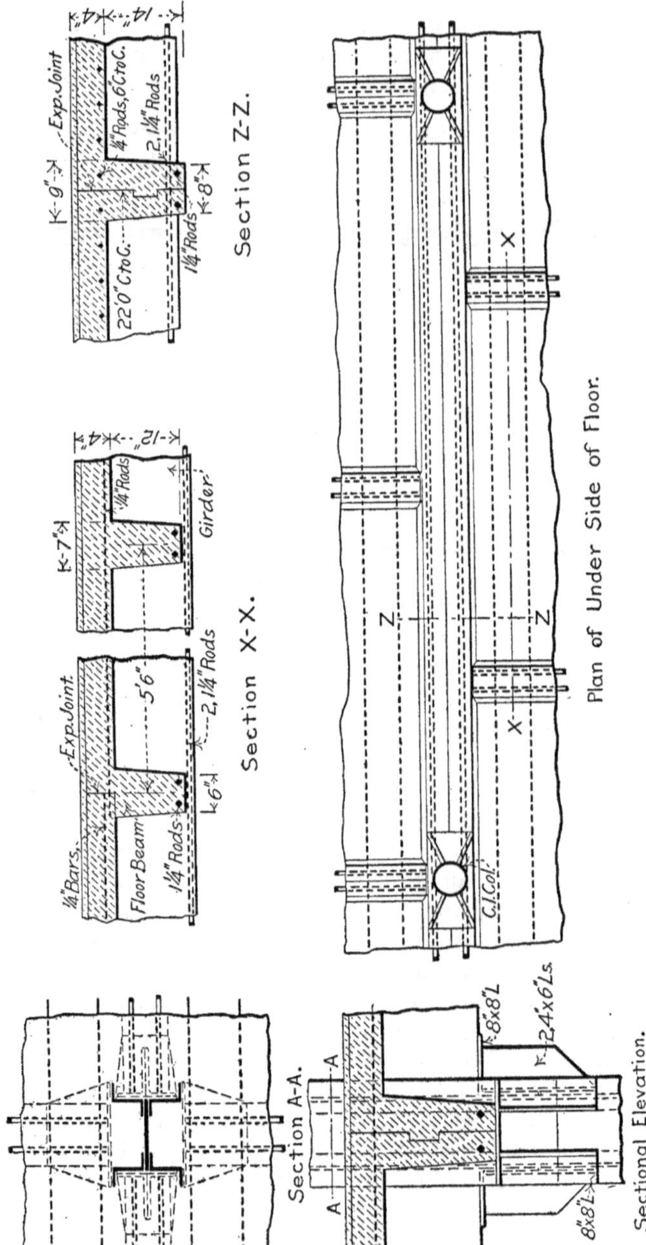

FIG. 75.—Ransome Floor for Foundry of Eastwood Co., Paterson, N. J.

a plane of weakness and not an open joint. The girders were built in two nearly equal halves separated by a vertical joint with an offset corresponding to a tongue and groove. These joints were not open, but, like those for the beams, were simply planes of weakness. The twisted rods in the floor-slabs were laid to project across the rebate in the expansion-beams and to lap past those in the adjacent slab so as to lock the two slabs together. In a similar way the rods in the beams were made to project a few inches at the ends, so that they would extend across the joint in the girder and lock the two halves together. The methods of attaching the floor-beams and girders to the steel supporting-columns and to the wall girders are shown clearly by the drawings of Fig. 75. The floor was proportioned for a load of 250 lbs. per square foot.

Another type of factory floor of Ransome construction is shown by Fig. 76. This form of floor was employed in the large works of

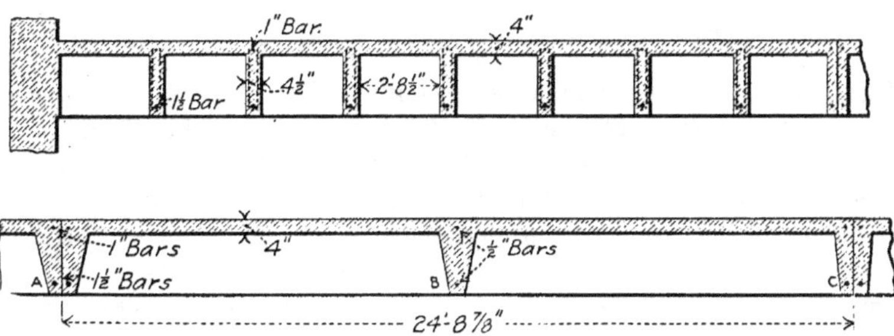

Fig. 76.—Ransome Floor for Pacific Borax Co.'s Factory, Constable Hook, N. J.

the Pacific Coast Borax Company, at Constable Hook, N. J., which were built in 1898. The main building, 200×75 ft., is divided into eight bays 25 ft. wide by seven transverse rows of columns set 12.5 ft. apart. These rows of columns carry transverse girders *A* and *C* 24 ins. deep, 9 ins. wide at the bottom, and 19 ins. wide at the top. Between girders *A* and *C* there is a secondary girder *B* of the same dimensions which divides the bay into halves. At right angles to the main and secondary girders there are floor-beams 24 ins. deep and 4½ ins. thick spaced 2 ft. 8.5 ins. apart in the clear. Covering the beams and giders and in one piece with them is a floor-slab 4 ins. thick. Each main girder and every eighth floor-beam is divided into halves by a vertical close joint; this divides the floor into 25-ft. squares separated by expansion-joints. The reinforcement of the main girders consists of upper and lower chord bars arranged as shown by the drawings, and the beams are similarly reinforced with the addition of stirrup bars at intervals. The reinforcing-bars of the main girders are made continuous across the building by means of turnbuckle connections. The floor was designed for a uniformly distributed load of 800 lbs. per square foot.

The drawings of Figs. 77 and 78 show the floor construction of the 16-story Ingalls Building at Cincinnati, Ohio. A row of interior columns runs lengthwise of this building dividing it transversely into two floor bays, one 17.5 ft. wide and the other 33 ft. wide. The columns are set 16 ft. apart. Fig. 77 shows this division and the arrangement

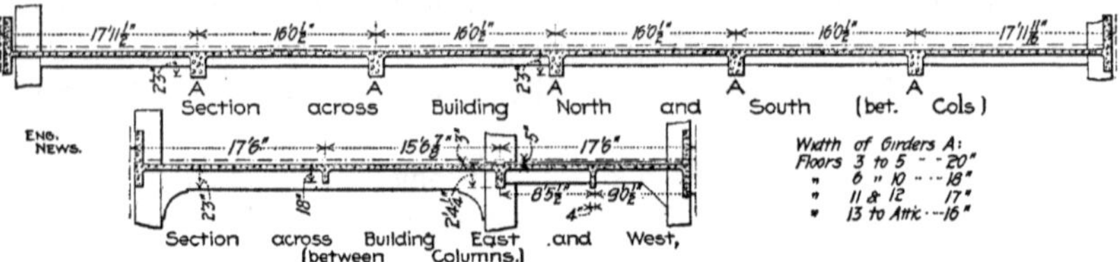

FIG. 77.—General Sections of Ransome Floors in Ingalls Building, Cincinnati, Ohio.

of girders and floor-beams planned to suit it. The main girders between columns are 36 ins. deep on the first floor, 34 ins. on the second, and 27 ins. on the floors above and they vary in width as shown by the legend on the drawings of Fig. 77. Between main girders there is a secondary girder running at right angles at the center of each transverse floor-bay. The main girder reinforcement is shown by the drawings of Fig. 78 and is of rather elaborate character. There is, as

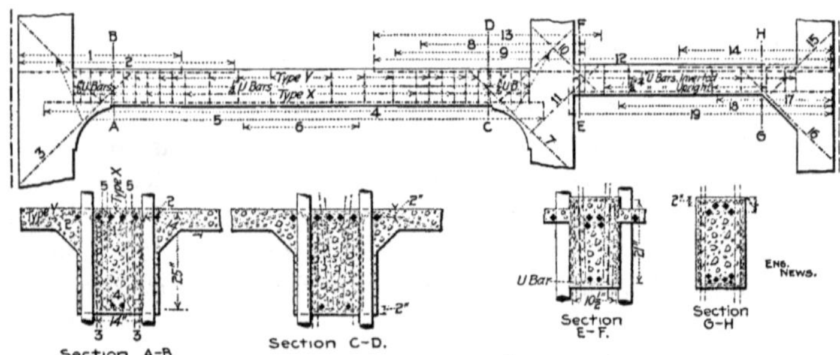

FIG. 78.—Details of Ransome Floors in Ingalls Building, Cincinnati, Ohio.

will be seen, a set of tension-rods and a set of compression-rods to strengthen the beam against flexure, a series of stirrups, and at each column a set of diagonal wind-bracing rods. The floor-plate is 7 ins. thick for the first floor and 5 ins. thick for the floors above. It is reinforced by a bottom network of two layers of bars laid at right angles. These floors are calculated for a load of 60 lbs. per square foot, except the first and second floors, which are calculated for loads of 200 lbs. and 80 lbs., respectively.

One of the most notable examples of Ransome floor and girder construction of which the writer knows is that of the brass works of Kelly & Jones Company at Greensburg, Pa. The essential details of this building are shown in the drawings of Fig. 79. The building is

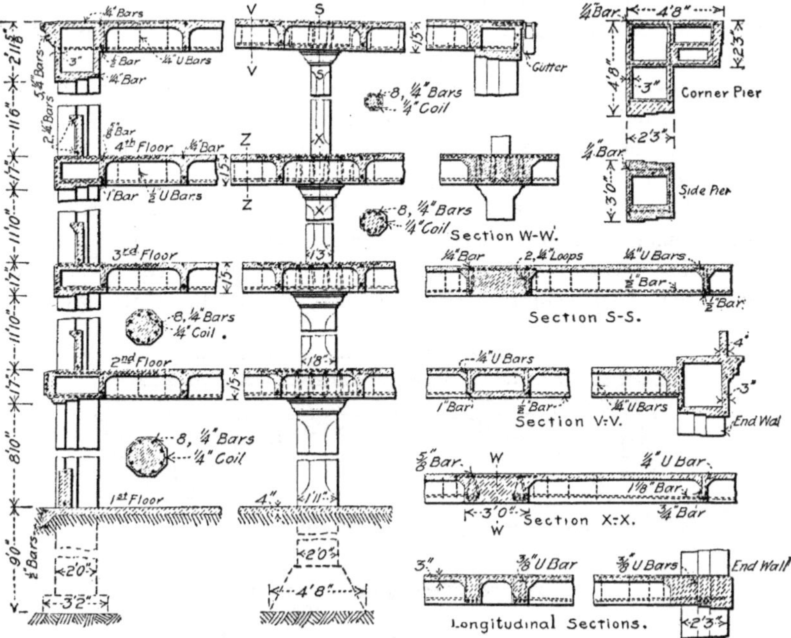

FIG. 79.—Ransome Floor and Girder Construction, Kelly & Jones' Factory, Greensburg, Pa.

four stories high and 60×300 ft. in plan, and it is built entirely of reinforced concrete. Each floor is a continuous slab of concrete 60 ft. wide and 300 ft. long separating the tiers of columns above and below. The roof is similar to the floors except that it has a slight pitch transversely. The floors are carried on four lines of girders resting on the wall piers and two rows of interior columns; these girders divide the floor into three 20-ft. bays transversely of the building. The columns are also spaced 20 ft. apart longitudinally of the building.

Considering only the girder and slab construction here, it will be seen from the drawings that the wall piers carry an outside box girder at each floor-level. Each transverse line of piers and columns carries a main transverse girder or double girder. These main girders are

20 ft. apart and midway between each pair is a secondary transverse girder of smaller section which takes support at the ends on the side-wall-girders. Over each interior column the two parts of the main double girder are filled between with concrete so as to make a solid slab 3 ft. square. The floor-beams at right angles to the transverse girders are spaced 3 ft. $8\frac{5}{16}$ ins. apart in the clear. All girders and beams are of the same depth, 14 ins.; the main girders are each 9 ins. wide, the secondary girders are 4 ins. wide, and the floor-beams are 3 ins. wide. The floor-slab is 3 ins. thick. The arrangement and dimensions of the reinforcement are shown by the drawings of Fig. 79. The floors and girders were made of a 1–2–4 concrete with $\frac{3}{4}$-in. broken stone. All floors were designed for a uniform load of 250 lbs. per square foot.

Roebling Construction.—The Roebling flat-top arch floor is a composite construction of concrete and metal which calls for particular description because of its extensive use. It is built in two styles, one having a paneled ceiling as shown by Fig. 80 and the other having a

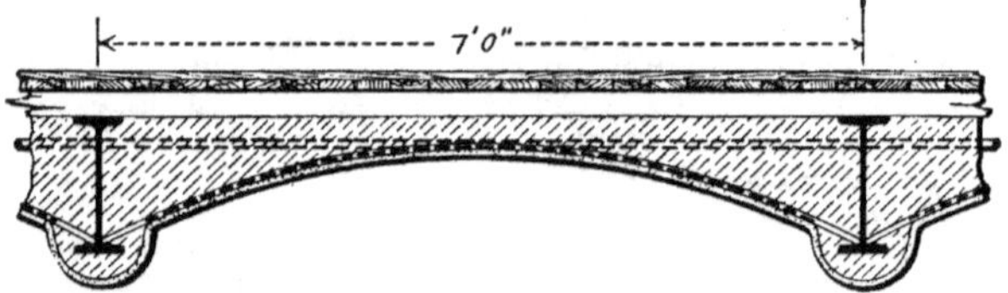

FIG. 80.—Roebling Arch Floor with Paneled Ceiling.

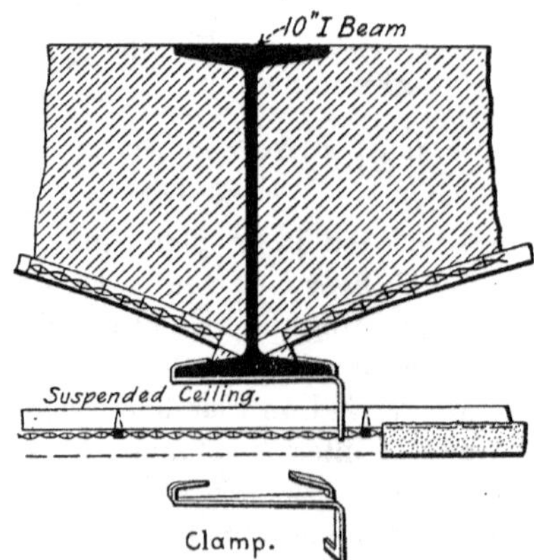

FIG. 81.—Detail of Flat Ceiling Construction for Roebling Floor.

suspended ceiling, as shown by Fig. 81, of wire lath and plaster. The metal element used consists of parallel round rods woven into a wire netting, the rods being $\frac{7}{16}$ in. in diameter for a 5-ft. span and $\frac{9}{16}$ in. in diameter for a 7-ft. span. This metal netting is cut into sheets and sprung between the lower flanges of the floor-beams, and the concrete is filled above it level with the tops of the beams. No centering is used in placing the concrete, the netting being strong enough to carry the concrete until it has set. By referring to the drawings it will be observed that the ends of the stiffening-rods project beyond the netting. This leaves an opening for the concrete filling to pass through the centering and form a firm seat on the lower flanges of the floor-beams. In the case of the paneled floor this bearing is further increased by a haunching around the lower beam flanges.

Modified De Vallière Construction.—One of the most important of the modified foreign constructions in use in America for building work is the De Vallière as adopted by Mr. J. O. Ellinger. This is a ribbed-plate construction and its typical form is shown by Fig. 82. Fig. 83

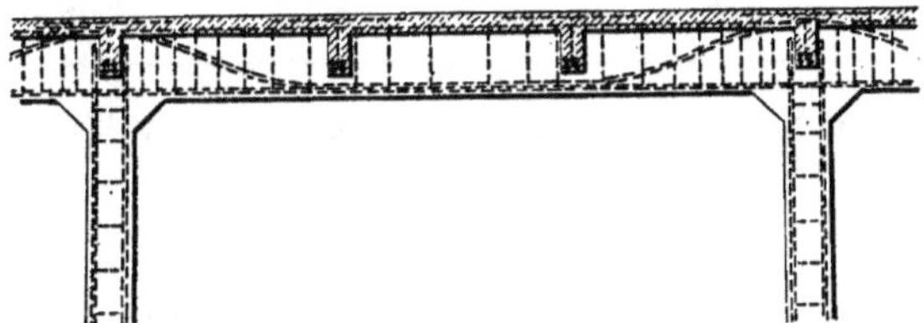

FIG. 82.—De Vallière Floor Construction.

shows the beam and slab reinforcement separated from the concrete. As will be seen from these illustrations the reinforcement consists, for the beam, of four tension-rods, two horizontal and two "catenary" rods of smooth round steel, and for the slab, of parallel tension-rods at right angles to the beam. The lower or horizontal rods in the beam are tied to the slab rods by twisted-wire hangers. The object of the "catenary" rods in the beam is to take up negative bending moment over supports, while the hangers tie together the beams and slab. The most representative buildings of this construction are the University of Pennsylvania Gymnasium and the Forrest Laundry at Philadelphia, Pa. The drawings of Fig. 84 show the floor construction for the Forrest Laundry. This building is 50 ft. wide and 160 ft. long and the floors are carried by the side walls and by a row of columns midway between the walls. There are four floors and a roof of reinforced concrete, the roof being similar in construction to the floors, but with a pitch to one side. The columns carry an 8×16 in. girder lengthwise of the building and this and the side walls carry 8×14 in. floor-beams. The floor-slab is 4 ins. thick and is in one piece with the girders and beams.

Fig. 83.—Reinforcement for Girder and Slab of De Vallière Floor.

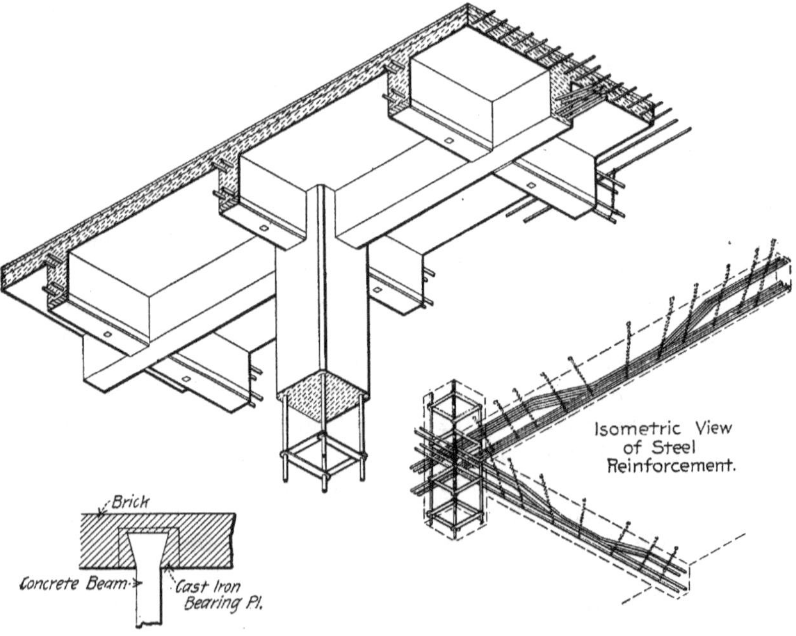

Fig. 84.—Details of De Vallière Floor Construction, Forrest Laundry, Philadelphia, Pa.

The roof has 8×14 in. girders, 8×12 in. beams, and a 3-in slab. The columns range from 25 ins. square to 8 ins. square. The beam-girder and column reinforcement are clearly shown by the drawings of Fig. 84. The floor construction in the University of Pennsylvania Gymnasium is similar to that of the building just described, but is particularly notable for the clear span of the 8×16 in. floor-beams, which are spaced 4 ft. apart in the clear and have a clear span of 32 ft. 5 ins. in one bay and of 30 ft. 7½ ins. in the other bay. The concrete used for the De Valliere floor is a 1–3–5 trap-rock mixture.

International Floor Construction.—Fig. 85 shows a form of ribbed-plate floor which has been employed in building construction by the International Fence and Fireproofing Company. The peculiarity of this floor is the use of steel-wire ropes in connection with a wire netting for reinforcing both slab and girders. The drawings show the arrangement of the reniforcement so clearly that no further description is necessary. It will be noted that the cables are anchored to the walls and are continuous from wall to wall; also that the metal fabric is carried on the tops of the cables.

Wire-fabric Construction.—Where a continuous netting or wire fabric is employed as reinforcement with steel-floor framing, the favorite mode of procedure is to carry the fabric over the tops of the beams, with a slight sag between beams. The drawings of Fig. 86 *a* and *b* furnish good samples of this type of construction. One of these drawings shows a long-span construction for a hospital building, and the other shows a short-span construction used in a hotel building at Baltimore, Md. The reinforcement used in these particular cases is the electrically welded wire fabric described on p. 332. This same reinforcement is employed for segmental-arch floors.

Long-span Girders.

It is seldom that girders exceeding 30 ft. span are required in floor construction, and in the majority of cases the span used is less than 20 ft. In roof construction occasionally and in constructing theater galleries, girders of much longer span are necessary. Two recent examples of such constructions are given here. The first is taken from the new College of Music at Cincinnati, Ohio. The recital hall of this building is 60 ft. 7 ins. wide and 80 ft. long, and has in the rear a balcony extending entirely across without intermediate support. This balcony is carried by one girder 12 ins. wide, 32 ins. deep, and 60 ft. 7 ins. span and another 48 ins. deep and of the same width and span. The details of the 12×32 in. girder are shown by Fig. 87,

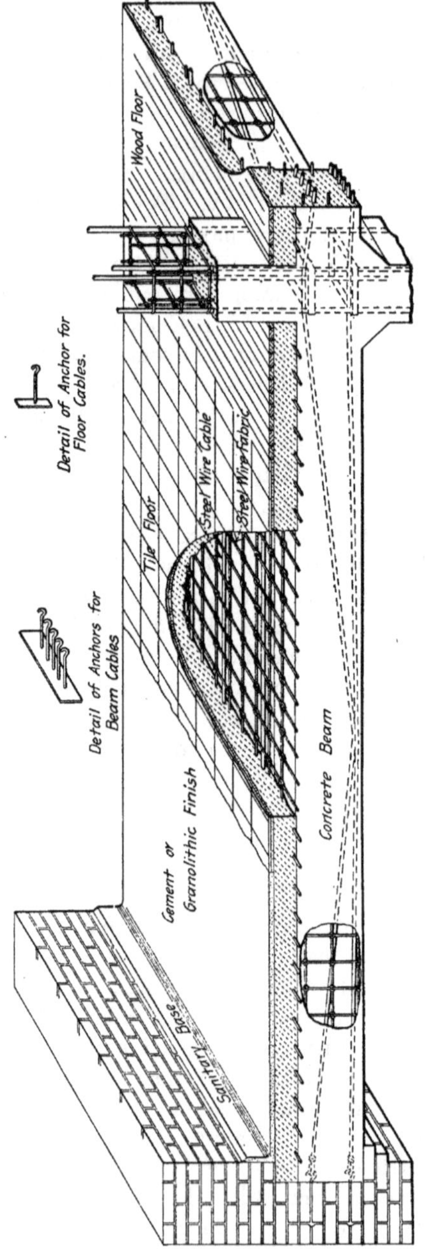

Fig. 85.—International Slab and Girder Floor Construction.

and those of the 12×48 in. girder are practically the same. The roof of the recital hall consists of a 4-in. slab stiffened by ribs 15 ft. apart supported by four girders 12×48 ins. ×60 ft. 7 ins. long. These roof girders are of substantially the same construction as the balcony girder shown by Fig. 87. The girder reinforcement consists of eight $1\frac{3}{8}$-in.

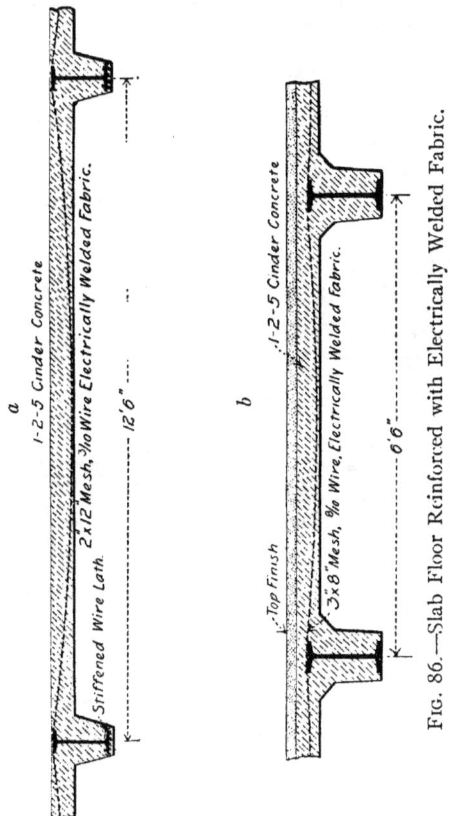

Fig. 86.—Slab Floor Reinforced with Electrically Welded Fabric.

rods on the tension side; four of these are straight and four are bent. From the straight rods, $\frac{3}{4}$-in. 16-gage strap-iron stirrups extend upward through the girder. As the balcony floor at the girder was not of sufficient section to take the compression, the girder was reinforced by four $1\frac{3}{8}$-in. compression-rods anchored down by $\frac{3}{16}$-in. annealed wire ties engaging with two $\frac{1}{2}$-in. anchor-rods placed about mid-height of the girder. The lower part of the girder was made of 1–2–3 concrete and the upper part of 1–2–4 concrete; the largest aggregate was $\frac{3}{4}$ in.

crushed gravel. The girder was designed for a floor load of 90 lbs. per square foot. Under test it deflected $\frac{3}{16}$ in. Under their own

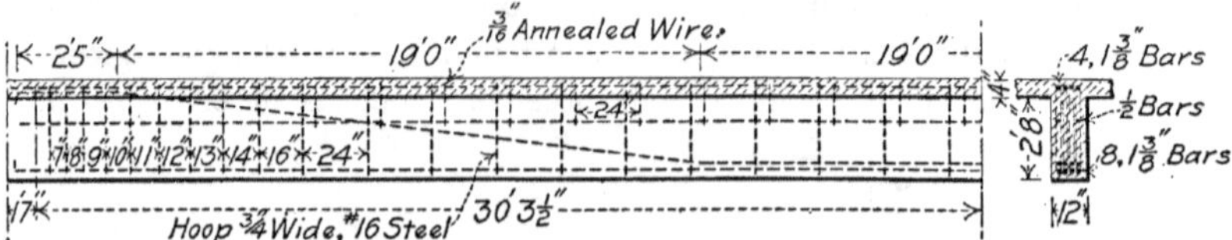

FIG. 87.—Girder Supporting Hanging Gallery, College of Music, Cincinnati, Ohio.

weight the roof girders deflected $\frac{1}{4}$ in.; the 12×48 in. balcony girder deflected $\frac{5}{8}$ in., and the 12×32 in. girder $\frac{1}{4}$ in.

The drawings of Fig. 88 show the details of the roof girders over the machine-room of the factory of the Central Felt and Paper Com-

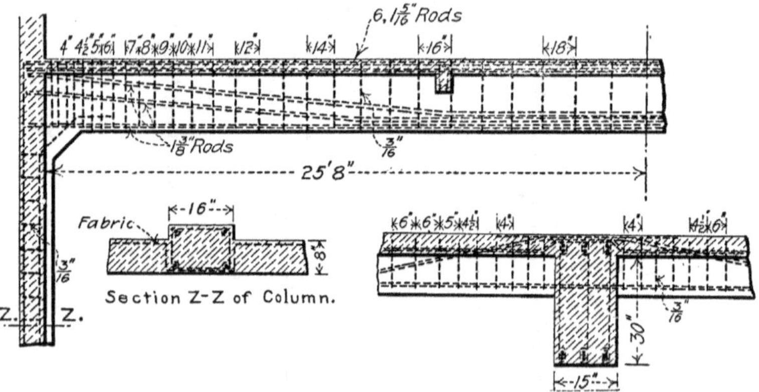

FIG. 88.—Roof Girder for Factory of Central Felt and Paper Company, Long Island City, N. Y.

pany, at Long Island City, N. Y. These girders have clear spans of 50 ft. 2 ins. and are 15×30 ins. in cross-section.

COLUMNS.

Current practice in column construction exhibits few vagaries in methods of reinforcement. A rectangular section, with straight reinforcing-rods arranged symmetrically and tied together at close intervals, is the normal construction. A few examples are sufficient to illustrate these features. It is quite common practice to use concrete columns without reinforcement in connection with reinforced floors. A number of American and English designers have also used concrete for filling and encasing columns built up of steel shapes, and some of these designs may be classed as composite structures of concrete and steel, but others provide for scarcely more than the fireproofing of the steel, which takes all the load and is, so far as the supporting power is concerned, structurally independent of the casing or fire-proofing.

Bank Building, Basel, Switzerland.—The columns illustrated in Fig. 89 were employed in a bank building at Basel, Switzerland, where

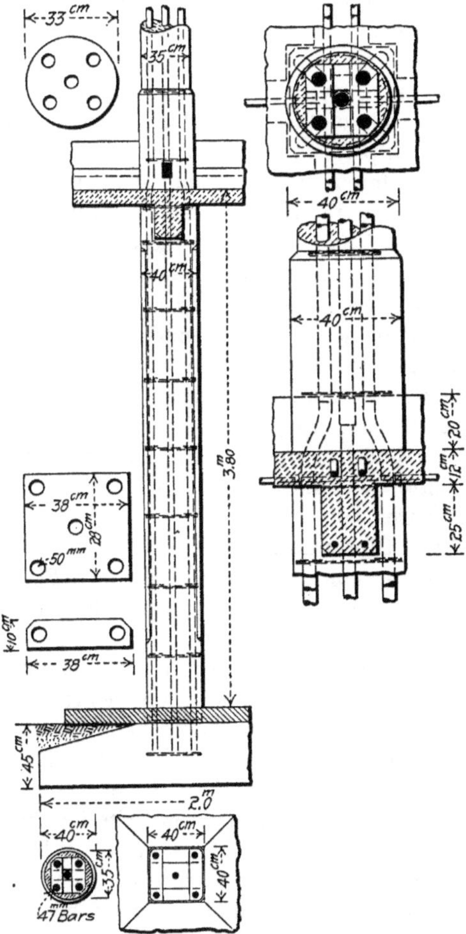

Fig. 89.—Hennebique Columns, Bank Building, Basel, Switzerland.

the basement columns are square and those of the upper floors are cylindrical in cross-section, and they will, with a general description of the type, illustrate the ordinary practice of the French engineer Hennebique in applying reinforced concrete to column construction. The standard reinforcement employed in columns of this type consists of four or more longitudinal bars of round iron bound together at regular intervals by straps or wires arranged in horizontal planes. Generally four bars are employed placed at the corners of square or rectangular columns and at the quarter-points of cylindrical columns, but a fifth bar at the center of the column is often added, and in large columns intermediate peripheral bars are added in numbers depending upon the size of the column and the load carried. The sizes of bars used vary from 8 mm. to 50 mm. (0.32 in. to 1.97 ins.) in diameter. At the base of the column the ends of the bars, which are cut square, rest on a horizontal sheet-metal plate embedded in the concrete. The horizontal ties between the reinforcing-bars are spaced about 50 cm. (19.7 ins.) apart vertically. There are two forms. The most common form consists of four straps of iron punched at the ends to slip over the rods and placed so as to form a peripheral strap within the column. Owing to the planes of weakness introduced by these flat-strap binders, it is the practice to employ wire binders much similar to those used in Hennebique piles (p. 139) in columns which have much bending load. The reinforcing-rods are made continuous from one floor to another, and where splices are necessary the rods are cut to square butt-joints and enclosed in a cylindrical sleeve. It is common in this construction also to mold hollow columns, the interior space being used for the passage of pipes, wires, etc.

Office Building, Cincinnati, Ohio.—Unlike the column described in the preceding paragraph, the column shown by Fig. 90 is a special rather than a standard design. It is an example of the columns used in the 16-story Ingalls Building at Cincinnati, Ohio. In this building there are 22 columns, all but 5 of which are wall columns, and they are spaced from 16 ft. to 33 ft. apart. They decrease in size from 34×38 ins. at the basement to 12×12 ins. at the roof. The footings are built independently of the columns, and each has a rectangular pedestal somewhat larger than the column. Upon this is set a cast-iron base-plate, having circular projections on top to form seats for the vertical round bars which act in compression only. The tops of the projections are faced to a true horizontal plane so as to form a true bearing for the faced ends of the bars. Each column has four, six, or eight of these bars from 2 ins. to 3½ ins. in diameter, the joints of which are made by faced ends in contact. In the lower part of the building these bars extend through one story only, but above the third floor they extend the height of two stories. At each stage of the work a sleeve is put around the top of the bar to form a socket for the next bar. These sleeves are filled with cement grout as the connection is made. The circular bars are intended to take compression only. To take the tension due to wind load each column has from four to ten smaller bars of twisted steel. The compression-bars are tied together diagonally through the column and the tension-bars are encircled every 12 ins.

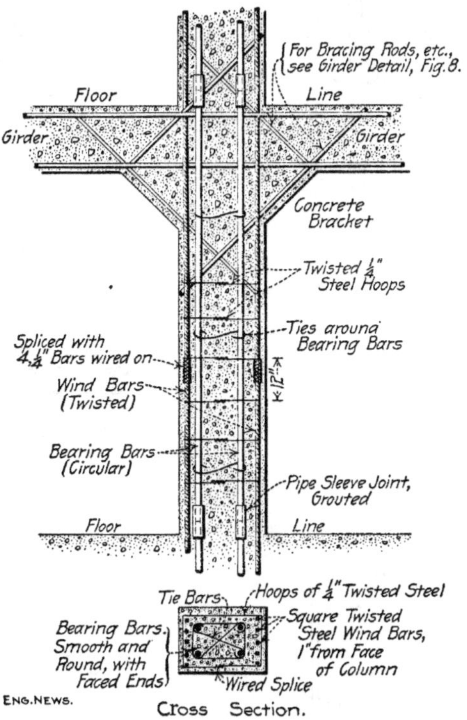

FIG. 90.—Column for 16-story Ingalls Building, Cincinnati, Ohio.

by hoop-bars of twisted steel. The column loads on the shoes range from 500 to 750 tons.

Factory Building, New York, N. Y.—Fig. 91 shows a cross-section of one of the columns used in constructing the one-story factory building of the Central Felt and Paper Company, at Long Island City. The columns were from 14 ins. to 20 ins. square and were made of 1–2–4 broken stone concrete reinforced by four vertical rods tied together at intervals. The vertical rods were from $\frac{3}{4}$ in. to $1\frac{1}{4}$ ins. in diameter, depending on the size of the column, and were circular in cross-section. The tie-rods were round and $\frac{3}{16}$ in. in diameter. The wall-column construction for this same building is shown by Fig. 97.

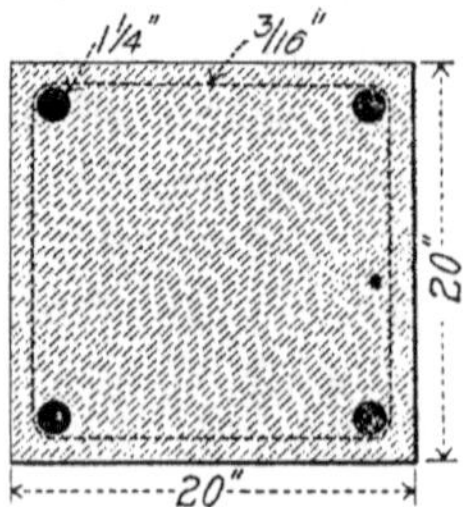

FIG. 91.—Column for Factory of Central Felt and Paper Company, Long Island City, N. Y.

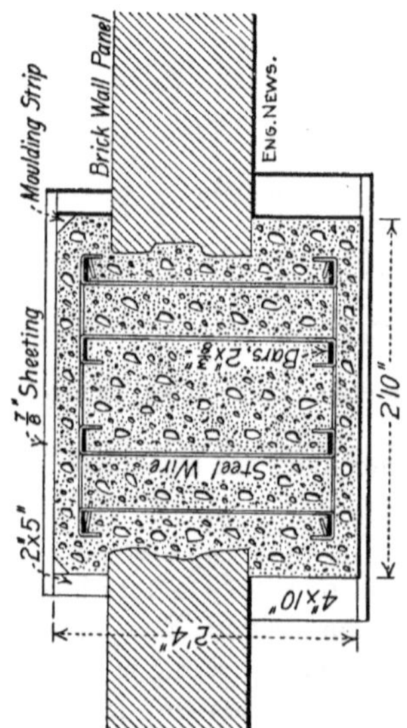

FIG. 92.—Wall Pier for Power House, at Louisville, Ky.

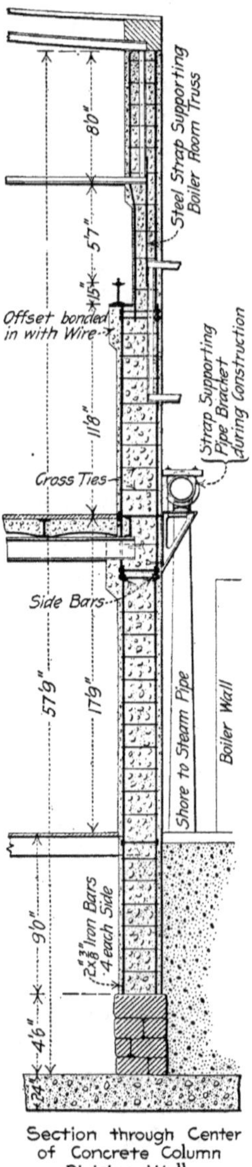

FIG. 93.—Vertical Section of Wall Pier Showing Floor and Roof Connections and Craneway Bracket.

Power House, Louisville, Ky.—A transverse section of one of twelve concrete-steel columns built at Louisville, Ky., in 1903, to strengthen the walls of a brick power house, is shown by Fig. 92. A vertical section of the same column drawn to smaller scale, but showing the floor girder and truss connections and the crane-track brackets, is shown by Fig. 93. The main reinforcing members of these columns consist of eight flat iron bars 2 ins. wide and ⅜ in. thick. These bars were made up of 15-ft. lengths spliced by two ¾-in. bolts and distanced by side-bars and cross-bolts at the joints. Between splices the bars were tied together by wires, as shown by the drawings. The concrete was composed of 1 part Portland cement, 3.8 parts sand, and 5.7 parts broken stone.

Factory Building, Greensburg, Pa.—In constructing the brass factory of Kelly & Jones at Greensburg, Pa., which is illustrated by Fig. 79, use was made for the first time in America of a hooped column such as is recommended by Mr. Considère. As shown by the drawings, each column is reinforced by eight vertical bars spaced equidistant in a circle and nowhere less than ½ in. from the surface, and these bars are enclosed in a helix of 4-in. pitch extending from end to end of the column. The bars and helix are both ¼-in. square twisted rods and are wired together at intersections. In the same building the wall columns or piers were made rectangular in section and hollow with 3-in. shells. They were reinforced by a ¼-in. rod at each corner and bv circumferential bars ⅛×½ in. spaced 12 ins. apart vertically.

WALLS AND PARTITIONS.

In actual practice wall and partition construction of reinforced concrete has taken a variety of forms. Some of the more notable of these forms are as follows: (1) Monolithic walls without a supporting metal framework; these walls may be either solid or hollow and when used for exteriors are often veneered with brick, stone, or terra cotta; (2) Walls consisting of a plastered metal network carried by a framework of steel; these walls may be either solid or hollow; (3) Walls composed of rectangular slabs of reinforced concrete laid up like masonry and commonly supported by a steel framework; (4) Walls consisting of reinforced concrete columns and lintels which are filled between with brickwork or terra cotta. The form of construction mentioned third is commonly limited to partition or interior wall construction, and the form mentioned last is exclusively an exterior wall construction. The other forms are employed for exterior or interior walls indifferently. The following examples indicate clearly the form and nature of the several constructions defined.

Monolithic Walls.—The drawings of Fig. 94 show the solid-plate monolithic wall construction extensively used by the French engineer Hennebique for both interior and exterior walls. This particular construction was employed for an interior wall of a bank building at Basel, Switzerland, and shows a door-opening. The reinforcement consists of a row of vertical rods in each face and a row of horizontal rods at the center of the slab. The vertical rods are tied into the slab transversely by stirrups. This partition was designed to resist a lateral

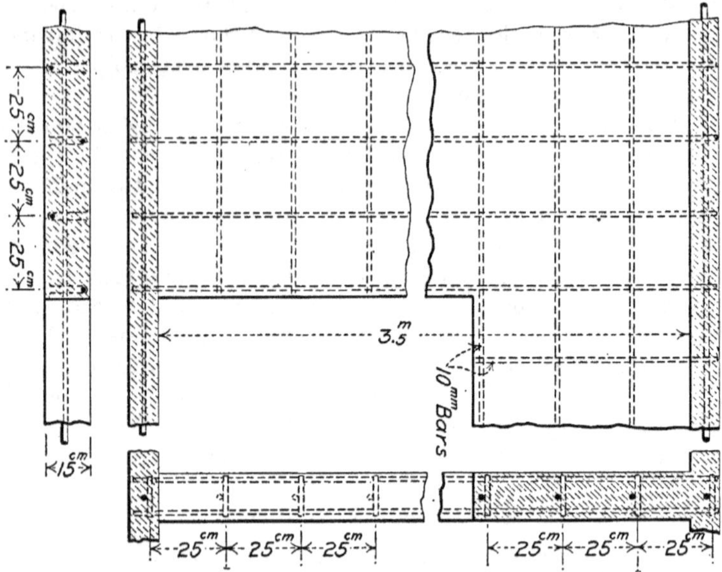

FIG. 94.—Monolithic Wall, Hennebique Construction.

thrust of 250 kgs. per square meter (51 lbs. per sq. ft.). For exterior walls practically the same plate construction is employed, the plates filling in the spaces between columns, and being in one piece with them.

The wall construction shown in Figs. 95 and 96 is of particular interest; it was designed for the 16-story Ingalls Building at Cincinnati, O., and is veneered on the outside with marble to a height of three stories and with brick and terra cotta above the fourth floor. Exclusive of the facing the walls are 8 ins. thick; the party walls were not faced and are only 3 ins. to 4 ins. thick. All the walls are reinforced by vertical and horizontal bars of twisted steel. The typical arrangement of the bars in the street walls is shown by Fig. 95. The vertical bars marked *A* are ½-in. bars 27 ft. long extending through two floors and having their ends lapped at the floor line. They are placed 2 ins. from the edges of the window-openings. The vertical bars marked *B* are the column wind-bars; joints are made at the middle of window-openings in each story, the ends being butted together. Over each joint is lapped a short piece of ½-in. bar or four ¼-in. bars of twisted steel. The horizontal bars marked *C* are pairs of ½-in. bars placed side by side near the outside and inside faces of the wall and 1½ ins. above or below the edges of the window-openings. These bars are in lengths of 27 ft., and the ends of adjoining bars are lapped at least 21 ins. When the joints occur over the window-openings, between the columns, the lapped ends are wrapped with wire or secured together with hog-rings. Fig. 96 shows the method of supporting the facing on the concrete wall. The brick facing is supported at every floor by a ledge formed in the concrete, and is also secured by wire anchors projecting from the concrete. The marble veneer and terra-cotta facing have grooves on the back, the top and bottom surfaces of which have an upward inclination. These grooves engage with ribs of similar section on the face of the wall and form a species of dovetail joint.

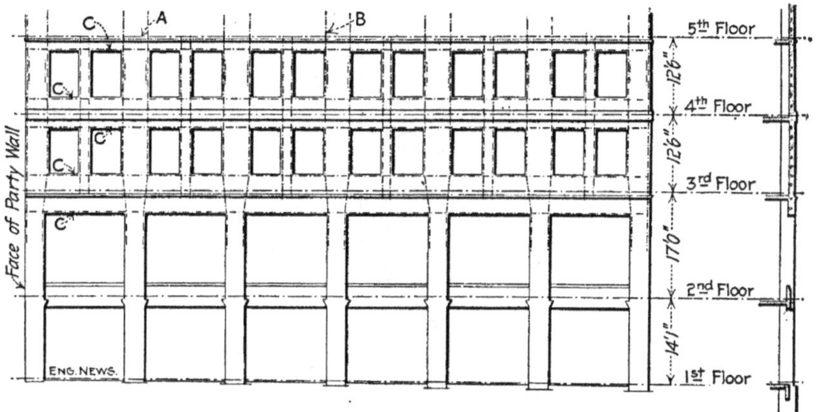

FIG. 95.—Elevation and Section of Wall of Ingalls Building, Cincinnati, Ohio.

The drawings of Fig. 97 show the monolithic wall construction of the factory of the Central Felt and Paper Company at Long Island City, N. Y. The walls of this building are 6 ins. thick between wall columns, and they carry no load except their own weight and the adjacent edges of the floor and roof-slabs. At the floor and roof levels and also at the bottom each wall bay is reinforced as a girder, as is clearly shown by Fig. 97. The wall-plate reinforcement consists of vertical

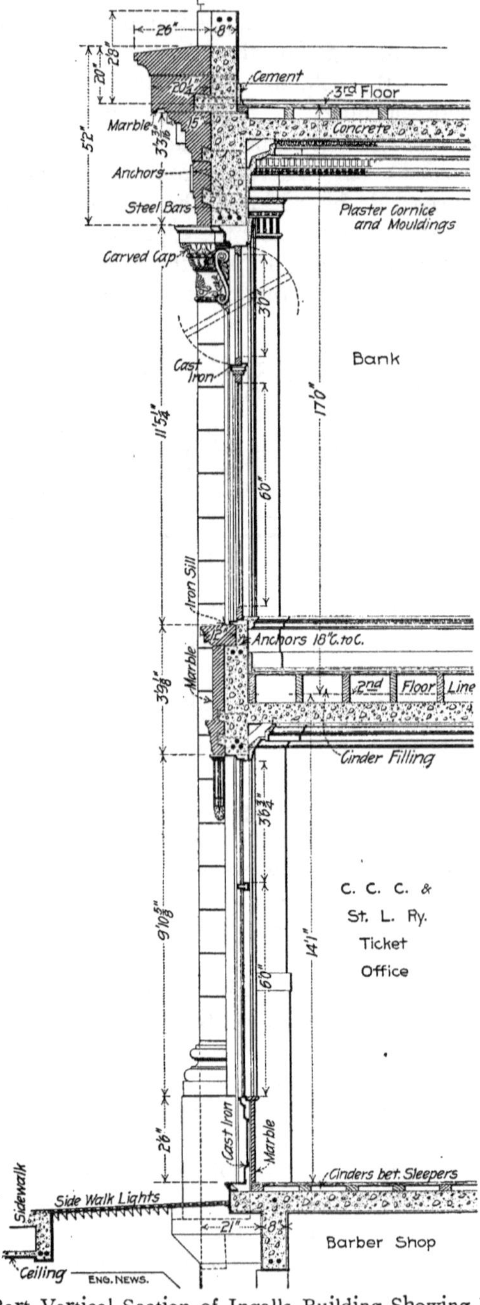

Fig. 96.—Part Vertical Section of Ingalls Building Showing Veneering

sheets of lock-woven steel fabric placed near the inner face. At window-openings this fabric is folded back so as to form a U-shaped net at the side of each opening. Sheets of the same material bent to U shape are also embedded in the wall concrete under the window-sill.

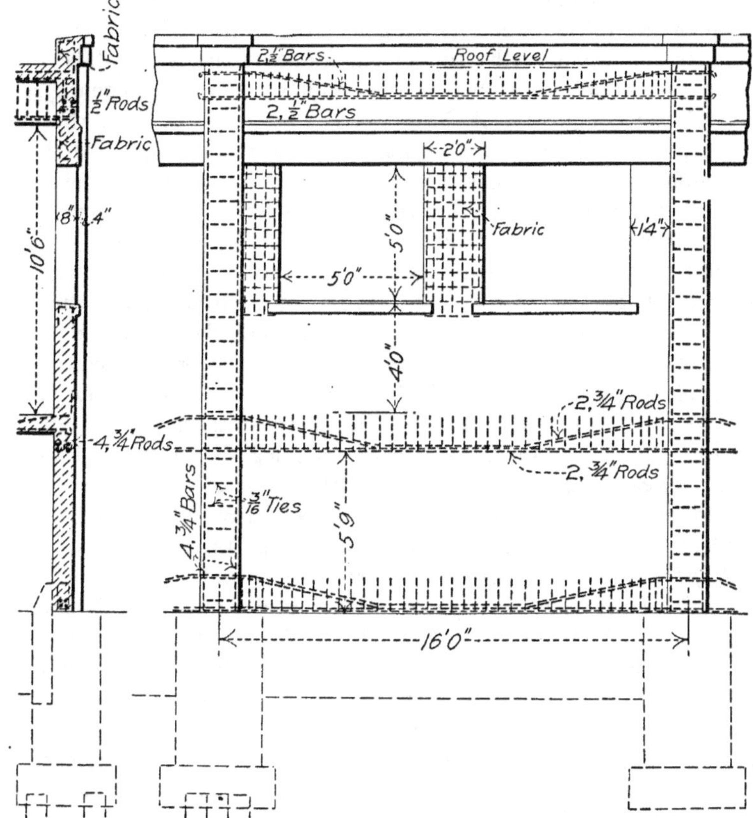

FIG. 97.—Details of Walls for Factory of Central Felt and Paper Company, Long Island City, N. Y.

The concrete for the walls was a wet mixture of 1 part cement, 2 parts sand, and 5 parts fine cinders.

Plastered Mesh and Steel-frame Walls.—The most notable forms of plastered mesh and steel-frame walls are the Monier and the expanded-metal constructions. The Monier construction has been very extensively employed in Europe, but only to a limited extent in the United States, while expanded-metal construction is particularly an American development, although it has gained considerable popularity in Europe.

Monier Construction.—Considering first the solid form of Monier wall, the Monier netting is stretched vertically between the points of support with the carrying-rods horizontal and the distributing-rods vertical. When openings for doors or windows pierce the metal network they are bound on all sides with timber or channel-iron casings to which the rods of the mesh are firmly attached. These rods are similarly connected to the supporting frame of the wall, which may be the main framing members of the building or an independent structure. Upon completion of the network it is plastered on both sides with cement or lime mortar, forming a plate 5 cm. (1.97 ins.) thick. A modification of the construction described is sometimes employed where it is desired to carry the weight of the wall-plate entirely on the vertical supports. This modification consists in arching the rods extending horizontally and inserting at intervals an extra-heavy carrying-rod. These curved rods act as arch ribs abutting against the vertical supports and carry the weight of the wall-plate free from the sill of the wall. For ordinary walls a single network placed at the center of the slab is the normal Monier construction. In places where a thicker wall is necessary two parallel sheets of netting are stretched and embedded in mortar, thus making a thick slab with a reinforcement near each face. Hollow walls of Monier construction are formed by two parallel plates with a space between them. The chief use of Monier construction is for interior walls and partitions.

Expanded-metal Construction.—Expanded metal and plaster wall construction has been developed in a number of forms for both exterior walls and partitions. The ordinary form of interior wall construction consists of channel-iron studding set crosswise of the partition and connected at top and bottom with the floor-beams, which is covered with expanded-metal lath on both sides and plastered. Door- and window-openings are cased on all sides with channels. This construction forms a hollow wall. For light solid partitions the practice is sometimes adopted of laying wood sills and ceiling-plates and stretching round bars vertically between them and about a foot apart. These bars are connected to the sill and ceiling-plate by means of hooks and screw-eyes. Between the bars sheets of expanded metal are laced alternately behind and in front. This mesh is plastered on both sides with cement or lime mortar. Door- and window-openings are cased with wood. This second construction is also used with metal framework, in which case the carrying-rods are connected at top and bottom to the floor-beams by means of hooks and clips.

Any of the interior wall constructions just described may be used for exterior walls, but ordinarily a special exterior wall framing is designed. This commonly consists of I-beam wall columns spaced about 15 ft. apart and connected at top and bottom and at the top and bottom window lines by horizontal angles and Z bars. Between the horizontal members are set vertical channels to which the expanded metal is attached. This sheeting is plastered on the outside with cement mortar and on the inside with lime plaster. To protect the columns from fire they are boxed around with expanded metal and plaster.

Walls of Cast Slabs.—Wall construction of especially cast slabs of reinforced concrete has been adopted in only a comparatively few buildings in Europe and is unknown in American practice. The concerns controlling Monier construction in Europe have employed two forms of cast-slab wall construction. In one form a frame work composed of horizontal and vertical members forming meshes about 1×1½ m. (3.28×4.92 ft.) is erected and the meshes are filled with cast Monier plates set like glass in a window-sash. These plates are cast so that the rods of the reinforcement protrude on the two horizontal edges and furnish a means of attaching the plates to the framework. The plates are made from 3 to 4 cm. (1.18 to 1.6 ins.) thick. In the second form of Monier wall construction with cast pieces the blocks used resemble burnt-clay hollow wall tile in form, and they are laid up in the wall with mortar joints, exactly as are building tile.

Reinforced Concrete Frame and Brick Filling.—A wall construction consisting of a concrete-steel framework composed of the wall-

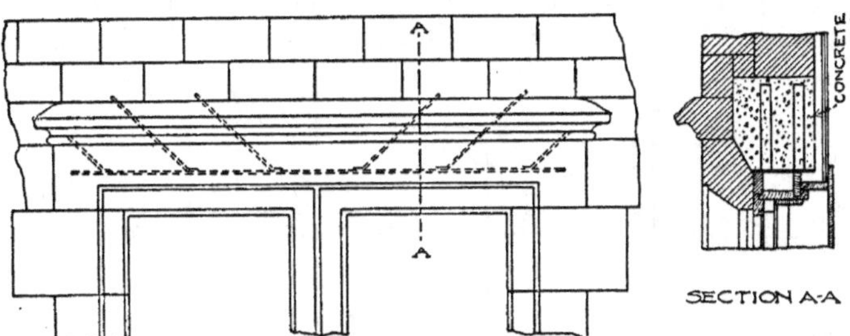

FIG. 98.—Cast Lintel with Kahn Bar Reinforcement.

columns, window-sills and lintels, cornice, water-table, etc., with the panels filled with brickwork, has been used occasionally by the French engineer Hennebique. In this construction the concrete-steel members are separately cast pieces laid up as similar members of cut stone would be laid. Walls of this construction afford opportunities for architectural treatment nearly equal to those furnished by veneered walls. The use of reinforced concrete lintels for carrying brick walls over window-

and door-openings is illustrated by Fig. 98. The form of reinforcement here used is the Kahn bar described on p. 337.

Roofs.

Two forms of roof construction in concrete-steel are employed. In the first form reinforced concrete plates are attached as a covering to the ordinary roof-framing of steel or timber. In the second form both frame and covering are of concrete-steel and are monolithic in construction.

Reinforced Plate and Steel-frame Construction.—Any of the flat-plate constructions employed for floors may be employed equally well for covering roofs having steel or wood framework, and Monier, expanded-metal, Columbian, and other floor-plates have been frequently so employed, they being sometimes molded in place on centers and sometimes cast separately in sheets or slabs which are attached to the roof-framing and joined to each other by mortar joints. The construction of these floor-plates has been described in a preceding section; the following are examples of plates designed specifically for roofing: In building the two-story $200 \times 88\frac{3}{4}$-ft. stable for the Anglo-Swiss Condensed Milk Company of Brooklyn, N. Y., the roof was divided into three spans each 28 ft. 11 ins., the trusses of which carry I-beam purlins. The concrete-steel roofing rests directly on these purlins and consists of slabs 4 ft. wide and long enough to reach from eaves to ridge, laid side by side, with scarfed joints, and plastered over with a $\frac{1}{4}$-in. mortar coat. The construction of the slabs is shown by Fig. 99. The

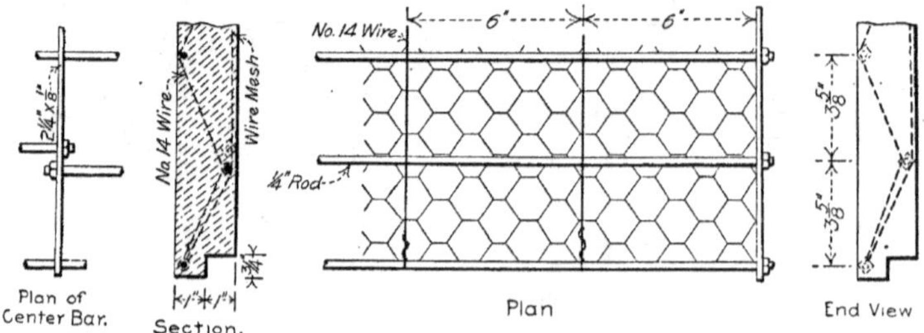

Fig. 99.—Roofing Slabs for Stable for Anglo-Swiss Condensed Milk Co., Brooklyn, N. Y.

main reinforcing-frame of each slab consists of a gridiron of $\frac{1}{4}$-in. rods spaced $3\frac{5}{8}$ ins. apart by three transverse spacing-plates containing staggered holes for the rods. Over and under these staggered bars No. 14 gage wires spaced 6 ins. apart are woven, and stretched over the bottom rods is a fine wire netting. This skeleton is filled flush with the edges of the spacing-plates with a 1–2–4 concrete made with very fine broken stone, to make a 2-in. plate. The edges of the slab are scarfed and overlap when laid, the joint being made with cement mortar. Voids are also left in the edges to permit thin flat steel bars or angle-clips to be bolted to the reinforcing-frame and locked around the roof-framing. Fig. 100 shows the concrete-steel-slab roof con-

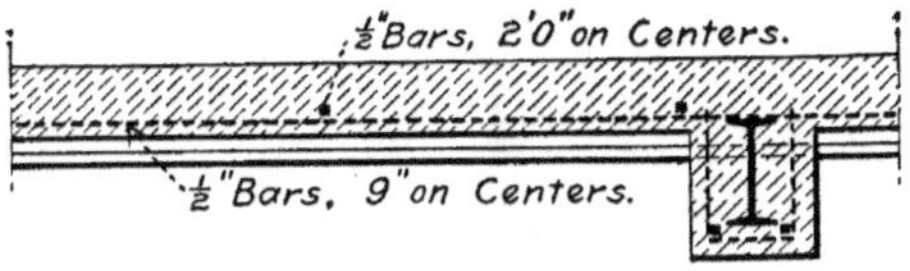

Fig. 100.—Roofing Slabs for Government Printing Office, Washington, D. C.

struction employed in the Government Printing Office at Washington, D. C. Slabs 12 ft. wide and reaching from eaves to ridge were molded in place. These slabs were 5 ins. thick and were haunched around the purlins; they were reinforced by a network of transverse and longitudinal twisted square bars. The joint at the ridge was filled with asphalt 1 in. wide and 3 ins. deep, and the joints between slabs were filled with ¾-in. pine strips bedded edgewise in the concrete. Another form of slab roofing is shown by Fig. 101. This is made of

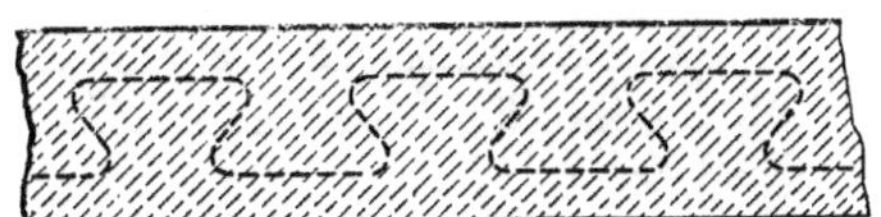

Fig. 101.—Roofing Slabs of Plastered Corrugated Steel Plate.

corrugated steel sheets which are laid on the roof-framing and then plastered on both sides with cement mortar. The corrugated sheets are made 10×20 ft.×½ in. deep. This roofing 1¼ ins. thick weighs about 15 lbs. per square foot and costs complete about $21 per 100 square feet.

Monolithic Construction.—In monolithic roof construction both the roof-framing and the roof-plate are of reinforced concrete. This construction has been applied to all forms of roofs from simple peaked roofs to arched roofs and domes of elaborate composition. European practice in particular exhibits roof construction of a very bold character in concrete steel.

Flat and Pitched Roofs.—Flat and pitched roofs of monolithic construction consist usually of ribbed plates and resemble closely plain ribbed plate floor construction. An example of both forms is given by Fig. 102. This roof was built for a New York City residence and consists of a mansard face on the street front and a flat roof extending backward from the mansard to the rear eaves. The flat roof consists of girders spanning the building transversely and carrying a roof-plate and a suspended ceiling. The mansard roof takes support on the front wall and on the flat roof and is braced by plate buttresses resting on the

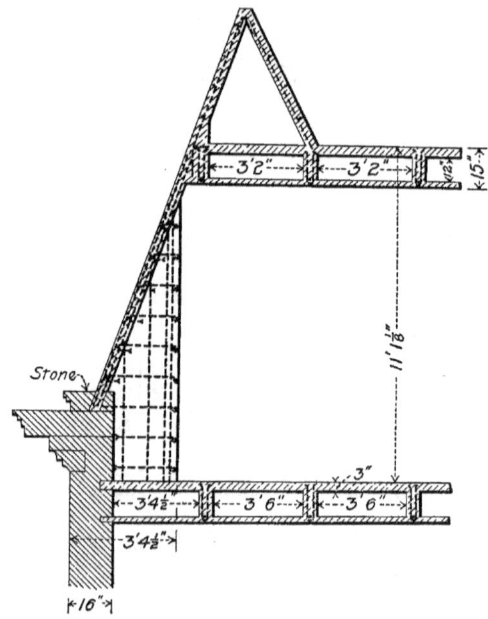

FIG. 102.—Monolithic Roof for a New York City Residence.

top or attic floor. The form of reinforcement used is the Hennebique. The roof covering used on this roof was asphalt bricks welded together. The flat roof construction here shown is exactly like the construction of the floor shown in the same drawing, and this is the normal flat-roof construction. The drawings of Figs. 79 and 88 also show flat-roof constructions. The roof shown by Fig. 103 is a notable example of pitched-roof construction because of its vaulted ceiling. It covers the court-room of the Nassau County Court-house at Mineola, Long Island. The roof-plate is 2 ins. thick and is stiffened by arched ribs spaced 3 ft. apart. These ribs have a thickness of 5 ins. and a span of 49 ft. The ceiling-plate, like the roof-plate, is 2 ins. thick and is connected to the arch ribs monolithically. The other details of the construction are shown by the drawings. In Europe practically all prevailing timber-roof forms are reproduced in concrete-steel by Hennebique, Bonna, Cotancin, and other prominent systems of construction, and examples of gable, mansard, and weaving-shed roofs are numerous.

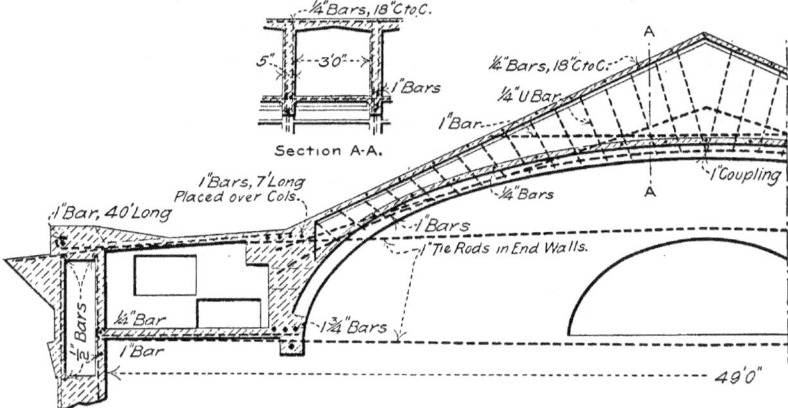

FIG. 103.—Pitched Roof with Vaulted Ceiling, Court-house, Mineola, N. Y.

Domes and Arched Roofs.—Both European and American practice furnish examples of domes and arched roofs constructed in reinforced concrete. Fig. 104 shows a European construction which has been

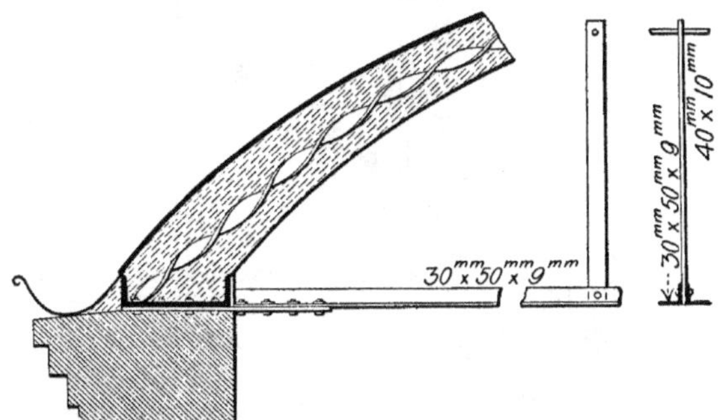

FIG. 104.—Details of Habrich Construction for Domed and Arched Roofs.

applied to both domes and arches. As will be seen, the arched plate is seated on a channel the opposite sides of which are connected across the span by tie-bars. The reinforcement consists of twisted straps arranged radially in domes and parallel in arched roofs. Domes of 22 m. (72.2 ft.) span of this construction have been built. At Basel, Switzerland, a railway shed 200 m. (656.2 ft.) long has an arched roof of 20 m. (65.6 ft.) span and 4 m. (13.12 ft.) rise of this construction. The twisted reinforcing-bars are spaced seven per meter, or about 5⅔ ins. apart. The roof was calculated for a snow load of 75 kgs. per square meter (15.4 lbs. per sq. ft.) and for a wind load of 40 kgs. per square meter (8.2 lbs per sq. ft.); the dead load is 225 kgs. per square meter (46 lbs. per sq. ft.). Almost the same construction as that just described is employed for domes and arched roofs, using Monier netting instead of twisted straps as reinforcement. Another form of dome construction consists of concentric spherical shells of Monier plates connected

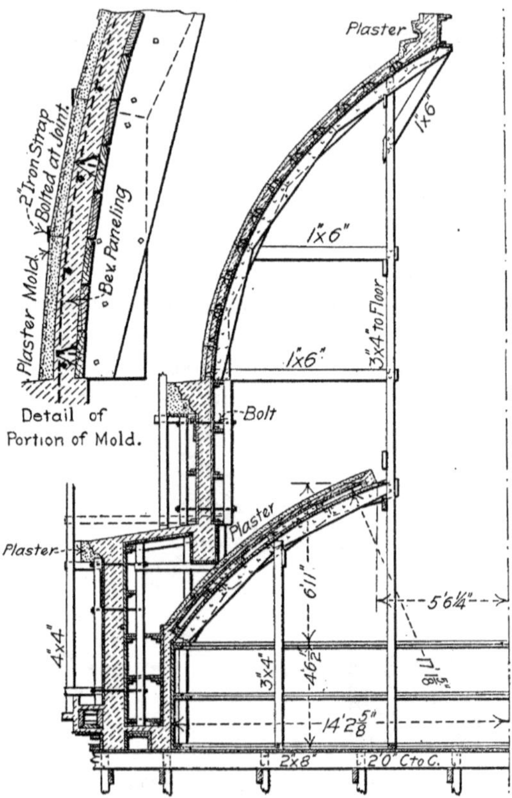

Fig. 105.—Domed Roof Construction for Court-house at Mineola, N. Y.

by vertical plates set radially. The lower part of the annular space between shells is filled with meagre concrete to stiffen the construction. A dome of this construction, with an outside diameter of 8.8 m. (28.9 ft.), covers the mausoleum of Frederick III. at Potsdam.

A notable example of dome construction is illustrated by the drawings of Fig. 105. This structure forms part of the Nassau County Court-house at Mineola, N. Y., and was built in 1901. It consists of an outer or roof dome and an inner or ceiling dome, the two being carried on a pair of concentric rings each supported by twelve columns. The ceiling dome is a segment of a sphere with an 11-ft. circular sky-light at the crown. Its shell is 2 ins. thick and is stiffened by radial and circular ribs on the under side. The reinforcement consists of radial and circular ¼-in. twisted rods in the shell and of four ¼-in. bars in each rib. This dome is supported on a 4-in. cylindrical wall or drum 4½ ft. high which has a 10×12-in. bottom ring or girder reinforced by two ½-in. rods. Outside of this drum and concentric with it is another 10¾ ins. thick and 9 ft. high. A flat annular roof 3½ ins. thick and reinforced with ¼-in. bars 18 ins. apart extends about 3 ft. inward from the top of this drum and is stiffened by 10×22-in. girder ribs on the under side. Each of these ribs has four 1-in. reinforcing-bars. These ribs carry most of the weight of the roof-dome drum 7 ins. thick and 6 ft. high. The oblique ribs under the circular annular roof are tangent at eight points to the drum wall on center lines and carry part of its weight directly to the inner row of columns and to the large circular girder, while the remainder of its weight is supported by the cantilever action of the circular roof. The roof dome has a solid shell 3 ins. thick without ribs and is reinforced by ¼-in. radial and circular rods. The horizontal or circular rods are 12 ins. apart and the radial rods are 12 ins. apart at the bottom of the drum. The radial rods are set ¾ in. from the outside surface for the first 7 ft. in height of the dome and then cross over to the same distance from the inner surface, which they follow to the apex. All rods lap 18 ins. at joints and are well wired together. The concrete of the dome was composed of Portland cement and unscreened ¾-in. broken stone.

Stairs.

Two general forms of stair construction are practiced with concrete-steel. In one the stairs are monolithic plates of concrete with notched tops to form the treads and risers and having the under side sometimes smooth and sometimes stiffened by ribs or string-pieces; in the other only the treads and risers are reinforced concrete and they are carried on metal framing. In the second form of construction the only change from ordinary metal construction is the substitution of reinforced concrete for the familiar stone and cast-iron treads and risers. The following examples illustrate some of the more simple staircases which have been built of reinforced concrete. Much more complicated forms have been constructed to meet unusual conditions or to illustrate the pliability of concrete-steel in adapting itself to bold and unusual structural forms.

Two examples of stair construction with Monier plates are shown by Figs. 106 and 107. In the first example the reinforced plate is

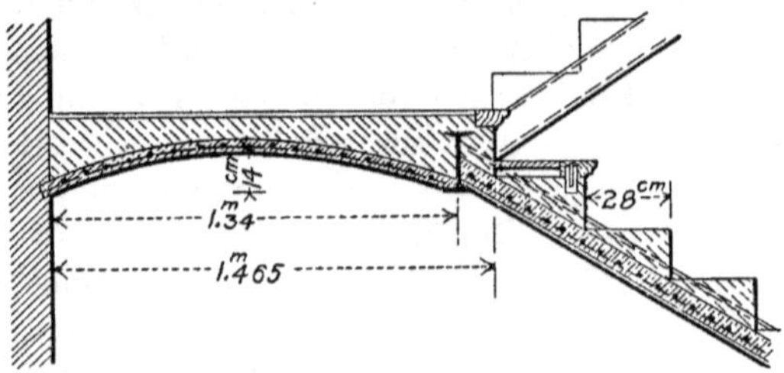

FIG. 106.—Stairs with Metal Strings Carrying Reinforced Plates.

carried on two channel-iron string-pieces supported at top and bottom on I beams. On top of this plate the steps are molded in plain concrete and covered with treads of wood, stone, or any other suitable material. The landing is formed by an arched Monier plate. In the second example shown there are no metal string-pieces, the Monier plate being arched instead and carried at top and bottom on beams.

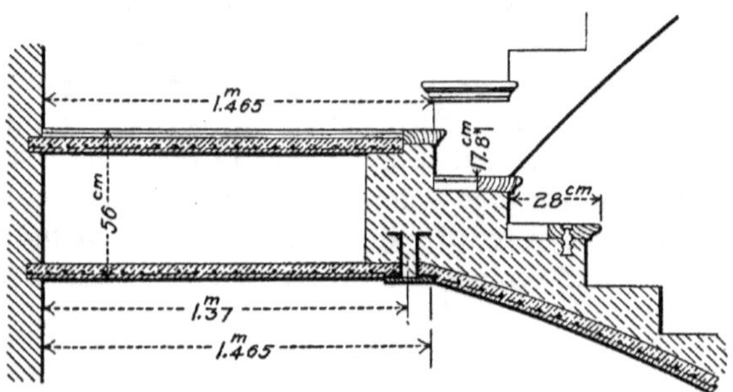

FIG. 107.—Stair Construction with Monier Plates.

The steps are molded in plain concrete as in the first example Several modifications of these two constructions are also employed. One of the most obvious of these is to carry one edge of the Monier plate on the building wall, using only an outside string-piece; another is to encase the plate in the wall so that it forms a bracket. The drawings of Fig. 108 illustrate a section of the main stairs built for a New York City residence. These stairs are in half-flights, with intermediate landings, and are seated at top and bottom on extra-heavy girders. They consist simply of a reinforced slab plain on the under

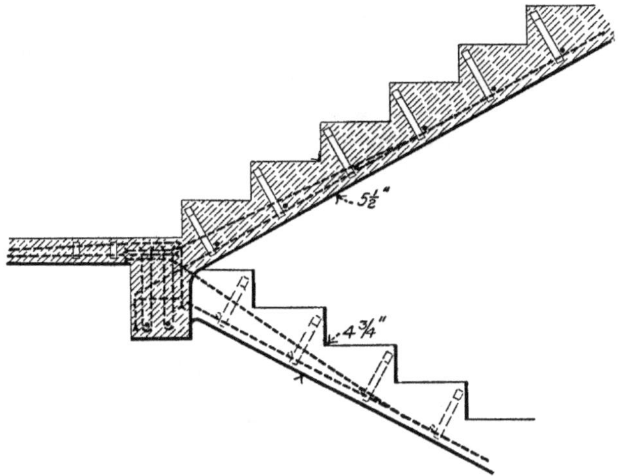

FIG. 108.—Reinforced Concrete Stairs for New York City Residence.

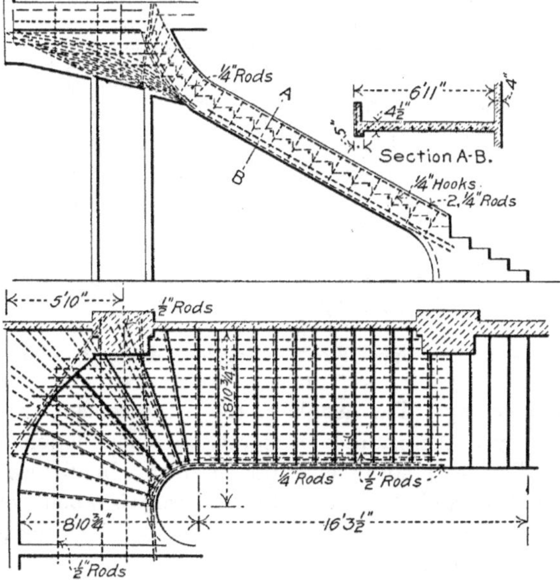

FIG. 109.—Main Floor Stairs for Ingalls Building, Cincinnati, Ohio.

side and notched on the upper side. Figs. 109 and 110 show the stair construction of the 16-story Ingalls Building at Cincinnati, O. The main stairs are shown by Fig. 109, and Fig. 110 shows the stairs for all floors above the third. As will be seen, these stairs are monolithic slabs of concrete-steel with their tops offset or notched to form the treads and risers, and their bottoms sloped and curved to form a smooth

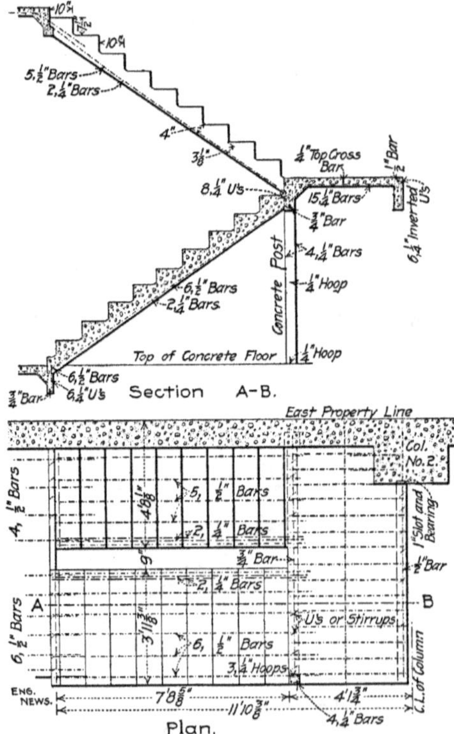

FIG. 110.—Stairs for Upper Floors of Ingalls Building, Cincinnati, Ohio.

and regular ceiling. The details of the reinforcement are given fully in the drawings. An example of spiral stair construction is shown by Fig. 111. This stairway ascends the shaft of a lighthouse tower constructed at Nikolaief, Russia. This stairway is about 108 ft. high and, as is shown by the drawings, is corbelled out from the tower walls. The reinforcement is entirely of round bars.

Corbels.

Reinforced concrete has been frequently and successfully used in constructing corbels, brackets, and cantilever beams, and several examples of such constructions have been noted in the preceding pages. The drawings of Fig. 112 show the cantilever construction by which the second floor of a mill building at Nantes, France, is corbelled out over the street. The brackets are spaced 4.27 m. (14 ft.) apart and are connected at their outer ends by a fascia girder 1.7 m. (5.58 ft.) deep. The brackets and girder carry the floor. The system of reinforcement employed was the Hennebique. Another example of cantilever construction designed by the same engineer is shown by Fig. 113. In double-tracking the Chemin de Fer Ceinture of Paris it was necessary to construct new retaining-walls set back from the original walls bordering the cut, and this narrowed the bordering roadways so that the sidewalks had to be carried on a cantilever platform over the edges of the cut. The drawings show the construction of this cantilever sidewalk, of which there were some 1,200 m. (3,936 ft.) The brackets were spaced 2.925 m. (9.59 ft.) and were all 0.5 m. (1.64 ft.) thick; their height is shown by Fig. 113 and their span varies, the maximum span being 3 m. (9.84 ft.). The reinforcement of the brackets and the manner of connecting them to the masonry retaining-walls are clearly shown by the drawings. The platform carried by the brackets consists of a wide shallow beam at the outer edge and of a narrow deep beam close to the wall; these beams carry the floor-slab proper, which is paved with asphalt and has a stone parapet. The drawings of Fig. 114 show the wall-bracket construction employed in the factory of the Central Felt and Paper Company at Long Island City, N. Y.

Shaft-hangers.

In mill-building construction it is necessary to provide the concrete beams and girders with some means by which the line shafting for operating the machinery may be attached to the ceiling structure. A variety of solutions for this problem have been presented by engineers. The one most commonly adopted has been to suspend from the ceiling girders by means of anchor-bolts a system of timbers to which the shaft-hangers could be attached by means of bolts or lag-screws in the usual manner. In this construction the anchor-bolts are built into the beam as it is molded, with their ends projecting below its soffit to receive the future timber. A preferable construction is to embed in the beam at intervals threaded sockets to which the timbers may be

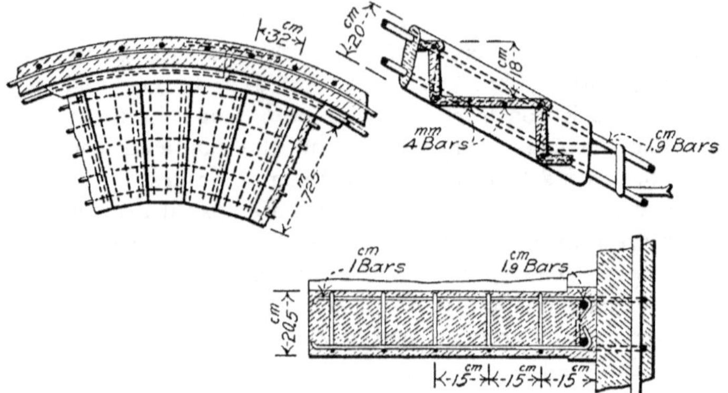

FIG. 111.—Spiral Stairway for Lighthouse Tower at Nikolaief, Russia.

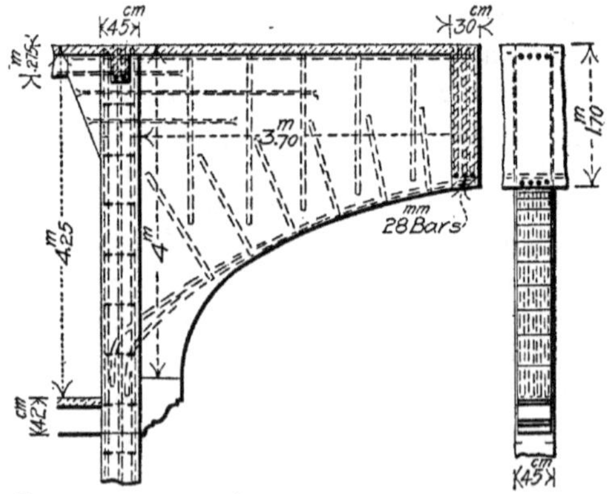

FIG. 112.—Cantilever Beam Carrying Corbeled Second Story of Mill Building at Nantes, France.

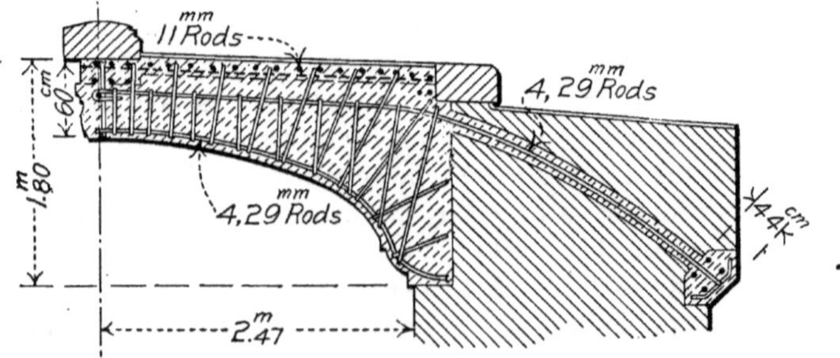

FIG. 113.—Cantilever Sidewalk Over Open Cut at Paris, France.

connected by suitable bolts. An example of such a socket construction is shown by Fig. 115. It consists of an inverted stirrup forged with a threaded socket and embedded at frequent intervals in the

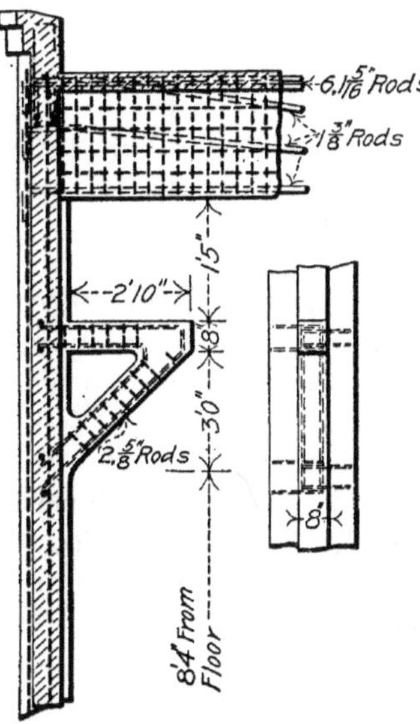

FIG. 114.—Wall Column Bracket for Factory Building at New York City.

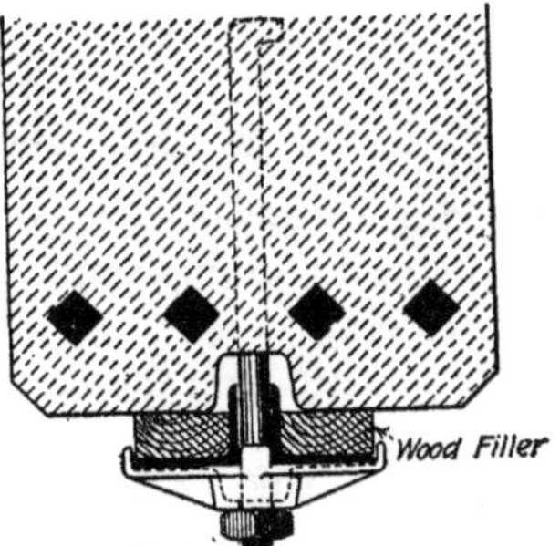

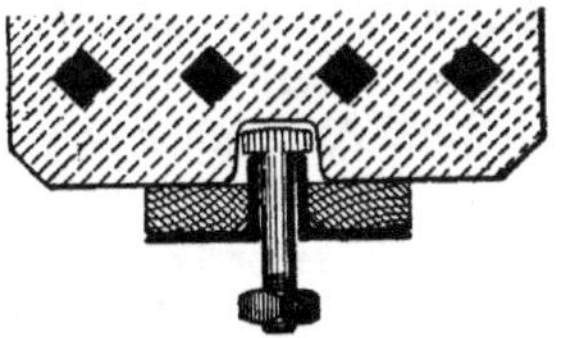

FIG. 116.—Slot Construction of Concrete Beams for Attaching Shaft Hangers.

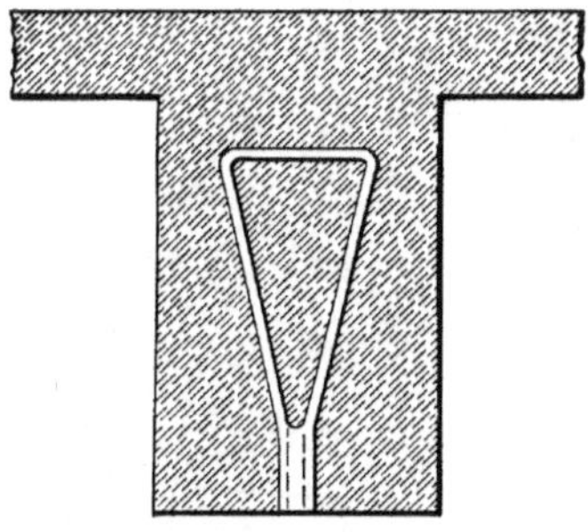

FIG. 115.—Socket Construction for Attaching Shaft Hangers to Concrete Beams.

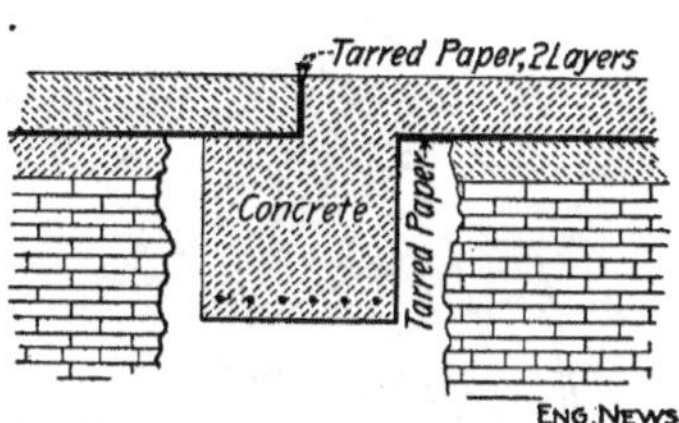

FIG. 117.—Detail of Expansion-joint for Concrete-steel Roundhouse Roof.

beam, as is indicated by the sketch. Wrought pieces of gas-pipe threaded inside at one end have been used in a similar manner. One of the most recent designs for attaching shafting to concrete beams is shown by the drawings of Fig. 116. As will be seen, the construction is very flexible; the hanger can be attached at any desired point along the beam, and can be shifted at will to another point or to another beam should the location of the machinery be altered. This shaft-hanger attachment has been patented by Mr. H. P. Jones of Yonkers, N. Y.

Expansion-joints.

There is no well-established practice in either locating or constructing expansion-joints for reinforced-concrete building work, and the following examples illustrate, therefore, individual ideas and preferences only.

In constructing the factory of the Pacific Coast Borax Company at Constable Hook, N. J., expansion-joints were introduced in both floors and walls. A portion of this building was four stories high and a portion was one story high, and it was rectangular in plan. Vertical planes passed through this building both longitudinally and transversely 25 ft. apart marked the lines in walls and floors at which expansion-joints were introduced. These joints, therefore, divided the floors into 25-ft. squares and the walls into sections 25 ft. wide, extending from foundation to roof. In the floors the joints in each case divide a girder or beam vertically into halves, as is shown by the drawings of Fig. 76. The joints in all cases are close joints formed by finishing one section of wall or floor against a vertical form and allow-

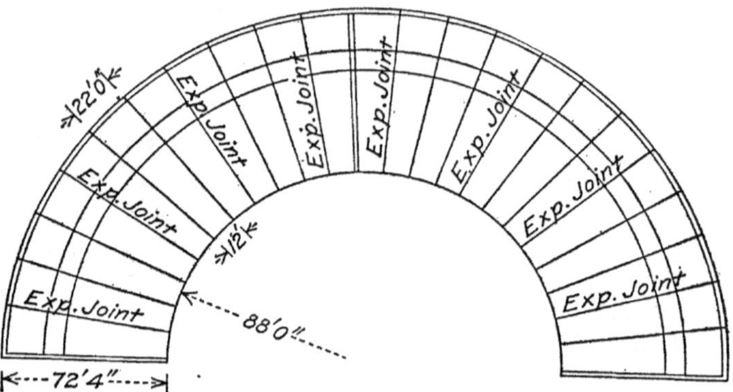

Fig. 118.—Sketch Plan of Roundhouse Roof Showing Arrangement of Expansion-joints.

ing the concrete to set before the succeeding sections were concreted. The joints are, therefore, planes of weakness simply. In building the wall joints the adjacent edges of the wall sections were reinforced by a vertical bar extending from top to bottom. A somewhat similar system of expansion-joints was employed in constructing the foundry floors of the Eastwood Company at Paterson, N. J. This construction is illustrated by the drawings of Fig. 75. Fig. 117 shows the construction of the expansion-joints for a roundhouse roof built in 1904 for the Long Island Railroad. A sketch plan of this roof showing the location of the joints and the dimensions of the areas bounded by them is given in Fig. 118.

CHAPTER VIII.—REINFORCED CONCRETE IN BRIDGE AND CULVERT CONSTRUCTION.

Bridge construction is one of the best-known applications of reinforced concrete. The most common form of concrete-steel bridge is the arch, and this is the only form which has been constructed to any extent in the United States. In Europe, however, girder bridges of concrete-steel are much used. These are sometimes plain flat plates or slabs resting on abutments, but the usual construction is a monolith consisting of a slab, with stiffening ribs or girders on the under side. The unstiffened slab is of course adapted only to very small spans such as meet the requirements of ordinary box-culvert work, but the ribbed construction is frequently applied to spans from 30 ft. to 50 ft. in length. This construction has met with particular favor in building foot-bridges and highway bridges across railway tracks, canals, and small streams, where the light weight, freedom from corrosion, and low cost of the structure are important items.

For longer spans than 50 ft., arch construction is almost universally employed, and concrete arches of 165 ft. span have been built and much longer spans of that material have been designed. The arch bridge of reinforced concrete occupies a position between the metal bridge and the stone arch; it possesses to a large degree the qualities of lightness and cheapness characteristic of the steel bridge, while having much of the beauty and durability of the masonry arch. Its use has been, therefore, chiefly for highway bridges and particularly, in America at least, for street bridges and bridges carrying park roadways. When first introduced the concrete-steel arch bridge made its principal appeal to American engineers as a cheap substitute for the stone arch; its claims as a type of construction having an individuality and structural merits peculiar to itself were only gradually realized.

In the earlier bridges of concrete-steel only the arch ring was reinforced, and this is yet the usual construction; but a number of such bridges have been built in recent years in which the supporting piers and the spandrel walls and parapet walls are also of concrete with a metal skeleton. The early form of arch was the plain arch or the arch ring with a smooth soffit, but ribbed arches are now being built in considerable numbers. With certain forms of reinforcement European engineers have also made use of hinged arches. The use of reinforced concrete for bridge floors is another for which the demand is increasing. A less familiar application is the reinforcement of steel girder bridges which have become weakened by rust or are in particular danger of such an attack on their strength.

Culverts of reinforced concrete are largely used in Europe, and recent American practice also shows a number of examples of such structures. In cross-section and in the arrangement of the reinforcement, culvert construction in concrete-steel approaches bridge practice at one extreme and conduit practice at the other extreme. This feature will be made plain by the description of representative structures which are given toward the end of this chapter. Previously, however, representative examples of girder bridges, arch bridges, and bridge floors will be described.

GIRDER BRIDGES.

Practically all forms of ribbed-plate construction in concrete-steel are applicable to girder-bridge construction. Only a few of them have, however, been so employed, and of these only those which have some claim to general use are considered here. From these few examples the engineer will readily see how other forms not commonly used may be adapted to the same purpose.

Hennebique Construction.—Hennebique construction has been employed in France for girder bridges having spans up to 50 ft. The construction used consists of a flat plate stiffened by and in one piece with longitudinal ribs or girders spaced from 3½ ft. to 6½ ft. apart. This platform is usually carried on masonry abutments or on abutments of unreinforced concrete. Fig. 119, which shows a transverse section of a canal foot-bridge at Yverdon, France, is a typical girder bridge of Hennebique construction. It will be observed that the standard floor-plate and joist construction employed for buildings (p. 156) is closely followed. The overhang of the plate beyond the girders is to be noticed particularly. Another and a more pronounced example of this detail is shown by the part section, Fig. 120, of a bridge carrying two 1.5-m. (4.92-ft.) railway tracks. This bridge has a span of 4.25 m. (13.94 ft.) and is built on a skew of 65° 40′; it has four girders, and the trough shape of the platform provides for the ballast on which the tracks are laid. A good example of still heavier construction

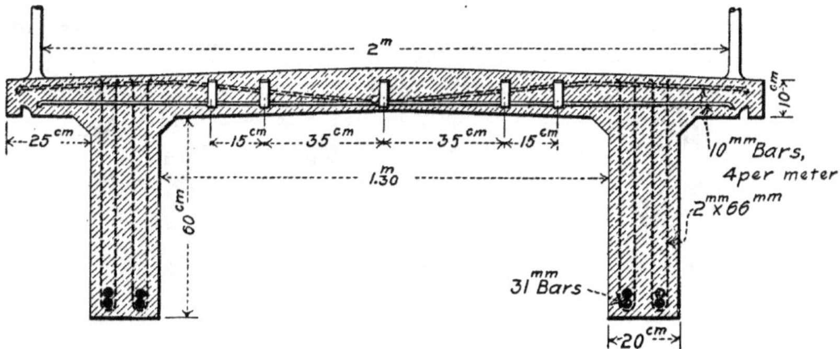

Fig. 119.—Cross-section of Girder Bridge at Yverdon, France.

is afforded by the highway bridge built in 1897 over the Flon, a small stream near Lausanne, France. This bridge has a span of 15 m. (49.2 ft.) and a width of 7.5 m. (24.6 ft.), with a floor-plate 7 ins. thick

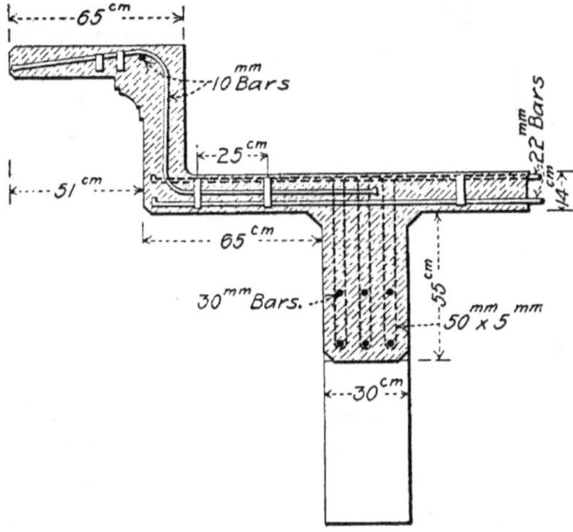

Fig. 120.—Part Section of Girder Bridge Showing Cantilever Sidewalk.

and girders 1 m. (3.28 ft.) deep. It was tested with a 20-ton road roller and cost for the platform alone about $1,700.

Hirdenheim Bridge, Germany.—The lateral overhang of the floor-plate referred to in the bridges illustrated by Figs. 119 and 120 is often extended to form sidewalks, as shown by Fig. 121. This is the cross-section of a three-span bridge built near Hirdenheim, Germany, for highway traffic. The center span is 14 m. (45.9 ft.) and the side spans are each 7.5 m. (24.5 ft.); the other dimensions are shown by the drawing.

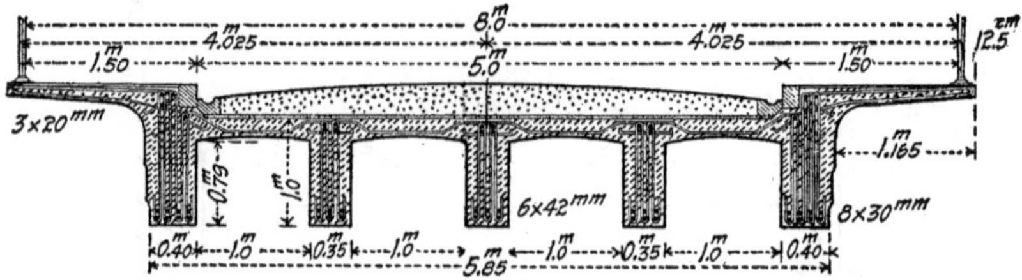

FIG. 121.—Girder Bridge near Hirdenheim, Germany.

This structure furnishes a good example of bracketed sidewalk construction.

Möller Construction.—In Germany the Möller ribbed-plate construction is a favorite one for girder bridges. Möller girder bridges consist, like the Hennebique, of a plate stiffened by and in one piece with vertical ribs spaced about 5 ft. apart. Fig. 122 is a drawing of a

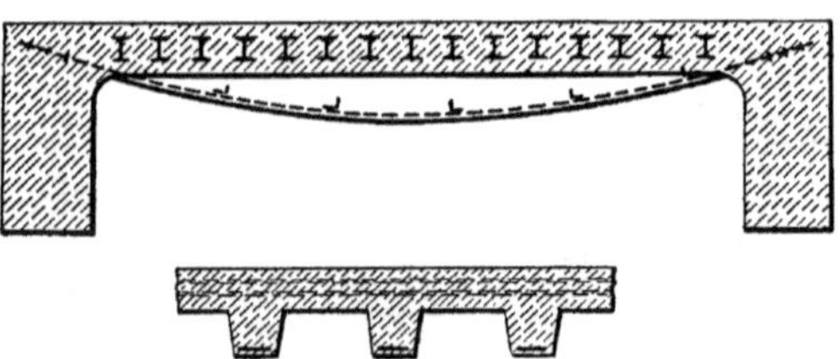

FIG. 122.—Typical Möller Girder and Slab Construction.

typical Möller plate and girder. It will be observed that the under side of the platform plate is haunched downward at the ends, giving it the appearance of a flat-topped arch, and that the ribs are of the fish-belly type. The reinforcement is peculiar, that of the plate consisting of small bars of I section placed parallel and transversely of the bridge, and that of the girders consisting of a flat strap near the bottom, which is anchored into the haunches of the plate and which has at intervals small angles riveted transversely to give a mechanical hold on the surrounding concrete. A representative structure of this type is the highway bridge over the Selke at Alexisbad, Germany, built in 1897, with a span of 15 m. (49.2 ft.). This structure withstood a test load consisting of a 22-ton road roller.

Embedded Steel-beam Construction.—A form of concrete-steel girder bridge which has been used in a number of important instances consists of steel I beams spaced from 2 ft. to 5 ft. apart spanning the opening and embedded in concrete. Fig. 123 shows a transverse section of a bridge of this type which was built in 1899 near Pittsburg, Pa.

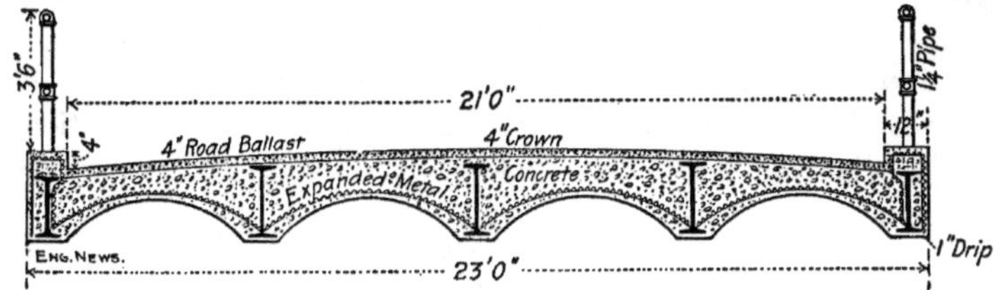

FIG. 123.—Embedded Steel Beam Bridge near Pittsburg, Pa.

This bridge has a span of 28 ft., with a 21-ft. roadway, and was designed to carry a 12-ton road roller. The outside girders are 18 ins. deep and the intermediate girders are 20 ins. deep. Between the beams

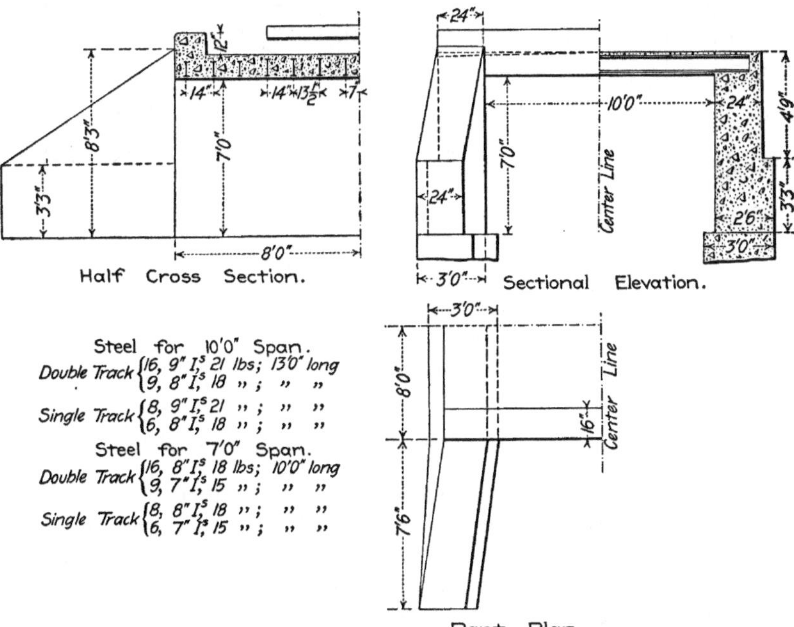

FIG. 124.—Embedded Steel Beam Cattle Pass, Aurora, Elgin & Chicago R.R.

and embedding them are concrete arches reinforced with expanded metal. An embedded-girder construction used for cattle-passes on the Aurora, Elgin and Chicago electric railway is shown by Fig. 124. This structure has a span of 10 ft. and the girders are 10-in. I beams spaced 14 ins. apart. The beams are entirely embedded in concrete, but are figured to carry the entire load. The claims made by the designing engineer for this construction is that it reduces the dimensions of the abutments so much that the structure is 20 per cent. cheaper than an open-floor bridge with I beams on concrete abutments. A second form of embedded-girder construction employed on the same road resembles that illustrated in Fig. 123. This construction is shown by Fig. 125. Concrete jack arches are turned between the

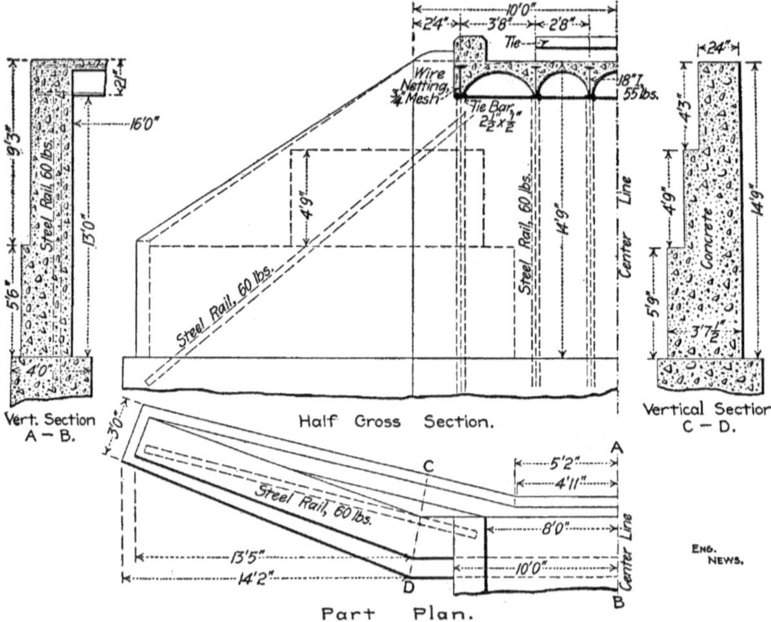

FIG. 125.—Embedded Steel Beam Construction, Aurora, Elgin & Chicago R.R

I-beam girders with a wire-netting reinforcement on their soffits. The end of each I beam rests on a vertical 60-lb. rail embedded in the abutment, and a diagonal rail is embedded in each wing-wall. This construction is said to give a saving in cost of 30 per cent over open-floor I-beam girder construction. A somewhat similar construction to these American examples has been used for a number of overhead highway crossings by the Chemin de Fer de l'Ouest of France. It consists of steel beams placed parallel between abutments and carrying thin brick arches between the lower flanges, with a concrete filling above these arches that completely embeds the beam. By this construction spans of from 50 ft. to 60 ft. have been built with girders from 30 ins. to 36 ins. deep.

Concrete Reinforced Steel-girder Construction.—A special application of concrete-steel which is deserving of brief notice is illustrated by an overhead highway bridge covering the tracks of the Orléans Railway at Perigueux, France. This bridge being in danger of destruction by the locomotive gases, it was decided to embed the steel members in concrete reinforced by bars or rods to provide for the extra dead load. The bridge has a span of 26 ft., and as reinforced is in all essentials a concrete-steel structure. The special importance of the example lies in the suggestion which it offers for the protection of some if not all of the members of metal bridges which are subjected to deterioration by locomotive gases.

ARCH BRIDGES.

Practically every form or system of reinforcement for concrete has been adapted to arch construction. It is impossible, however, to give within reasonable space an example of each, and it would be to a large extent a useless consumption of space to do so. Only a few of the numerous types or systems of concrete-steel have gained prominence in arch-bridge construction and, therefore, really demand description. These are all included in the succeeding examples.

Monier Construction.—In Europe Monier construction for arch bridges takes precedence of all others in age and in the number of representative structures for which it has been employed. The first Monier arches were built in 1867, and there are now several hundred bridges of this type located principally in Germany and Austria. In practically all Monier bridges the use of reinforced concrete is confined to the arch ring, the abutments and spandrel walls being of stone or of brick masonry or of unreinforced concrete. The use of reinforced spandrel walls and spandrel arches is, however, not unfamiliar. The reinforcement employed for bridge arches is the standard Monier network of round rods and wires described on p. 331, and it is usually embedded in a 1 Portland cement and 3 to 4½ sand mortar and not in concrete in the usual meaning of that term. The arrangement of the reinforcement in the arch ring differs in different bridges and is best illustrated by the structures whose descriptions follow. The great majority of the bridges which have been built of Monier concrete-steel have had spans of less than 100 ft., but a number have greater spans, and the longest span so far built, that of the Waidhofen Bridge in Austria, is 44 m. (144.2 ft.). The following bridges are fairly representative examples of European practice in Monier arch-bridge construction.

Fig. 126 shows a Monier arch of 17.3 m. (86.7 ft.) span built to carry a park roadway at Nymphemburg, Bavaria. This arch is of the basket-handle type and is reinforced both at the extrados and the

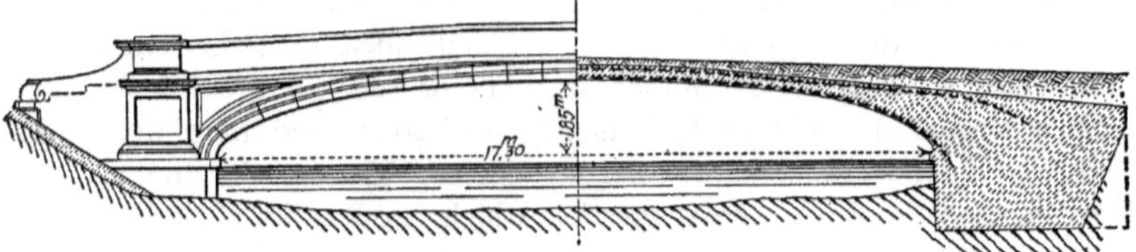

FIG. 126.—Monier Arch Bridge at Nymphemburg, Bavaria.

intrados of the ring as is shown by the half-section. The bridge has a total width of 10 m. (32.8 ft.), with a 6.3-m. (20.7 ft.) roadway, and was designed to carry wagons of 26 tons. The roadway is paved with wood blocks resting directly on the arch ring at the crown and on a concrete filling over the haunches. The double, or top and bottom, reinforcement shown by Fig. 126 is less frequently used in Monier arches, however, than the form illustrated by Fig. 127, which shows

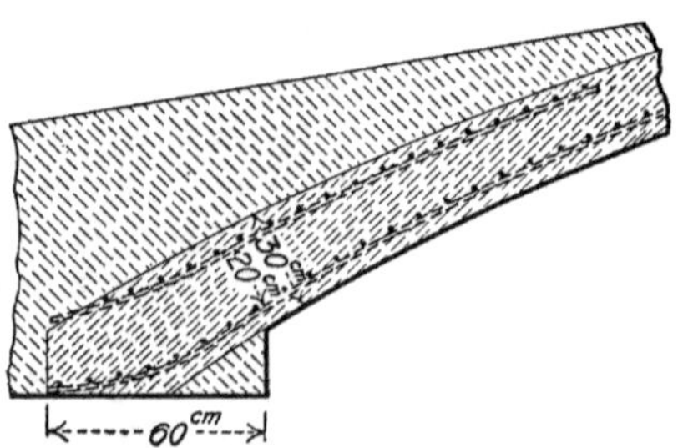

FIG. 127.—Monier Arch with Extradosal Reinforcement at Haunches only.

a highway bridge over a canal at Draulitten, Germany. This is a polycentric arch of 26.3 m. (86.3 ft.) span and 4.3 m. (14.1 ft.) width, and was designed for a load of 90 lbs. per square foot. The abutments are masses of concrete, the spandrel walls are of brickwork, and the spandrel filling is concrete. The reinforcement of the ring consists of a continuous intradosal net and an extradosal net reaching from each skew-back part way up the haunch. In many Monier arches the upper ends of these extradosal nets are brought down and tied to the intradosal net. These two forms of reinforcement have been preferred in Europe to the double reinforcement or continuous top and bottom net. One of the longest and boldest Monier arches ever built, however, has a double reinforcement. It is the Wildegg,

Switzerland, Bridge built in 1890, with a span of 37.22 m. (122 ft.), a width of 3.9 m. (12.8 ft.), and a ring thickness of 0.17 m. (6.69 ins.) at the crown and of 0.25 m. (9.85 ins.) at the springing-lines. The remarkable thinness of arch ring employed by the European builders of Monier bridges is most notable in this structure. The Wildegg Bridge has a skew of 45° and was designed for a load of 112 lbs. per square foot.

Wünsch Construction.—The flat-topped arch construction of reinforced concrete designed by Robert Wünsch of Budapest, Hungary, in 1891, has been employed in building several notable structures in the country of its origin. The Wünsch reinforcement consists of a horizontal top member of T section placed with its flange upward and of a curved intradosal member of the same section placed with its flange downward. At the crown of the arch the webs of the two members overlap and are riveted together, and at the ends of the arch a vertical tie connects the two members and extends downward into the pier or abutment masonry, where it is anchored to a girder extending along the pier transversely of the bridge. These frames, constructed as described, are placed parallel to each other longitudinally of the arch and from 18 ins. to 2 ft. apart. The ratio of the cross-sections of the reinforcement and of the concrete at the crown of the arch is generally about 2 to 100. The concrete embedding the curved

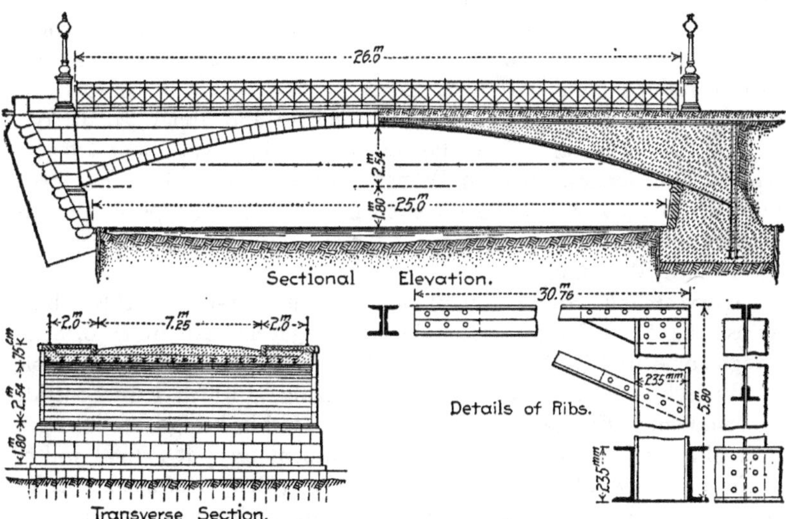

FIG. 128.—Wünsch Arch Bridge at Sarajevo, Bosnia.

member of the frame is much richer than that at the top of the arch which embeds the horizontal member. To illustrate Wünsch construction as applied to bridges, the Emperor Bridge at Sarajevo, Bosnia, and the Maryborough Bridge, Queensland, have been selected.

Emperor Bridge, Sarajevo, Bosnia.—The Emperor Bridge at Sarajevo, Bosnia, crosses the river Miljącka, with a clear span of 25.36 m. (115.8 ft.) and with a rise of one-tenth the span. Fig. 128 shows the details of the construction. The abutments have a stone facing, but the entire remainder of the structure is concrete. The construction and arrangement of the reinforcement are clearly shown by the drawings. The concrete is composed of Portland cement and unscreened gravel in the following proportions: Foundations, 1 to 10; abutments, 1 to 8; arch ring for a distance of 3 dcm. (9.8 ins.) above the soffit, 1 to 6; arch filling, 1 to 8. The bridge was tested with a uniform load of 108 lbs. per square foot, and there were used in its construction about 865 cu.m. (1,031 cu. yds.) of concrete and about 33,392 kgs. (73,325 lbs.) of metal. The total cost of the bridge was about $16,500. This is the longest span of Wünsch concrete-steel which has been built up to 1904, and is an excellent example of the type in all particulars except that in smaller spans the made-up T section is replaced by a rolled T section.

Maryborough Bridge, Queensland.—The low-level bridge built across the Mary River at Maryborough, Queensland, in 1896, after designs by Alfred B. Brady, M. Inst. C. E., is the most notable example of Wünsch construction, as applied to bridges, which exists outside of continental Europe. The dimensions of this bridge are: length over all, 613 ft.; waterway between abutments, 395 ft.; total width, 22 ft. 8 ins.; width of roadway, 20 ft. 8 ins.; length of spans, 80 ft.; rise of spans, 4 ft.; thickness of arch at center, 18 ins. at curbs, crowned to 20 ins. at center; thickness over piers, 5 ft. 8 ins. There are eleven spans carried on two concrete abutments and ten piers founded on pneumatic caissons.

Fig. 129 shows the arrangement of the arch reinforcement and other structural details of the bridge. The concrete used in the arches consists of 1 part Portland cement, 1½ parts sand, and 4 parts stone broken to pass a 2-in. ring. The reinforcing-frames are constructed according to the standard Wünsch construction, except that the members are railway rails of Vignoles section weighing 41.25 lbs. per yard. The horizontal rail is placed base downward and is continuous from end to end of the bridge. It is connected to the segmental member by a vertical tie over each abutment, but there is no connection between the two members at the piers. The segmental members are placed

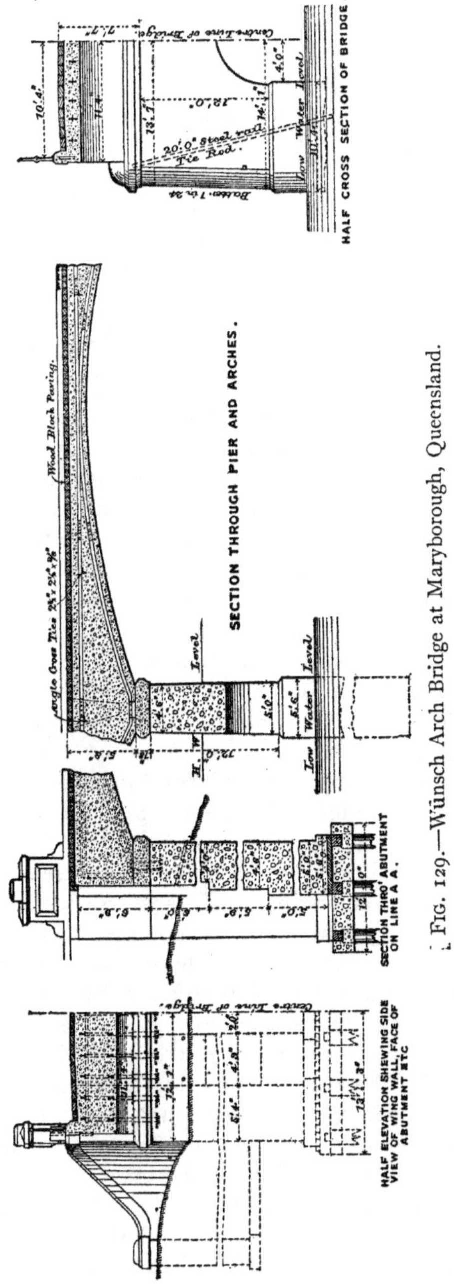

FIG. 129.—Wünsch Arch Bridge at Maryborough, Queensland.

base up and are distinct for each arch; they are spliced together at each pier and anchor-bolted to the masonry. There are eleven reinforcing-frames spaced 2 ft. apart, and they are braced together over each pier and at two places in each span by a $2\frac{1}{2}\times2\frac{1}{2}\times\frac{3}{8}$ in. angle riveted to the flanges of the horizontal members. The principal purpose of these cross-braces was to hold the frames in place while the concrete was being filled around them. All exposed concrete surfaces were finished by floating them with a 1 to 2 Portland cement and sand mixture in two coats aggregating not less than $\frac{5}{8}$ in. in thickness and by giving this facing a 6 to 1 cement and sand fluid wash. The total cost of the bridge in round figures was $75,000. A full description of the Maryborough Bridge is given in Trans. Inst. C. E., Vol. CXLI, p. 246.

Hennebique Construction.—The ribbed-arch construction of Hennebique has been employed in building a number of bridges, one of them being the longest-span concrete-steel arch bridge ever built. For short spans the typical construction is a flat plate stiffened by longitudinal arched ribs spaced from 5 ft. to 10 ft. apart. In some cases transverse ribs stiffen the floor-plate between the main arch ribs. This general form of construction is employed for spans of from 30 to 60 ft. For spans longer than 60 ft., an arch ring stiffened intradosally by arch ribs is the usual construction. An example of each form of construction is given in the following paragraphs:

Aisne Bridge, Soissons, France.—The three-span highway and tramway bridge built over the river Aisne at Soissons, France, in 1903, is an excellent example of the Hennebique flat-plate and arch-rib construction. This bridge is built on a skew of 30° and has three spans of 24.25 m., 24.48 m., and 24.25 m. (about 80 ft. each) carried by two abutments and two piers of reinforced concrete. The drawings of Fig. 130 show a longitudinal and transverse section of one of the shore spans and the supporting pier and abutment. As will be seen, each span has seven ribs spaced unequal distances apart which carry and are in one piece with a floor-slab stiffened by joists running transversely between ribs. These transverse joists are not shown by the drawing, but they are all 0.2 m. wide and 0.3 m. deep, and are spaced 3.5 m. apart under the roadway and adjacent sidewalk and half this distance apart under the tramway and adjacent sidewalk. The reinforcement of the arch wing and the floor-slab is shown clearly by the drawings. The arch ribs, with the exception of the two outside ones, which are of the same thickness throughout, are made thicker near the piers and abutments, through which they are carried so as to be continuous from end to end of the bridge. This thickening of the rib is indicated by the transverse section in Fig. 130.

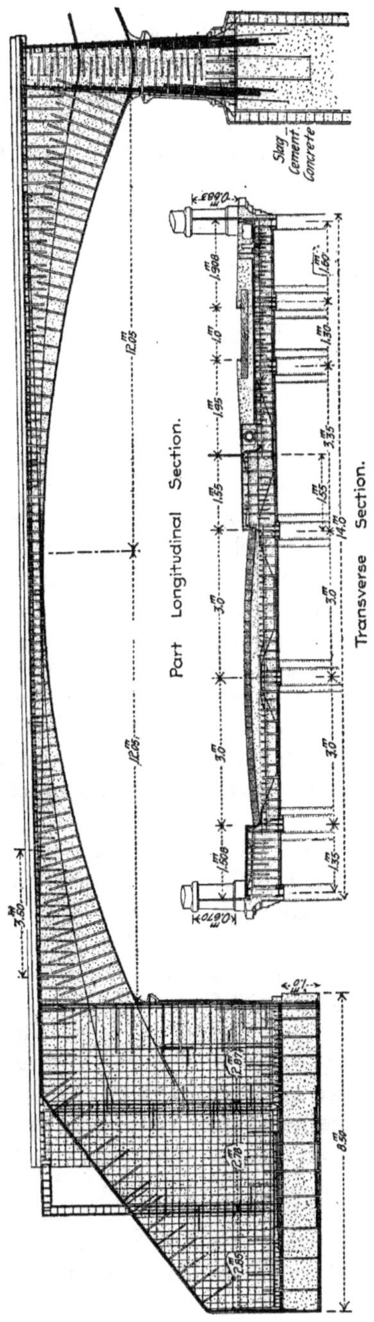

FIG. 130.—Hennebique Arch Bridge over the Aisne at Soissons, France.

Chatellerault Bridge, France.—The longest-span concrete-steel arch bridge ever built is the Chatellerault Bridge over the river Vienne at Chatellerault, France, constructed in 1899. This bridge has a center span of 50 m. (164 ft.) and two side spans each 40 m. (131 ft.) long. The center span has a rise of 15 ft. 8 ins. and the side spans have each a rise of 13 ft. The total length of the bridge is 443 ft Each span is composed of an arch ring stiffened by four parallel arch ribs and carrying vertical columns supporting a roadway platform of slab and girder construction. The arch rings are 6 m. (19.68 ft.) wide and the roadway platform is 8 m. (26.24 ft.) wide. The arch rings increase in thickness from crown to skew-backs and the stiffening-ribs have a similar variation in depth but a constant width. The reinforcement of each arch rib consists of intradosal and extradosal. rods connected by stirrups. At the piers and abutments this reinforce ment is carried into the substructure masonry as shown by Fig. 131, and both abutments and piers are carried up above the arches to the level of the roadway platform which is anchored to them. The piers and abutments consist of shells of concrete-steel backed with concrete. The roadway platform and the arch ring of each span merge into each other at the crown, and here the total thickness of the concrete-steel structure is 0.5 m. (1.64 ft.) for the 164-ft. span and 0.44 m. (1.44 ft.) for the 131-ft. spans. The bridge was designed for a moving load consisting of two files of four-wheeled carts weighing 16 tons and a static load of about 90 lbs. per square foot on the sidewalks, the roadway being 5 m. (16.4 ft.) wide and each sidewalk 1.5 m. (4.92 ft.) wide. The arch was tested by a dead load of 165 lbs. per square foot for the roadway and 123 lbs. per square foot for the sidewalks, each being loaded over the total length, then on each half-span, and then over the middle third of the span. Under this loading the maximum deflection of the right-shore span was 17/32 in., of the left-shore span, ¼ in., and of the center span, 13/32 in. The arches resumed their original positions upon removal of the load The moving-load test consisted of a procession of first a 16-ton roller, two four-wheeled 16-ton carts, and six two-wheeled 8-ton carts. After this the steam-roller was run over the bridge so as to pass over 2-in. sticks of wood so strewn as to produce a series of shocks. No permanent deformation resulted from these tests. The centering and forms being placed and the foundation beds ready, concreting was begun on Aug. 15 and finished on Nov. 5, 1899. The centers were removed on Dec. 15 1899. The total cost of the bridge was about $35,000.

Arch-rib Constructions.—A number of concrete-steel bridges, have been constructed in Europe in which isolated arch ribs carry

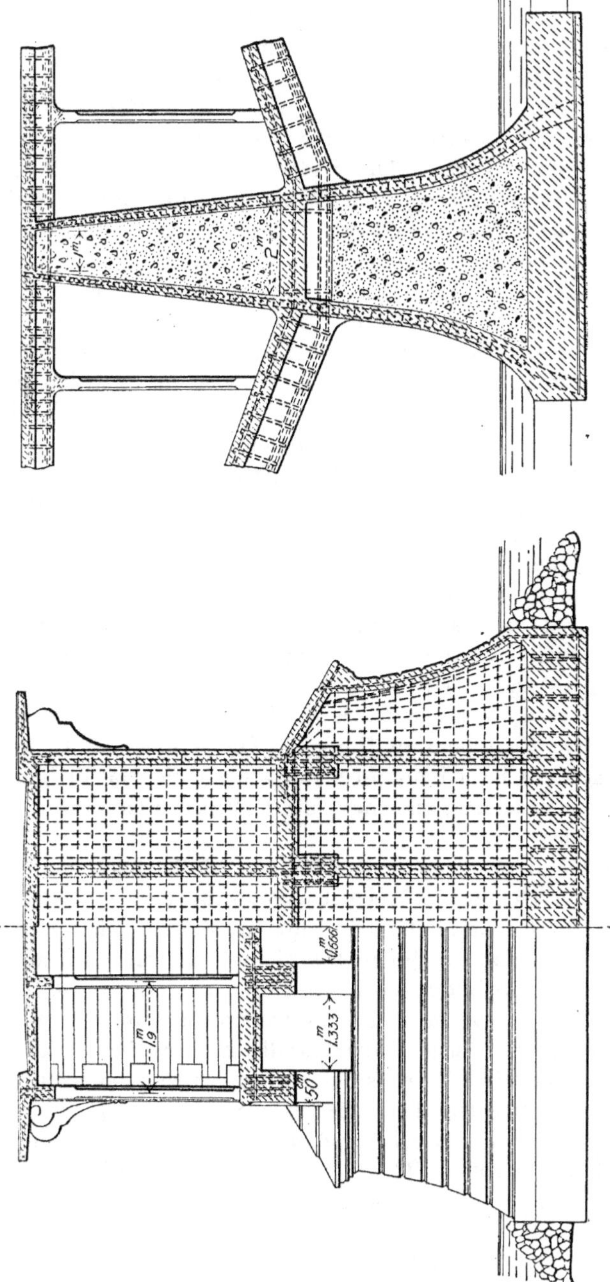

FIG. 131.—Hennebique Arch Bridge at Chatellcrault, France.

the roadway, the construction being closely analogous to steel-arch construction. In fact, in the construction of a bridge of this type at Auch, France, Mr. Bonna first erected self-supporting steel-arch ribs carrying vertical posts supporting a roadway platform all in steel skeleton and self-supporting, and then embedded the ribs and other steel members in concrete. This bridge has four ribs of 21 m. (69 ft.) span and a rise of 1 in 10½, each rib being 0.4 m. (1.31 ft.) wide and 0.3 m. (0.98 ft.) deep at the crown. These ribs carry vertical columns of reinforced concrete on which rests the roadway platform 4.5 m. (14.76 ft.) wide. On each side of the roadway platform there is a cantilever sidewalk 0.75 m. (2.46 ft.) wide.

Melan Construction.—The form of concrete-steel arch construction invented by Prof. Joseph Melan, of the Polytechnic School at Brunne, Austria-Hungary, has had its principal development in the United States, where the construction was patented in 1893. In this construction the reinforcement consists of parallel arch ribs spaced from 2½ to 3¼ ft. apart, and having a depth and longitudinal profile closely corresponding to the depth and profile of the arch ring. Usually the arch ring is the only portion of a Melan arch which is reinforced, the abutments and spandrel walls being of ordinary concrete masonry. Steel is almost universally employed for the reinforcing members. These are usually rolled I beams bent to the curve of the arch ring for short spans and of latticed arch ribs for longer spans. A recent development of the construction is the hinged arch, several examples of which have been built in Europe. The number of Melan arch bridges that have been built precludes the mention of all but a few representative structures, and these are described briefly in the following paragraphs:

Stockbridge Arch, Massachusetts.—One of the finest examples of foot-bridge construction in Melan concrete-steel in the United States is furnished by the bridge built in 1895 by Mr. Fr. von Emperger to give access from Laurel Hill to the picturesque Ice Glen at Stockbridge, Mass. This bridge has a space of 100 ft. with a rise of 10 ft., and is 7½ ft. wide over all. Its construction is shown by Fig. 132. The reinforcement consists of four parallel I beams bent to the curve of the arch and with the ends resting against and bolted to angles crossing the abutments. The I beams were 7 ins. deep and weighed 15 lbs. per foot; they were spaced 28 ins. apart and raised about 1½ ins. to 2 ins. above the soffit. The arch ring is 18 ins. thick at the crown and 30 ins. deep at the haunches. The concrete used consisted of 1 part Portland cement, 2 parts sand, and 4 parts broken stone, except that at the crown of the arch this mixture was enriched by reducing the aggregates to 4½ parts of the mixture. The placing of the concrete for the arch ring, 22 cu. yds., was done in one day with a force of 24 men. In constructing the arch ring it was necessary for the sake of economy to use a rough-board lagging on the center, and to secure a smooth soffit finish. Under this condition the following resort was had: The lagging was plastered with a layer of puddled clay which was covered with a layer of heavy paper. This layer of

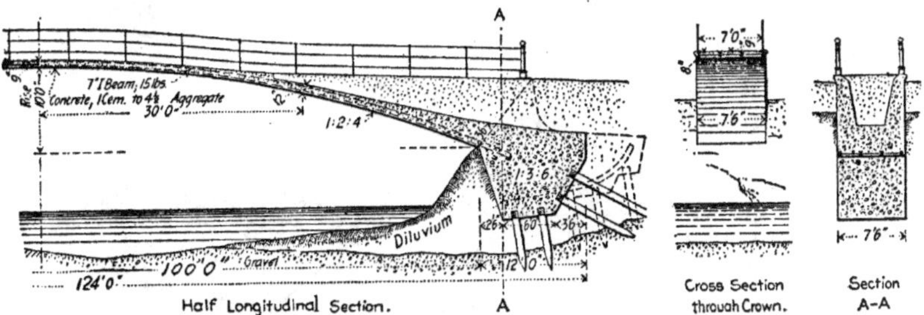

FIG. 132.—Melan Arch Bridge at Stockbridge, Mass.

wet clay served the second purpose of keeping the lagging wet and thus preventing it from shrinking away from the concrete as often happens. The false works were removed six weeks after the completion of the arch ring, and upon their removal the arch deflected ¾ in. at the crown and ½ in. at the haunches. When the arch ring was seven weeks old it was tested by a uniform load of 75 lbs. per square foot over an area of 7×100 ft., and then by half of this load concentrated over one-half of the arch. The arch showed no appreciable deflection under these loads. The total cost of the bridge was $1,450.

Topeka Bridge, Kansas.—The most important highway bridge of Melan construction in the United States is the one built across the Kansas River at Topeka, Kan., by Keepers & Thacher in 1896–7. This bridge consists of five spans, one 125 ft. long, two 110 ft. long, and two 97½ ft. long. The bridge has a 26-ft. roadway and two 7-ft. sidewalks. The piers and abutments are of concrete on pile foundations. Figs. 133 and 134 show their construction and also that of the arches. The reinforcement consists of twelve arch ribs of lattice construction, spaced 3 ft. apart. Each rib is spliced at several points, and the abutting ribs of adjacent arches are connected by a riveted joint. The drawings of Figs. 133 and 134 show clearly the rib construction and the nature of the splices and connections. The material used in these ribs was steel meeting the following specified requirements: Tensile strength, 60,000 lbs. to 68,000 lbs.; elongation in 8 ins., 28 per cent.; reduction of area at fracture, 40 per cent. The concrete used for the arch rings was composed of 1 part Portland cement, 2 parts sand, and 4 parts limestone broken to from ½-in. to 1-in. sizes for all parts except exposed faces where a 1-in. facing of 1 part Portland cement and 2 parts sand mortar was employed. The finish

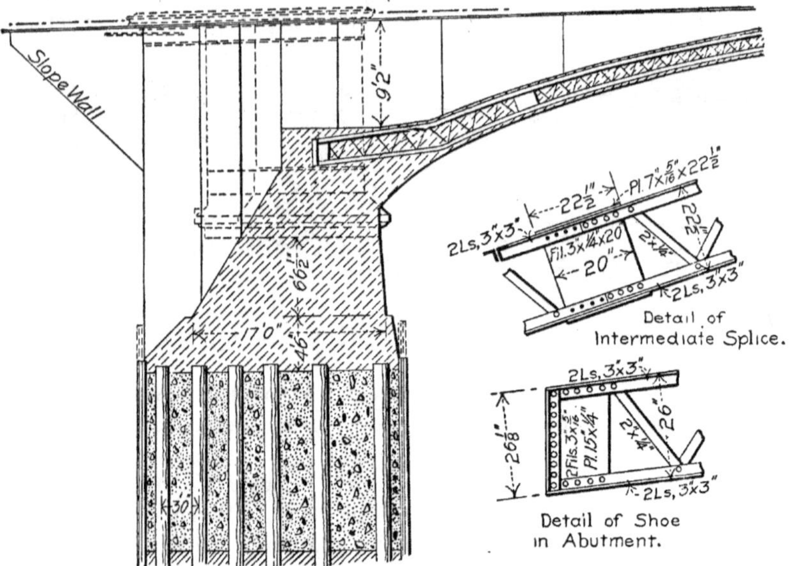

FIG. 133.—Abutment Construction of Melan Arches at Topeka, Kan.

of the exposed surfaces consisted of sifting on sand and cement and rubbing the surfaces hard with a float. The filling above the arch ring was earth tamped in layers. The cost of the bridge complete was $125,000.

Hinged Arches, Melan System.—All the Melan arches constructed in the United States have rigid arch rings. In Europe a number of hinged arches of this construction have been built. The most notable of these is the 42.4-m. (139.1 ft.) arch built in 1898 at Steyr, Austria, with a rise of 2.617 m. (8.56 ft.) or one-sixteenth of the span. This arch has three hinges located at the crown and the springing-lines. Fig. 135 shows a transverse section and a half longitudinal section of the span. By referring to the longitudinal section it will be seen that each half of the arch ring is a separate piece or segment, and that it is thicker at the haunches than at the ends. The exact depths at the three points are: crown, 0.6 m. (1.97 ft.); springing-lines, 0.7 m. (2.29 ft.);

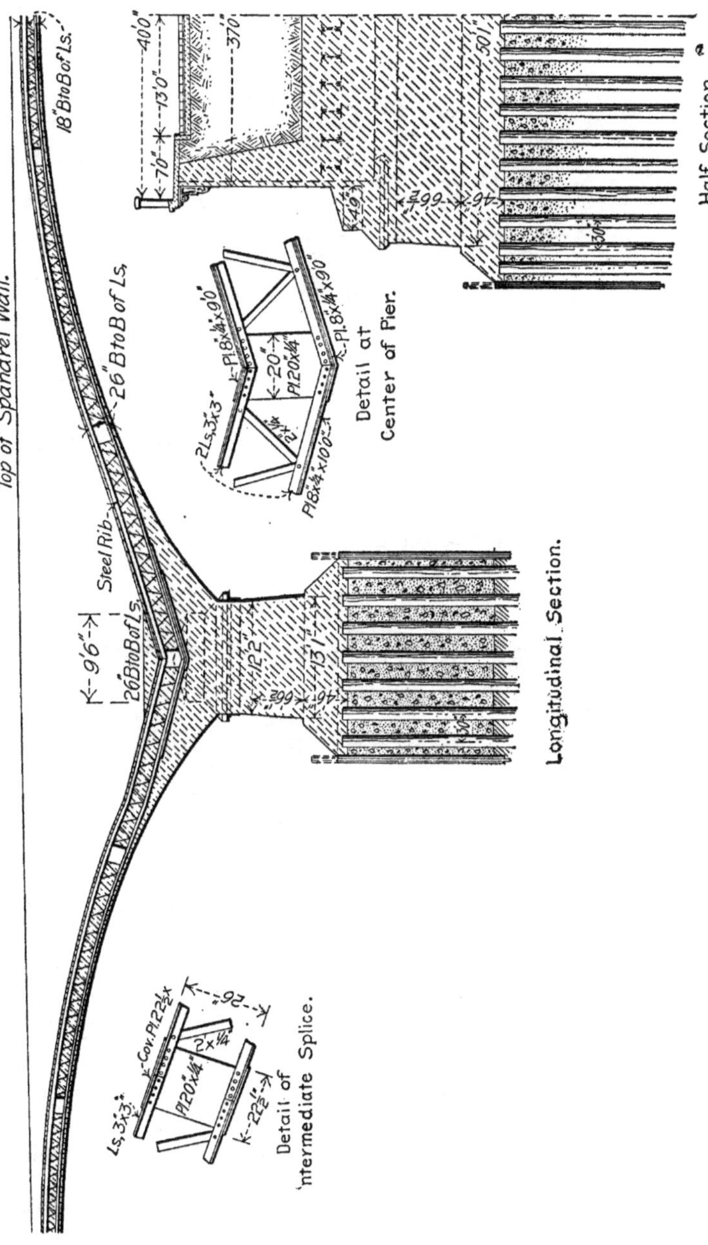

FIG. 134.—Pier and Arch Construction of Melan Arches at Topeka, Kan.

haunch, 0.8 m. (2.62 m.). The half-ribs which reinforce the arch ring are similar in outline and are 0.5 m. (1.64 ft.) deep at the haunches.

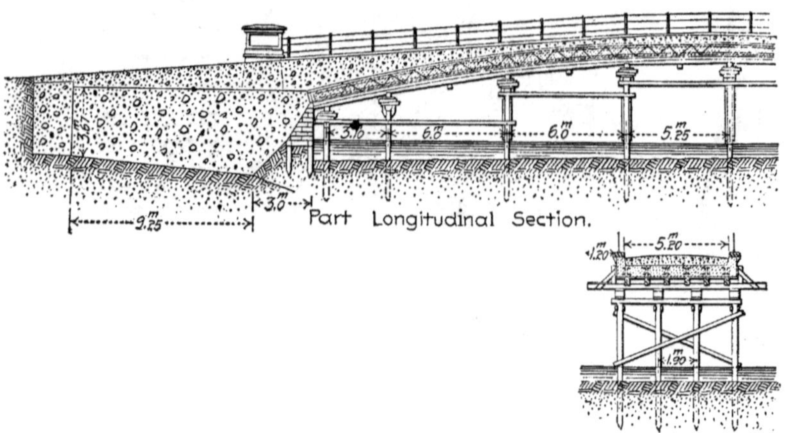

FIG. 135.—Hinged Melan Arch Bridge at Steyr, Austria.

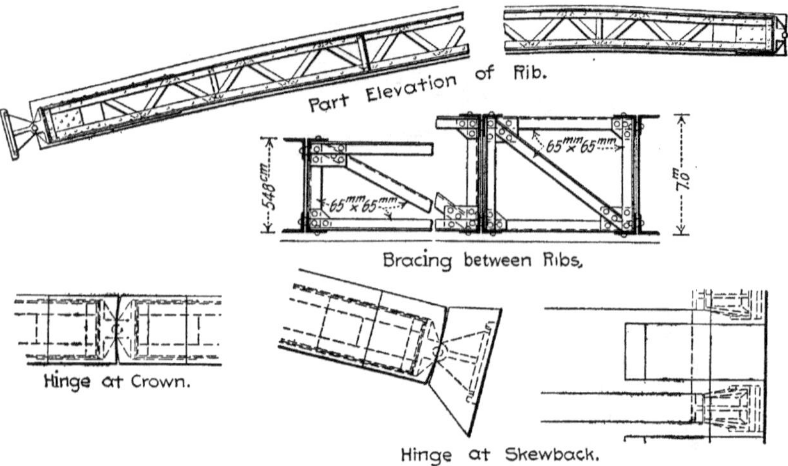

FIG. 136.—Steel Girder and Hinge Construction for Bridge at Steyr, Austria.

The cross-section of the reinforcement is 2.2 per cent. of that of the arch ring at the crown.

The reinforcing-ribs have T-section chords made up of angles, and a lattice-web as shown by Fig. 136. At three points in each half-arch the ribs are braced together by cross girders constructed as shown by Fig. 136. There are six reinforcing-ribs, each having a hinge at the crown and at each abutment. These hinges are quite similar in construction to those employed in steel arches, and are sufficiently explained by the drawings of Fig. 136. The articulations of the concrete ring are separate from those of the ribs and between them. At the ends they consist of a skew-back block, with a concave bearing surface, and a ring-block, with a convex bearing surface. These adjoining concave and convex bearing surfaces have different radii. The skew-back blocks and the ring-blocks were cast concrete of 1 part Portland cement and 4 parts aggregate. The articulation at the crown is of similar form and arrangement. Between the concrete bearing-blocks there were placed sheets of asbestos board to distribute the pressure. These cushions have proven to be too compressible, and have caused some trouble for this reason. By referring to the plan of the abutment hinges given in Fig. 136 it will be observed that there is an open space between the rib hinges and the arch-ring hinges on each side. These spaces were left open during the construction of the arch but were afterwards filled with concrete.

The concrete of the arch ring is composed of 1 part Portland cement, 2 parts sand, and 4 parts broken stone. The arch ribs are of soft steel, the skew-back shoes of cast iron, and the other hinge castings of steel. Openings are provided in the spandrel wall and the roadway over the crown hinge and both skewback hinges. The other details of the construction are evident from the drawings of Figs. 135 and 136.

A hinged Melan arch of exactly similar construction to that at Steyr just described, but with a span of 33 m. (108.3 ft.) and a rise of 4.4 m. (14.4 ft.), was built in 1900 at Laibach, Austria. A more unusual example of hinged construction is presented by the 26 m. (82 ft.) span built at Payerbach, Austria, in the same year. This arch is hinged at the crown only. The arch ring is 0.45 m. (1.47 ft.) deep at the crown and 1.46 m. (4.79 ft.) deep at the springing-lines, and the lattice-girder reinforcing-ribs have the same outline. The chords of these ribs extend into the abutments and there converge to a connection with a transverse beam, which in turn is connected by tie-rods to a transverse anchor-beam buried beneath in the abutment masonry.

A view of the Laibach Bridge is shown by Fig. 137, and as the method of construction will call for attention in a succeeding chapter, the main structural details will be briefly outlined as follows: The bridge is built on a skew of 9° 14′, and has a roadway 33 ft. wide and two sidewalks each 6½ ft. wide. The outside width of the main arch is about 50 ft. The loading for which the structure was calculated consists of two wagons weighing each 13 tons, and a uniformly distributed live load of 96 lbs. per square foot. The axle loads were assumed distributed over a width of 8 ft. The main arch has a clear span of 33 m. (108.3 ft.) and a span center to center of hinges of 109.34, with a rise of 14.34 ft. Its intrados is three-centered, following as nearly as possible the curve of pressure of half the load. On the middle 33-m. arch the intrados has a radius of 123 ft., while the radii of the end arcs are 98 ft. The thickness of the main-arch ring is 20 ins. at the crown, 27½ ins. at the haunches, and 25½ ins. at the skew-backs. Its reinforcement consists of 14 steel lattice ribs spaced from 3.3 ft. to 3.8 ft. apart, whose flanges are everywhere about 2 ins. within the concrete. These ribs are connected transversely by four sets of steel

Fig. 137.—View of Hinged Melan Arch Bridge at Laibach, Austria.

cross-frames and are parallel with the axis of the bridge, i.e., are skewed with respect to the abutments. Figs. 138 and 139 show the hinge construction. The concrete hinge-blocks were molded several months before construction was begun. Full and uniform contact of the curved abutting faces was secured by placing in the joint a strip of hard lead 4 ins. wide and $\frac{1}{16}$ in. thick. The center hinges of the steel ribs are simple abutting pin-joints, as will be seen from the drawing, but the skew-back hinges of the ribs have wedges for accurate adjustment of the span. After the arch was completed and the centering removed, the hinges of the metal ribs were encased in concrete. As all the permanent deflections of the arch had probably taken place before that time, it was thought that this concreting would not interfere with

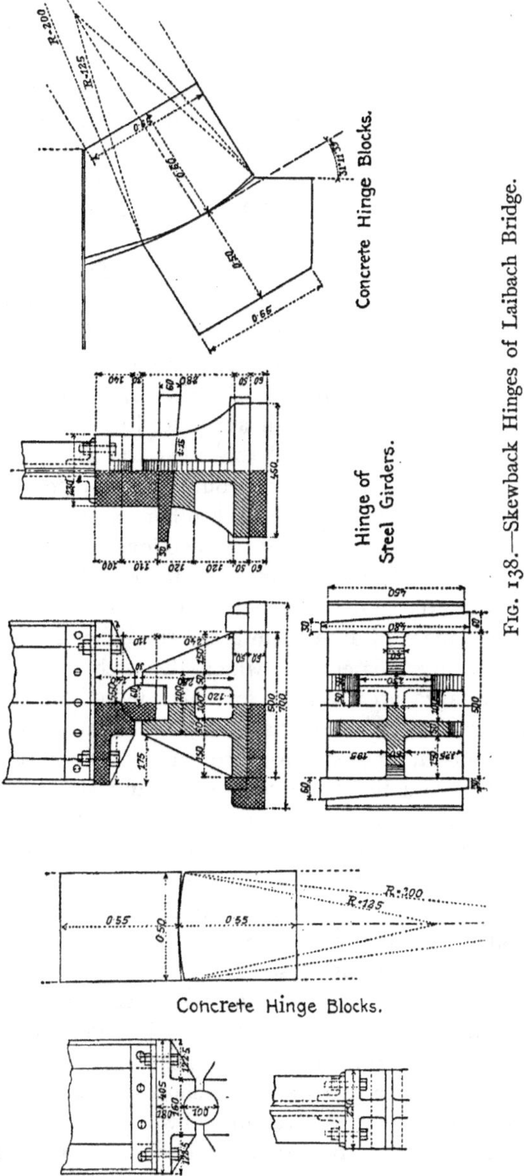

FIG. 138.—Skewback Hinges of Laibach Bridge.

Hinge of Steel Girders.

FIG. 139.—Center Hinges of Laibach Bridge.

the proper action of the hinges; as a matter of fact, no cracks had appeared three years after the bridge had been put into service. The back of the arch ring was covered with a 1-in. layer of asphalt to shed water, and drain-holes were formed through the ring at all points of depression. The weight of metal in the bridge was about 67 tons, of which 16 tons are in the hinge-castings. The total cost of the bridge was $32,000, which includes $2,000 for ornamentation and a large sum for the heavy foundations. The bridge was designed by Prof. Melan.

Thacher Construction.—In 1899 Mr. Edwin Thacher, M. Am. Soc. C. E., the designer and builder of the Topeka, Kan., bridge and other bridges of Melan construction, patented a form of concrete-steel arch construction. In this construction the reinforcement consists of flat steel bars in pairs spaced at proper distances apart transversely of the bridge. The two bars of each pair are placed one above the other, so that one follows the extrados and the other the intrados of the arch ring, while both extend well into the abutments or piers. The bars of each pair have no connection with each other except through the concrete, but each bar is provided with projections, generally rivet-heads of extra height, spaced at short intervals to provide mechanical reinforcement of the adhesion between the steel and the concrete. The bars act as the flanges of a beam to assist the concrete in resisting the thrusts and bending moments to which the arch is subjected. The following advantages are claimed for this method of reinforcement: It gives a larger amount of inertia and consequently greater strength for the same amount of steel; a more reliable connection is secured between the steel and the concrete than in any system that depends upon adhesion alone; the bars can be shipped straight in any convenient form and bent cold to the curve desired. A number of important bridges have been built of the construction described. More recently Mr. Thacher has designed a reinforcing-rod consisting of a round bar rolled with flat spots at intervals. This bar is used in all respects like the flat bars just described.

Y-Bridge, Zanesville, Ohio.—The most notable bridge of Thacher construction in many respects is the so-called Y bridge at Zanesville, O., built in 1900–1. This bridge is built at the junction of the Muskingum and Licking rivers, and to reach all three shores it has three arms radiating from a triangular pier as a center. Two of these arms have three spans and the other has two spans. The main dimensions of the several spans are:

Arch, No.	Span, Ft.	Rise, Ft.	Arch, No.	Span, Ft.	Rise, Ft.
1	122	11.5	5	99	6.28
[illegible]	122	14.6	6	81	14.6
3	122	14.6	7	81	10.85
4	120.6	11.5	8	81	6.06

The arches were built first and the spandrel walls later, the arches and spandrels being tied together with steel anchor-rods or dowels as shown by Fig. 140. The thickness of ring at the crown of the arch varies from 18 ins. for the 81-ft. spans to 30 ins. for the 122-ft. spans. The dimensions and construction of the spandrel and parapet walls are shown by Fig. 140. Each arch is reinforced by 15 pairs of steel bars $\frac{3}{8}$ in. thick and of varying widths; 3 ins. for the 81-ft. spans, 5 ins. for the 122-ft. spans, and $4\frac{1}{2}$ ins. for the other spans. These bars are placed, as shown by the cross-section, Fig. 141, about 3 ft. apart transversely of the bridge and about 2 ins. from the intrados and extrados. Each bar has rivets driven in it at 8-in. pitch, and splices are formed by double-splice plates with six rivets. The bars of adjacent arches are made continuous over the piers and the ends are simply embedded in the concrete skew-backs of the abutments and the center pier. As shown in the cross-section, Fig. 141, they are 43 ft. wide over the arches and 42 ft. wide between parapet walls. This width is divided into a 30-ft. roadway and two 6-ft. sidewalks. The spandrel-filling is earth compacted by flooding. This filling is crowned for the roadway and covered with 6 ins. of concrete, on which a brick pavement is laid on a sand-cushion.

Jacaguas River Bridge, Porto Rico.—The Jacaguas River Bridge carries the military road leading from San Juan to Ponce, Porto Rico, over the Jacaguas River at Juana Diaz. It has three spans: a center span with a length of 120 ft. and a rise of 12 ft. and two end spans with a length of 100 ft. and a rise of 11.38 ft. each. The total length of the bridge is 404 ft., and it has a width of 20 ft. The reinforcement of each arch ring consists of seven pairs of flat bars, arranged as shown by Figs. 279 and 142. The concrete of the arch ring was composed of 1 part Portland cement, 2 parts sand, and 4 parts broken stone, and the abutment and pier concrete had the proportions of 1:3:6. The arch rings were built on centers in three longitudinal sections, a center section 9 ft. wide and two side sections $3\frac{1}{2}$ ft. wide. The exposed surfaces were mortar-faced, and a considerable number of cracks and shells developed subsequently in this facing.

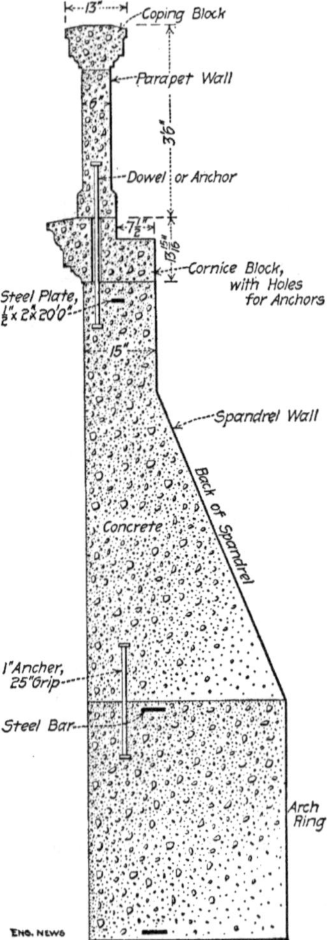

FIG. 140.—Section of Spandrel and Parapet Wall, Y Bridge, Zanesville, Ohio.

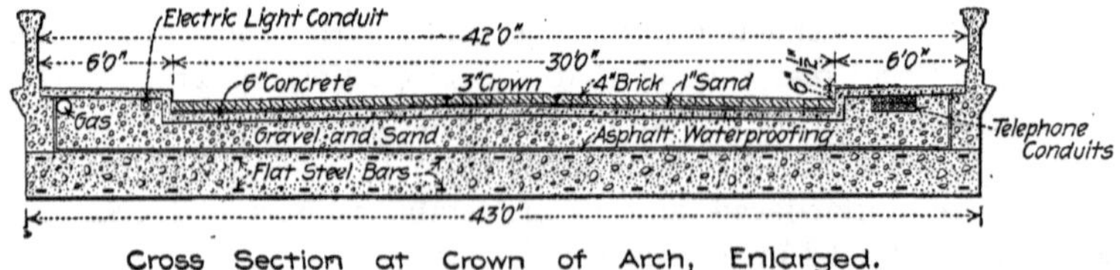

FIG. 141.—Cross-section of Arch at Crown, Zanesville Y Bridge.

The centering of the west-end span was lowered in two months and the deflection of the arch was $\frac{5}{16}$ in., which gradually increased to

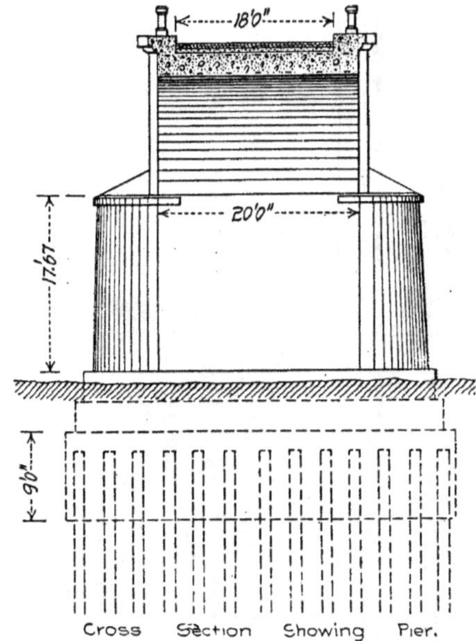

FIG. 142.—Cross-section of Jacaguas River Bridge.

$\frac{3}{4}$ in. in sixty days after which no further deflection took place. The centering of the center span was lowered after forty days and deflected $\frac{5}{8}$ in., which increased to about $\frac{7}{8}$ in. in one month. The centering of the east-end span was lowered after thirty days, and the arch deflected $\frac{3}{4}$ in., which increased to about $1\frac{1}{8}$ ins. in two months. The bridge was tested by the Board of Public Works as follows: 107 bags filled with sand and from 30 to 60 men giving a total weight of from $21\frac{1}{2}$ to $23\frac{1}{2}$ tons were concentrated at the center and intermediate points of the east 100-ft. span and at the center of the 120-ft. span. The deflections of the spans were 0.012 in. and 0.015 in., respectively.

Green and Goat Island Bridges, Niagara Falls, N. Y.—During 1900 there were constructed at Niagara Falls two concrete-steel arch bridges, one connecting the mainland with Green Island just above the brink of the American Falls and the other connecting Green Island with Goat Island. The first named of the two bridges is the larger, and it alone will be described. This bridge is shown by Fig. 143. It has

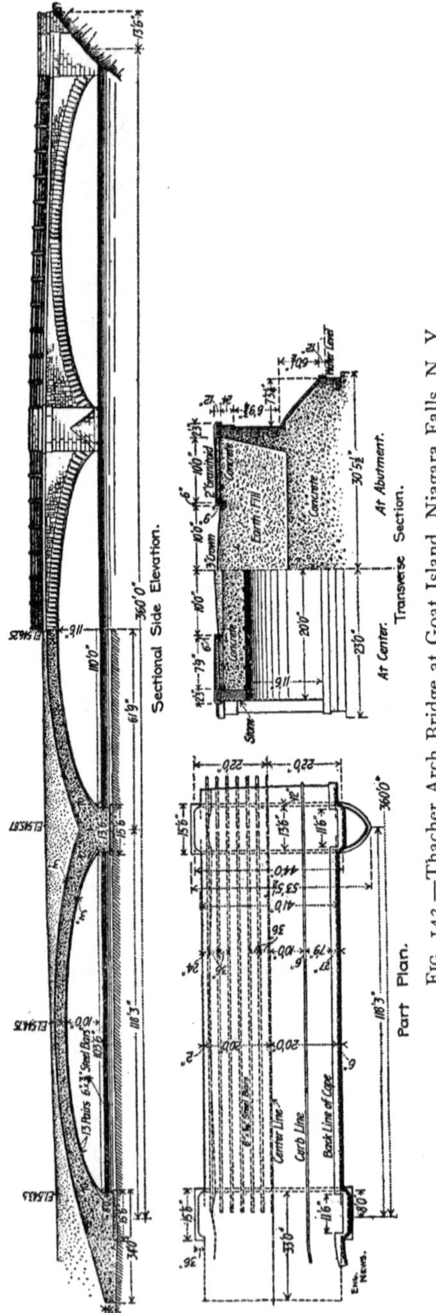

FIG. 143.—Thacher Arch Bridge at Goat Island, Niagara Falls, N. Y.

three spans: a center span of 110 ft. and a rise of 11½ ft., and two end spans each 103½ ft. long with a rise of 10 ft. The width of the bridge over all is 44¼ ft. and its total length is 371 ft. The reinforcement consists of twelve pairs of Thacher bars in each arch ring. The arch rings are of concrete composed of 1 part Portland cement, 2 parts sand, and 4 parts broken stone (run of crusher); the concrete of the abutments had the proportion of 1:3:6. All exposed faces except the soffits of the arches are faced with cut stone. The arch-ring concrete was laid in longitudinal sections. The contract price for the bridge was $102,070.

Expanded-metal Construction.—Arch bridges with expanded-metal reinforcement are comparatively rare, the concerns controlling the use of this material having devoted their attention more toward extending its use for other classes of structures. As a consequence no definite method of arch construction with expanded-metal reinforcement has been developed, and each bridge which has been built has been of a construction peculiar to itself. Of these structures, the one shown by Fig. 144 is a particularly good example of concrete-steel arch construction. This bridge was constructed on a private estate at Oconomowoc, Wis., in 1899. It has a clear span of 21 ft. and a rise of 6 ft. 8 ins. The total width is 15 ft. and the arch ring is reinforced on the under side by three ribs. The reinforcement consists first of two flat ⅜×3½ in. bars on each side, one being near the top of the parapet and the other directly beneath in the arch ring proper. These bars follow the curve of the arch ring and reach down into the abutments, where they connect with separate anchor-plates. The lower bars are connected across the bridge by ½-in. round rods hooked over them and spaced 18 ins. apart. Across these rods was drawn expanded metal of No. 16 gage and 2½-in. mesh, which was lapped and wired at all edges. The concrete for the entire structure was composed of 1 part Portland cement, 3 parts bank torpedo sand, and 4 parts of ¾-in. broken limestone. The finishing-coat, 1-in. thick, was composed of 1 part Portland cement, 1 part granite screenings, and 1 part torpedo sand.

Luten Construction.—In concrete-steel arches as ordinarily designed the horizontal thrust of the arch ring is resisted by heavy abutments. A number of small arches have been built by Mr. Daniel B. Luten, in which a concrete-steel tie-plate between the ends of the arch is introduced to receive this thrust, the abutments being largely done away with. Fig. 145 illustrates a highway bridge built at Pontiac, Mich., in 1902. This bridge, which is shown in longitudinal and part transverse section has a span of 30 ft. and a width of 60 ft. The reinforcement of the arch ring consists of parallel steel rods so curved as to be near the intrados of the arch at the crown and to approach the extrados at the haunches. Between these rods at three points near the crown

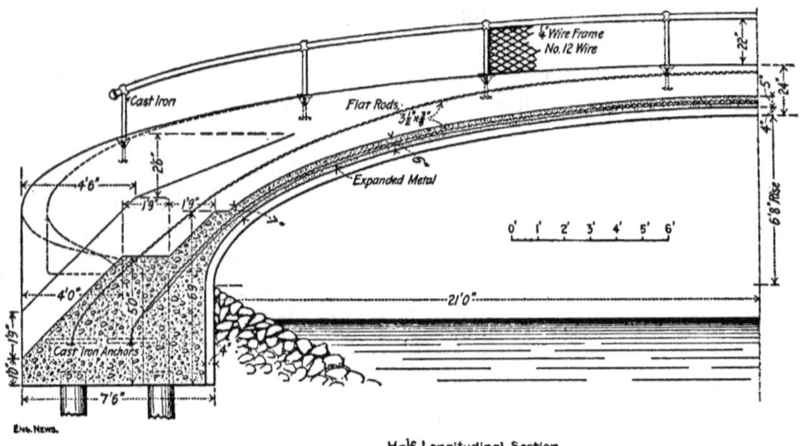

Half Longitudinal Section.

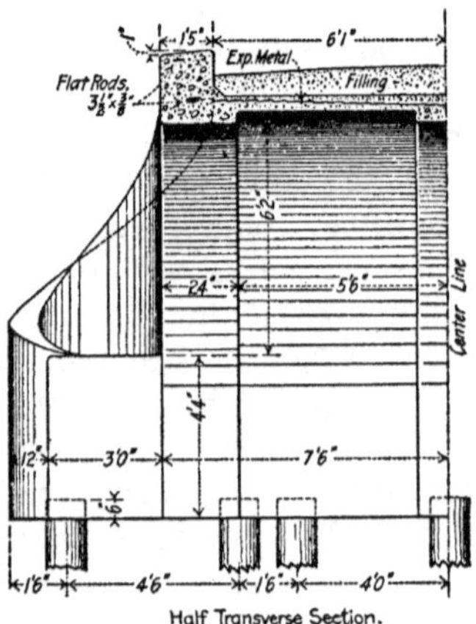

Half Transverse Section.

FIG. 144.—Arch Bridge with Expanded Metal Reinforcement, Oconomowoc, Wis.

are looped transverse webs which, at the ends, turn upward into the parapets. These rods are clearly shown by the cross-section of Fig. 145. Between the lower ends of each arch-ring rod there is stretched a horizontal rod to serve as a tie. These horizontal rods are embedded in a plate of concrete which serves as a paved stream-bed underneath the arch. The pavement is in one piece with the arch ring, the whole structure being monolithic. The concrete was made of 1 part Portland cement, 3 parts sand, and 5 parts broken stone. The tie-rods and pavement have been spoken of as horizontal; as a matter of fact they are dished at the center to a depth of 8½ ins. below the ends.

Corrugated-bar Construction.—One of the newest forms of construction for arch bridges of reinforced concrete is that in which the reinforcement consists of corrugated bars. The new bridge built over the Des Peres River in Forest Park, St. Louis, Mo., is a typical

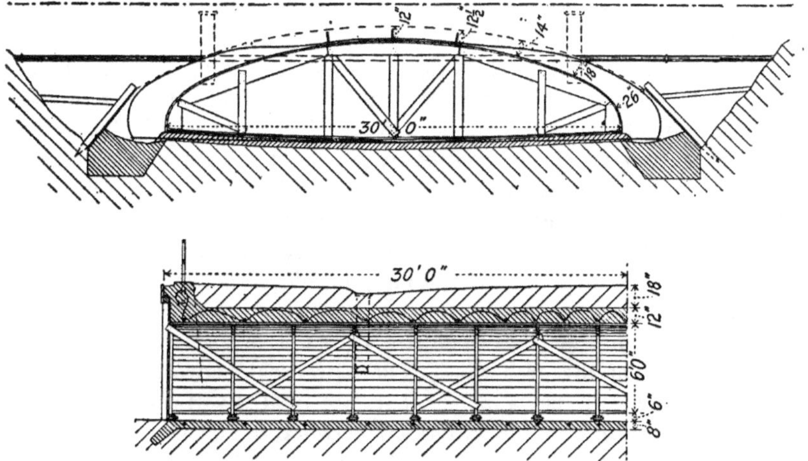

FIG. 145.—Luten Arch Bridge at Pontiac, Mich.

example of this construction. This bridge is illustrated in Fig. 146. It has a span of 45 ft. and is 45 ft. wide. The concrete used for the arch ring and spanded walls was composed of 1 part Portland cement, 2 parts sand, and 4 parts limestone broken to 1½-in. size. The reinforcement consisted of corrugated bars ¾ in. square placed parallel with the axis and 16 ins. apart, one layer close to and following the curve of the intrados and another layer close to and following the curve of the extrados. The corresponding bars of the two layers were in the same vertical plain. For the haunch portions of the arch the extradosal bars are doubled, as shown by the drawing. In some designs the extradosal bars are doubled the full length of the arch and in others transverse bars are used with each layer of longitudinal bars so as to form a square-mesh network. In a bridge of 100-ft. span designed for a three-track railway by the patentees of the corrugated-bar reinforcement, the reinforcement was arranged as follows: 1¼-in. longitudinal bars spaced 5½ ins. apart crossed by ½-in. transverse bars spaced 24 ins. apart for the intradosal net; for the extradosal net, two layers of 1¼-in. longitudinal bars, one with 8-in. and the other with 5¼-in. spacing, and between them a layer of transverse ½-in. bars spaced 24 ins. apart. The depth of the arch ring at the crown was 36 ins.

A notable example of corrugated bar and stirrup reinforcement is shown by the illustrations comprised by Figs. 147 to 149. The bridge illustrated carries Seeley Street over Prospect Avenue, in the Borough of Brooklyn, New York City, and is a skew structure with a span of 85 ft. 4 ins. and a rise of 8 ft. 6⅜ ins. The drawings of Fig. 148 show a transverse and a longitudinal section of the bridge.

The reinforcement of the bridge consists of corrugated steel bars. Tests of these bars showed an elastic limit of 66,630 lbs. per square inch and an ultimate strength of 106,270 lbs. per square inch for the large bars, and an elastic limit of 65,880 lbs. per square inch, and an ultimate strength of 105,170 lbs. per square inch for the small bars. These bars are arranged in the form of an intradosal and an extradosal network of intersecting bars which are connected through the arch ring by vertical tie-bars. The intradosal network consists of longitudinal bars 1¼ ins. square spaced 8 ins. apart transversely of the arch ring, and of transverse bars ½ in. square spaced 18 ins. apart longitudinally of the arch ring. The extradosal network is identical in construction and its component bars are located vertically above the corresponding bars in the intradosal network. The vertical tie-bars are located at the intersections of the longitudinal and transverse bars, but they connect the upper and lower longitudinal bars only as indicated by Fig. 148. These tie-bars are ¼ in. square and are attached by being hooked under the lower bars and hooked over the upper bars. In the intradosal netting the transverse bars are beneath the longitudinal bars and in the extradosal netting the transverse bars are above the longitudinal bars. The transverse and longitudinal bars in each netting are wired together at intersections at intervals only, and only for the purpose of holding them in their proper relative positions during construction.

By referring to the illustration, Fig. 147, it will be seen that at the springing-lines the intradosal netting bends downward and enters the abutment in a vertical plane. The extradosal netting, however, is discontinued at these points by the omission of the transverse bars, while the longitudinal bars are continued back into the abutments as anchor-bars. The abutments themselves are of rather novel design, since they are corbeled out at the rear and top, and this corbel is reinforced to withstand cantilever action, as shown by the illustration.

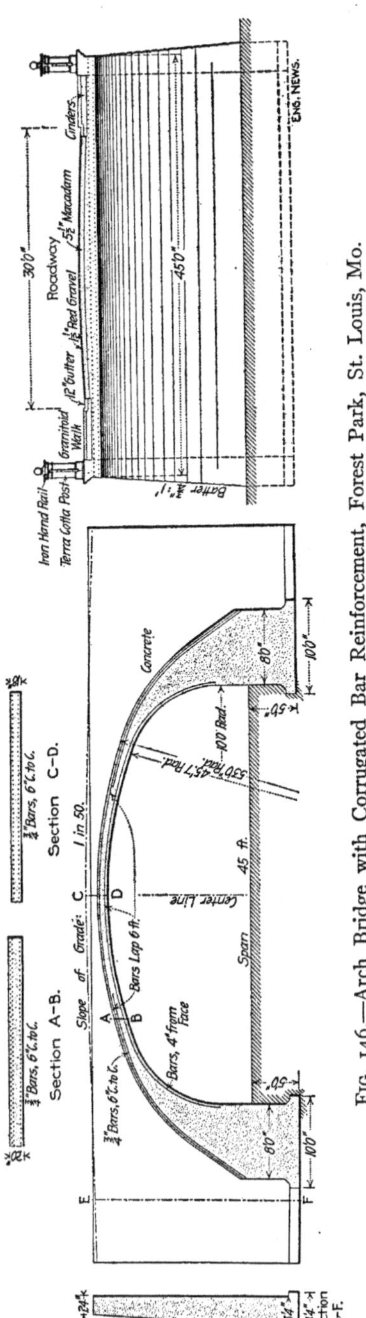

FIG. 146.—Arch Bridge with Corrugated Bar Reinforcement, Forest Park, St. Louis, Mo.

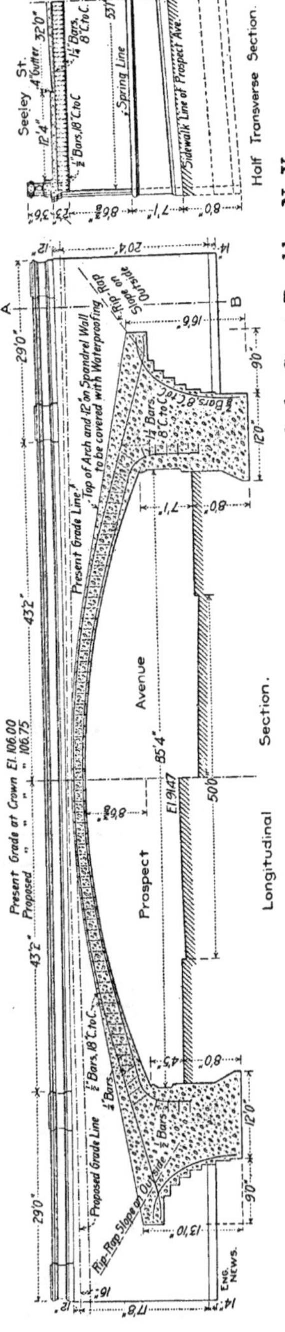

FIG. 147.—Arch Bridge with Corrugated Bar and Stirrup Reinforcement, Seeley Street, Brooklyn, N. Y.

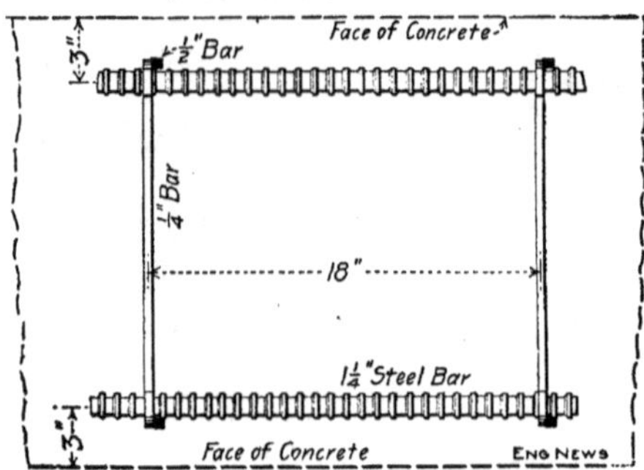

Fig. 148.—Detail of Bar and Stirrup Connections, Seeley Street Bridge.

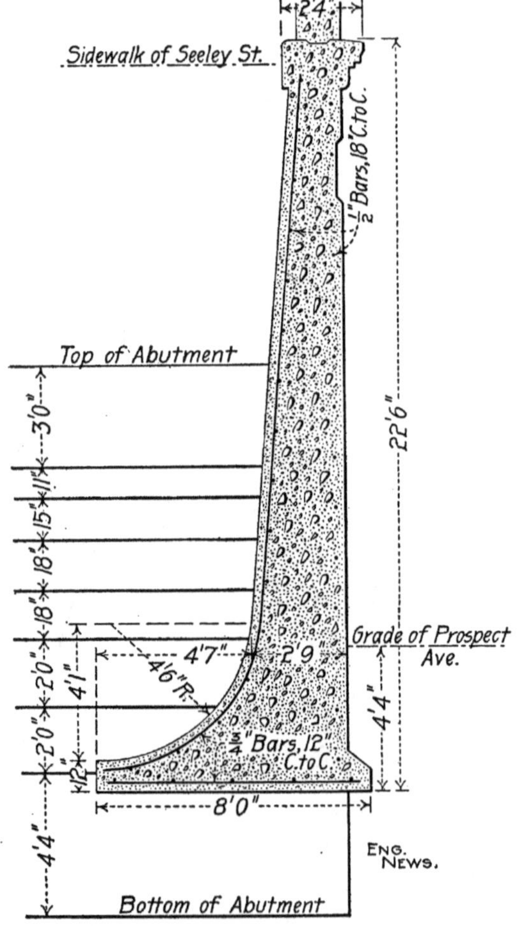

Fig. 149.—Reinforced Spandrel Wall, Seeley Street Bridge.

The abutment-wing walls and the spandrel walls are continuous and are reinforced with corrugated bars, as shown by Fig. 149.

Portland-cement concrete was employed for all parts of the bridge. For the abutments the mixture used was 1 part cement, 2½ parts sand, and 5 parts 1½-in. broken limestone, and for the arch ring the mixture was 1 part cement, 2 parts sand, and 4 parts 1½-in. broken trap-rock.

The drawings of Fi . 150 show some of the special details of a 75-ft. reinforced-concrete arch built for the Chicago, Burlington & Quincy R.R. in 1903. This arch was designed for a load of 1,000 lbs. per square foot, to be carried by the concrete alone, and has a depth of 3 ft. at the crown. The arch-ring reinforcement consists of corrugated steel bars arranged as shown by Fig. 150. The concrete for

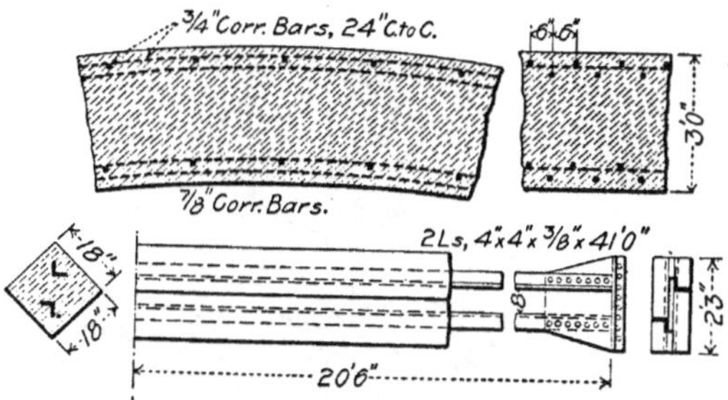

FIG. 150.—Details of Arch with Corrugated Bar Reinforcement, C., B. & Q. R.R.

the ring is a 1–2–4 mixture. The spandrel walls are not reinforced but are anchored to the arch ring by ⅞-in. corrugated bars spaced 2 ft. apart along the arch and acting as dowels. Where the spandrel walls join the abutment-wing walls, which are in the same plane, expansion-joints are introduced. These joints are ¾ in. wide and are packed with asbestos board and covered on top and inside the walls by cover-plates bolted to the concrete with ½-in. bolts. The top of the ar h is waterproofed with ¼ in. of California asphalt. The abutment-wing walls are tied together across the road bed by means of reinforced-concrete ties of the form shown by Fig. 150.

Combination Constructions.—In Europe several arch bridges have been built in which a combination of two forms of concrete-steel construction has been employed. Two of these structures are the Skodsburg Bridge in Denmark and the Mieres Bridge in Spain.

S! o lsborg Bridge, Denmark.—A foot-bridge built in 1897 at Skodsborg, Denmark, has an arch ring reinforced with the railway rails of T section and carrying transverse walls on which rest the spandrel arches of the platform. The walls and the arches which they carry h ve a Monier reinforcement. Fig. 151 is a section showing the arch-ring, wall, and platform construction. The arch has a span of 21.85 m. (71.5 ft.) and a rise of 2.57 m. (8.43 ft.), and the area of the ring reinforcement is 2 per cent. that of the total sectional area at the crown. The concrete, except the filling above the platform arches, was com-

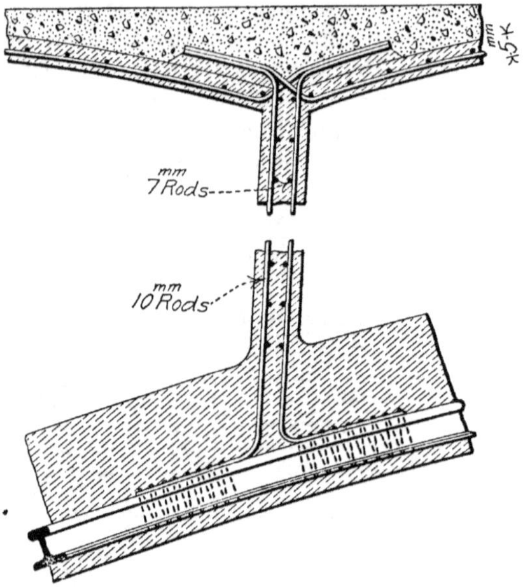

Fig. 151.—Detail of Skodsborg Bridge, Denmark.

posed of 1 part Portland cement and 3 parts gravel. The bridge cost about $2,000.

Mieres Bridge, Spain.—A bridge built across the River Caudal, at Mieres, Spain, has a three-hinged Melan arch ring carrying spandrel columns and roadway arches reinforced by round rods arranged after a system devised by the engineer of the work, Mr. J. Eugenio Rebira. The bridge has three girder spans of 10.5 m. (34.4 ft.) and two arch spans of 35 m (114.8 ft.), with a rise of one-tenth the span. Fig. 152 is a half-transverse section of one of the arch spans and Fig. 153 is a half-section of one of the girder spans. The rod and wire reinforcement devised by Mr. Rebira for the girder spans and the roadway platform of the arch spans is clearly shown by the drawings. The total width of the bridge is 7 m. (22.96 ft.), and is made up of a 5-m. (16.4 ft.) roadway and two 1-m. (3.28 ft.) sidewalks. The total cost of the bridge was about $30,300.

Long-span Concrete-steel Arches.—Designs for concrete-steel arches of considerabl longer span than any which have been constructed up to the present time (1904) have been worked out by a number of engineers. These designs are of interest as showing the

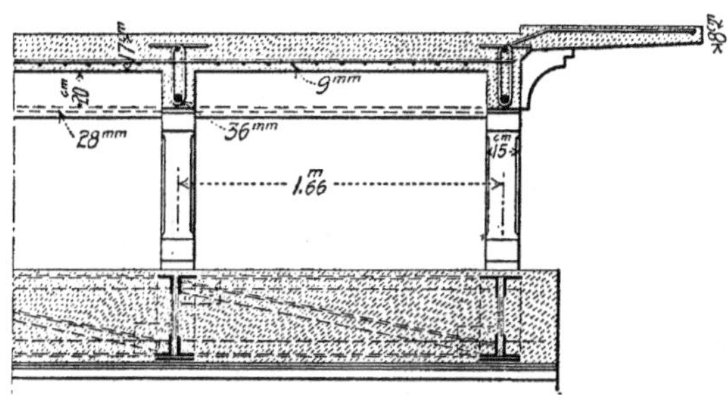

FIG. 152.—Half Cross-section of Arch Span, at Mieres, Spain.

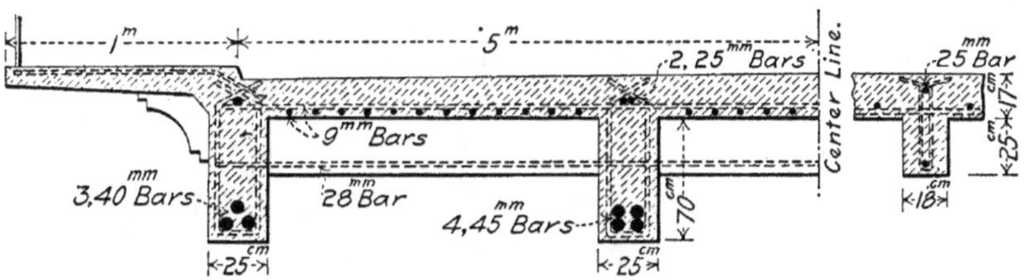

FIG. 153.—Cross-section of Girder Span, Mieres Bridge, Spain.

possibilities of this form of construction as conceived by well-known engineers, and they deserve brief mention for this reason.

Connecticut Avenue Bridge, Washington, D. C.—In 1897 the Commissioners of the District of Columbia invited competitive designs from four different engineers for a bridge of monumental character to carry Connecticut Avenue over Rock Creek in the National Park at Washington, D. C. Two of the designs submitted called for one or more spans of concrete-steel arch construction. One of these designs was by Mr. L. L. Buck, M. Am. Soc. C. E., and it called for a bridge with seven arch spans of Melan construction. The middle arch had a span of 190 ft. (Fig. 154) and a rise of 50 ft.; the two end arches had spans of 105 ft. and rises of 29 ft., and the intermediate arches had spans of 175 ft. and 160 ft., with rises of 50 ft. All the arches had segmental arch rings. The second design was by Mr. Wm. H. Breithaupt, M. Am. Soc. C. E., and called for a middle arch of steel with a span of 544 ft. 8 ins. flanked by two 225-ft. concrete-steel arches and by a 130-ft. arch at one end and a 115-ft. arch at the opposite end, both also of con-

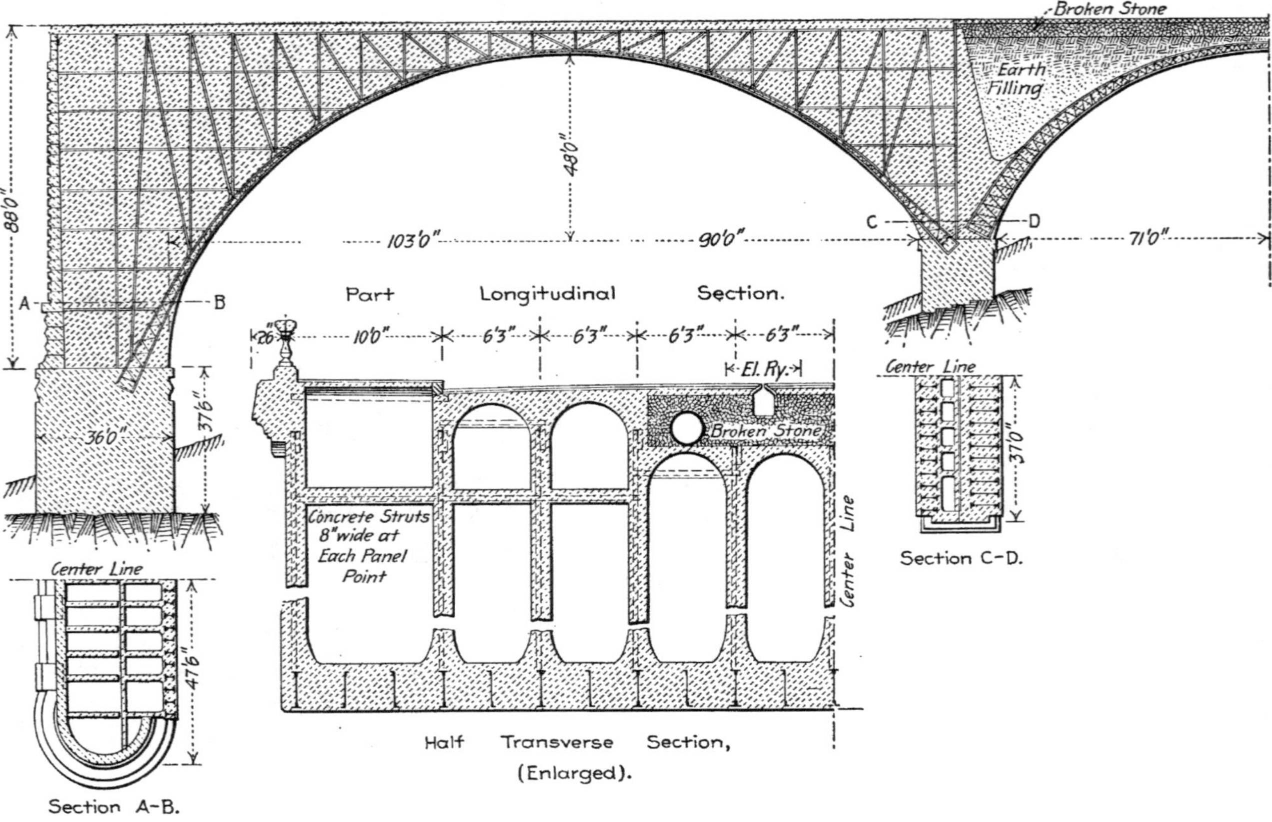

Broken Stone
Earth Filling
88'0"
48'0"
103'0"
90'0"
71'0"
A
B
C
D
36'0"
37'6"
Part Longitudinal Section.
26"
10'0"
6'3"
6'3"
6'3"
6'3"
El. Ry.
Broken Stone
Concrete Struts 8" wide at Each Panel Point
Center Line
Half Transverse Section, (Enlarged).
Center Line
47'6"
Section A-B.
Center Line
37'0"
Section C-D.

crete-steel construction. The reinforcement for the 225 ft.-arches consisted first of a series of parallel arch ribs, with horizontal top chords and curved bottom chords connected by a vertical post and diagonal tie-web system. Between these main ribs, which occupied the full spandrel height o′ the arch, were curved girders occupying the height of the arch ring only, as in the ordinary Melan construction. The concrete work consisted of the arch ring, vertical curtain walls, parallel vertical spandrel walls, and horizontal concrete struts. The spandrel walls and the curtain walls enclosed one main-arch rib each; the concrete struts enclosed the horizontal bracing between the arch ribs, and the arch ring enclosed the curved soffit members. Concrete arches between the spandrel and curtain walls carried the floor. The 130-ft. and 115-ft. end arches were of the usual Melan construction, with latticed ribs.

Memorial Bridge, Washington, D. C.—In 1900 the Chief of Engineers, U. S. A., received competitive designs from four engineers for a memorial bridge across the Potomac River at Washington, D. C. Two of these designs, both submitted by Prof. Wm. H. Burr, M. Am. Soc. C. E., called for concrete arches for a number of the spans, and on of these was awarded the first prize in the competition. In the prize design the bridge proper composed a center-draw span of steel 159 ft. long flanked on each side by three 192-ft. spans of concrete steel. The 192-ft. spans are segmental, with a rise of 29 ft., and are ribbed arches, with the exterior ribs of granite masonry and the intermediate ribs of concrete-steel (Fig. 155). The concrete-steel ribs were 30 ins. deep at the crown and 7 ft. 3 ins. deep at the springing-lines. The second concrete-steel design submitted by Prof. Burr called for a center-draw span of 213 ft. flanked on each side by two 283-ft. concrete-steel arches of elliptical form and a rise of 70 ft. 9 ins. As in the prize design the arch has outside ribs of granite. In these arches the concrete-steel ribs were 8 ft. deep at the crown and 23 ft. deep at the abutments. In connection with this last design it is of interest to note that one of Prof. Burr's designs called for a steel structure identical in all respects, excluding the architectural and ornamental features. The comparative estimated costs of the steel structure and of the reinforced-concrete structure were $1,000,000 and $1,450,000, respectively, or an excess of reinforced concrete over steel of $450,000.

Schenley Park Bridge, Pittsburg, Pa.—In 1876 Mr. Edwin Thacher, M. Am. Soc. C. E., designed two long-span concrete-steel bridges for erection in Schenley Park at Pittsburg, Pa. The smaller of these structures, planned to cross St. Pierre Hollow, is shown by Fig. 156. The main features to be noted are that a Melan arch ring carries spandrel bents in steel which support a concrete-steel roadway. The larger of the two bridges at Schenley Park was designed to cross Junction Hollow and to have a span of 300 ft., with a rise of 66 ft. 2 ins. flanked by 70-ft. and 61-ft. spans, and to have a width of 80 ft. The main-arch ring had a depth of 4 ft. at the crown. This bridge, 590 ft.

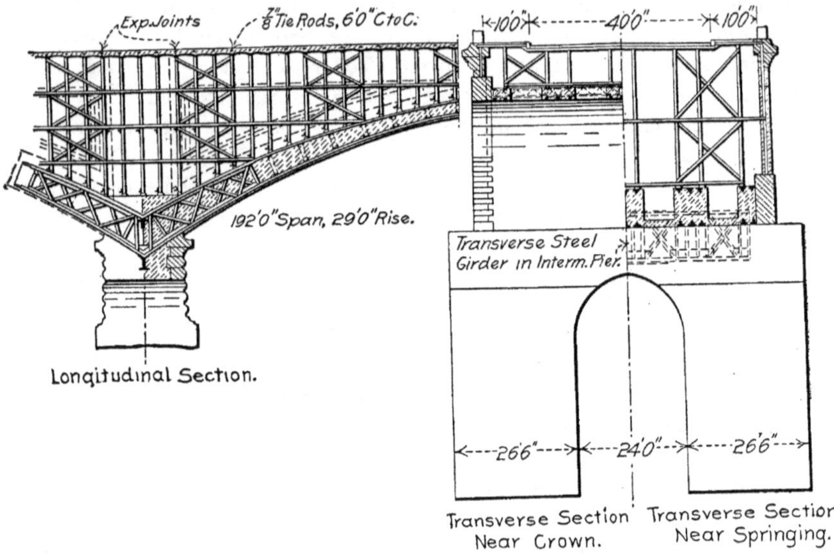

FIG. 155.—Proposed Arch Span for Memorial Bridge, Washington, D. C.

long, was estimated to cost about $7 per square foot of area covered. Mr. Thacher considers that there should be no insuperable difficulty in constructing arch spans of 500 ft. according to a similar design.

BRIDGE FLOORS.

Any of the forms of reinforced-concrete floor construction previously described can be employed for bridge floors, and there are a number of examples of their use in this manner. Fig. 157 shows a bridge flooring composed of Monier arches sprung between the floor-beams and filled above with concrete to give a flat surface for the support of the paving. This construction is employed for spans between beams of from 2.34 m. (7.67 ft.) to 5 m. (16.46 ft.). The reinforcement consists of a single network.

Fig. 158 shows the construction of one of the earliest bridge floors of concrete-steel to be built in the United States. This floor was invented by Mr. M. F. McCarthy of Chicago, Ill., and was employed

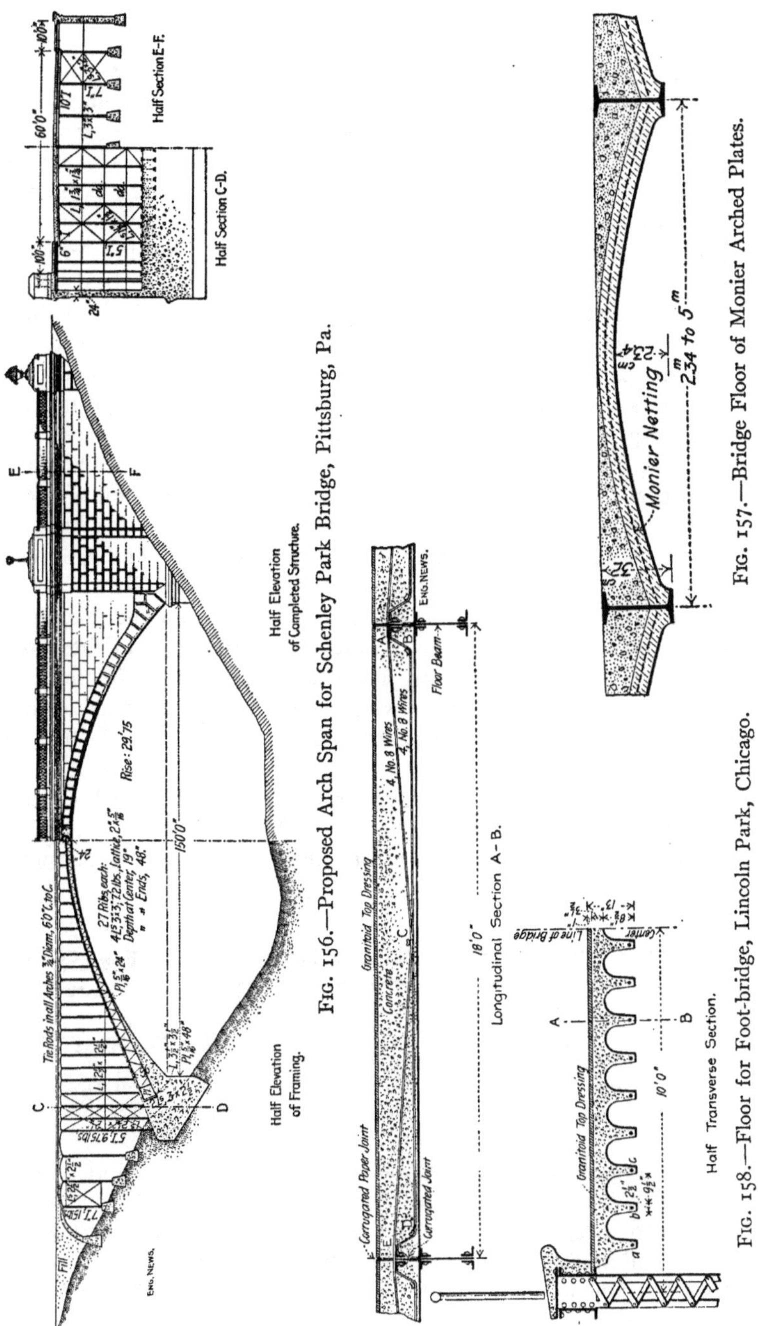

FIG. 156.—Proposed Arch Span for Schenley Park Bridge, Pittsburg, Pa.

FIG. 157.—Bridge Floor of Monier Arched Plates.

FIG. 158.—Floor for Foot-bridge, Lincoln Park, Chicago.

in a cantilever foot-bridge built in Lincoln Park, Chicago, by Mr. W. L. Stebbings, as engineer. As will be seen from the drawings, the floor beams are placed at the panel-points 18 ft. apart, there being at the center of the span two beams placed near together, with the expansion-joint between them. Beginning at each of these center beams a series of two groups of four steel wires each is carried back over the intermediate floor-beams toward each end of the span. One group of wires is stretched along the line *A*, *B*, *C*, *D*, *E*, while the other group is stretched directly above, along the line *ACE*. The vertical planes passing through these groups are 1 ft. apart horizontally in a direction transversely across the bridge, as shown in the transverse section. These groups of wires are filled around with concrete, composed of 1 part Dyckerhoff cement, 2 parts fine torpedo sand, and 4 parts blast-furnace slag, which was rammed into place as soon as the wires were adjusted. The method of construction was as follows: The false-work had a longitudinal joist directly under each of the points *a*, *b*, *c*, etc. These joists were covered with boarding, and on this boarding were placed the forms for making the corrugation. The groups of wires were first stretched into place under a tension of 60 lbs., and fastened into their proper positions by hook bolts fastened underneath the wooden joists. These hook bolts were so constructed that they could be removed when the concrete filling was firmly set. In designing this floor it was assumed that the wires would take all the tensile strain, while the concrete in the top part of the beam would take the compression.

A number of overhead highway-bridge crossings have been built by the Erie R.R., Mr. C. W. Buchholz, M. Am. Soc. C. E., Chief Engineer, with the general form of concrete-steel floor construction

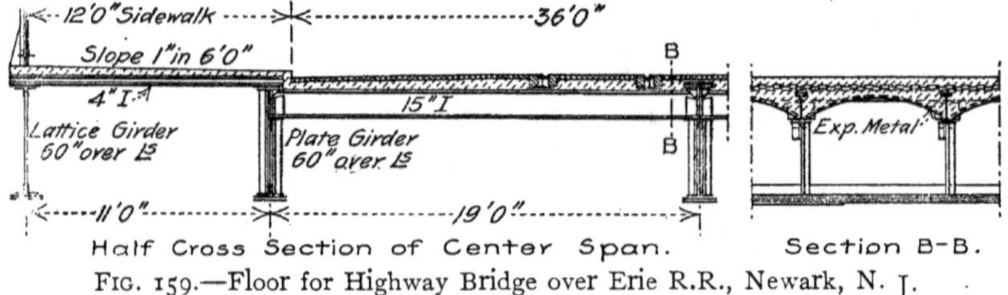

Fig. 159.—Floor for Highway Bridge over Erie R.R., Newark, N. J.

shown by Fig. 159. The particular floor illustrated is that of a 134-ft. plate-girder bridge on Kearney Avenue, Newark, N. J., which was designed to carry two lines of electric-railway tracks, a carriageway, and two sidewalks. The main-floor beams are 6½ ft. apart and the sidewalk beams are 4 ft. apart. The lower flanges of the floor-beams were wrapped with expanded-metal lath, and sheets of 3-in. No. 10 expanded-metal were sprung between them to form the soffits of segmental arches of 14 ins. rise. Concrete arches were built on these soffits, enclosing all the upper portions of the floor-beams and reaching several inches above their tops. The concrete was made of 1 part Portland cement, 2½ parts sand, and 6 parts cinders, well rammed. After the concrete had set and the centers were removed, the lower flanges of the floor-beams were plastered and the under sides of the arches were floated. The concrete filling was paved with asphalt. The other details of the construction are shown by the drawings of Fig. 159.

Fig. 160 shows the reinforced-concrete floor construction employed

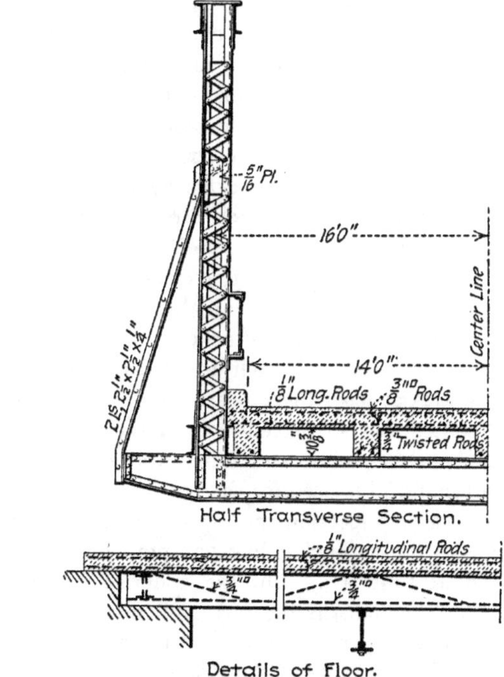

Fig. 160.—Floor for Highway Bridge, Portage County, Ohio.

for a riveted-through truss highway bridge of 100-ft. span built in Portage County, Ohio, in 1903. In this bridge the floor-beams are spaced 12½ ft. apart, and they carry a plate and joist floor reinforced by square twisted steel rods. The joists run lengthwise of the bridge and rest directly on the top flanges of the beams. The details of the reinforcement are shown by the drawings.

Culverts.

Culverts of both arch and box form are built of reinforced concrete. Molded pipe of concrete-steel has also been employed in exactly the same manner as vitrified clay and cast-iron pipes are used by many American railways. Culvert construction approaches bridge practice at one extreme and aqueduct practice at the other extreme. A few examples will illustrate this fact and will also be sufficient to illustrate recent practice in this class of concrete-steel structure.

Illinois Central R.R. Construction.—A type of reinforced-concrete construction, which bears a close resemblance to the embedded steel-beam girder bridge construction described in pp. 199, has been extensively employed during the last few years by the Illinois Central R.R., and is by that company termed an arch construction. This construction has been adopted by other railways, following the example of the Illinois Central. It has been used mostly for culvert work and is therefore included here in the section devoted to culvert work. The first example given, however, is a bridge structure which is illustrated by Fig. 161. This shows a concrete bridge 77 ft. long, consisting of four arch spans. This is a double-track structure, replacing a light 82-ft. single-track iron bridge over Ames Creek, near Buckley, Ill. The width over the copings is 36 ft., the tracks being spaced 15 ft. center to center. The spans are 15 ft. in the clear, with a radius of 20 ft. for the crown and 2 ft. for the haunches. The concrete is 18 ins. thick at the crown, and over each arch are laid 9-in. I beams, 17 ft. 6 ins. long. There are seven of these beams under each track, spaced 18 ins. center to center and so placed that at the center of the arch there are 3 ins. of concrete below and 6 ins. above the beams. Concrete made of small material is used for the bottom 3 ins. and for 1½ ins. above the beams. In cross-section the top of the concrete is flat, but slopes at each side to the coping or low parapet, through which 4-in. drain-tiles are laid at intervals. On the concrete is laid the stone ballast 18 ins. thick, in which the tiles are embedded as shown. The piers are 3 ft. thick at the springing line, and the abutments 4 ft. thick at the same level. The east side of the bridge was first built, parallel with the old structure. When this was completed, traffic was diverted to it, the old bridge was removed, and the west half of the concrete structure built. This bridge contains about 725 cu. yds. of concrete. The cost of the double-track concrete structure was less than that of a single-track steel structure of modern design and weight

Figs. 162 and 163 show a similar construction adopted for culverts, cattle-passes, private roads, drainage-ditches, etc., ranging from

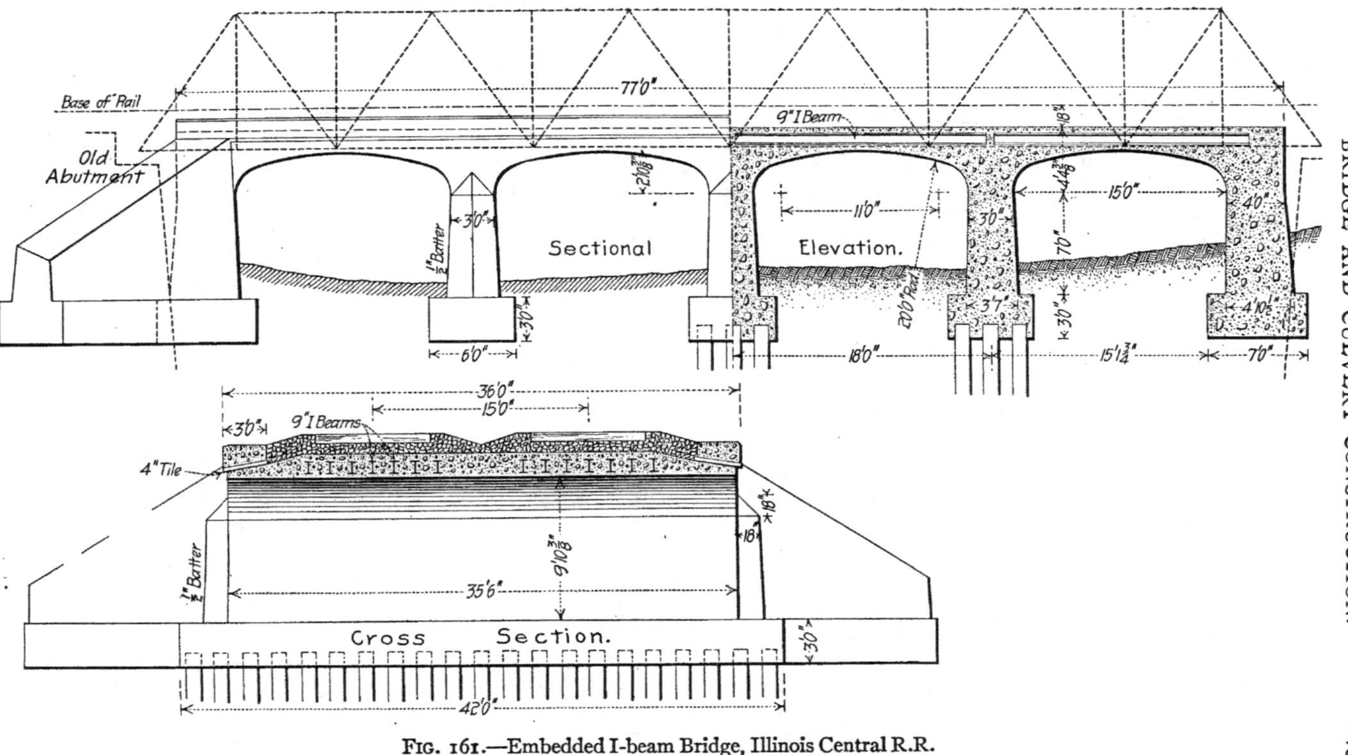

FIG. 161.—Embedded I-beam Bridge, Illinois Central R.R.

10 to 15 ft. span. The arches are as a rule quite flat. The 12-ft. arch has a radius of 16 ft. for a chord of 12 ft., and corner curves of 2 ft. radius to the face of the abutment, which has a batter of 1 in 24. For single track the length of the culvert is 19 ft., or 0 ft. 6 ins. over the parapet walls. The thickness of concrete at the crown of the arch is 8 to 23 ins., and five lines of 10- to 12-in. I beams, 16 ft. long, 2 ft. apart, are placed under each track, embedded in the arch-work, with about 3 ins. of concrete below and 8 ins. above at the crown. The upper surface of the arch forms, with the parapets, a basin or trough in which the ordinary stone ballast is laid, so that the roadbed can be continued unbroken over the structure. The wing walls may be straight or flaring, and are finished off in steps or with a sloped coping, as desired. The invert has a radius of 26 ft The amount of concrete in such a 12-ft. culvert is about 200 cu. yds. In concrete culverts with a steep fall, scouring under the 8-in. invert is prevented by apron walls at the ends, and occasionally at intermediate points. These walls are 18 ins. thick and 2 to 3 ft. deep below the bottom of the culvert. Small concrete culverts are also made, having openings $3 \times 5\frac{1}{2}$ ft., or 3 ft. diameter.

The culvert shown in Fig. 162 is typical of a great number of the

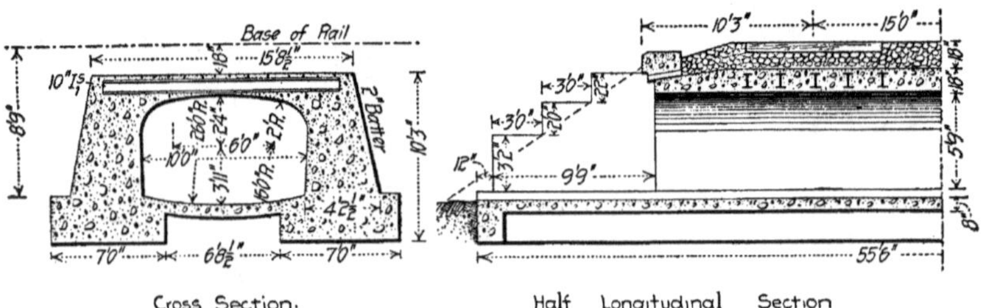

FIG. 162.—Embedded I-beam Single Culvert, Illinois Central R.R.

smaller structures, some of which, however, are without the invert. This is a 10-ft. culvert, with side walls 3 ft. 6 ins. thick at the springing line. Under each track are five 10-in. I beams, 14 ft. long, spaced 2 ft. apart. The invert is 8 ins. thick, with an 18-in. apron wall 3 ft. deep at each end. This culvert contains about 190 cu. yds. of concrete. A somewhat similar class of structure, but having greater height, and used in replacing short trestles, is shown in Fig. 163. This has two 12-ft. arches, and 12-in. I beams, spaced as before.

Some of the most interesting examples of reinforced-concrete culvert construction of recent development are illustrated in Figs. 164 and 165 from standards adopted by the Oregon Short Line R.R. The first form shown is a box culvert and the second is an arch culvert. The side walls are battered and rest on wide footings, and the arch and cover-plate are in each case reinforced by steel rails. In the

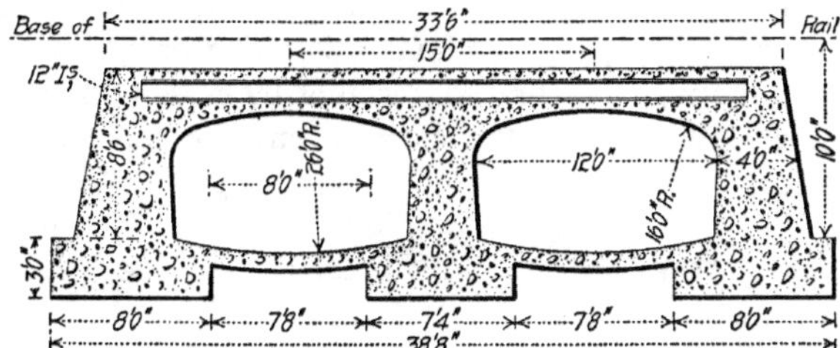

FIG. 163.—Embedded I-beam Double Culvert, Illinois Central R R

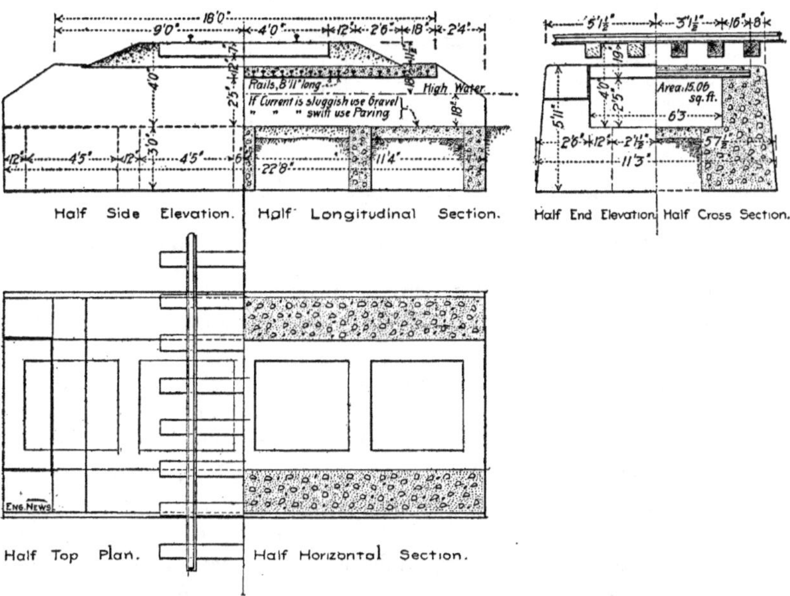

FIG. 164.—Embedded Rail Box Culvert, Oregon Short Line R.R.

box culverts the rails are straight and are spaced close together under the tracks, and further apart toward the ends of the walls. The rails are all set base downward and extend well over the abutments at the ends. In the arch culverts the rails are set alternately base and head downward and are bent to approximately the curve of the arch ring. The spacing of the rails is similar to that noted for the box culverts, and under the center where they are closest together there are embedded two parallel intradosal strips of expanded metal. The bottoms of both box and arch culverts between the side and wing walls are provided with a gravel bed where the flow is sluggish, and with stone paving where the current is rapid. To hold this filling or paving the side- and wing-wall footings are braced together by cross-walls of concrete. Two of these walls connect the extreme ends of the wing walls on

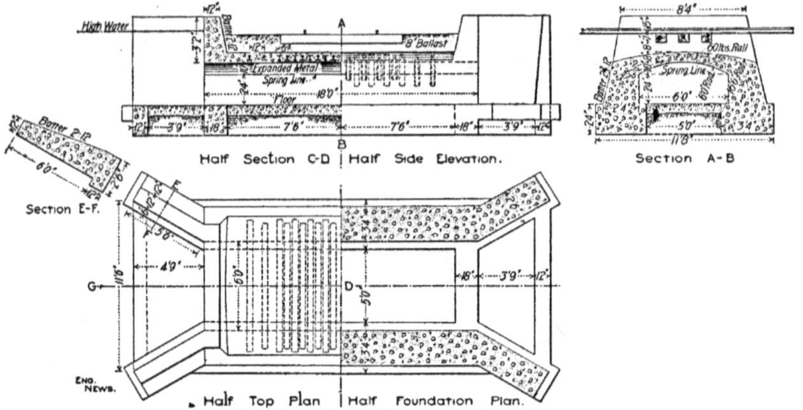

FIG. 165.—Embedded Rail Arch Culvert, Oregon Short Line R.R.

each side of the roadway and serve to prevent underscouring. The concrete used for the culvert-work is a 1–3–6 Portland cement and broken stone mixture.

New York Central & Hudson River R.R.—Fig. 166 illustrates a reinforced-concrete culvert built on the Pennsylvania Division of the New York Central & Hudson River R.R. in 1902. This culvert carries the roadbed over a deep gulley known as Gravel Run and is 300 ft. long with a clear span of 12 ft., and contains 3,000 cu. yds. of concrete. The reinforcement consists of No. 8 galvanized-wire netting in the arch-ring and of embedded steel rails in the parapet and wing walls. The arch ring netting has a 1×2-in. mesh. One of the notable features of this culvert was the method employed to secure a smooth surface finish which is described on p. 413.

Chicago, Burlington & Quincy R.R.—The drawings of Fig. 167 show a box-culvert construction of concrete reinforced with corrugated bars. This culvert is 147 ft. long and passes through an embankment about 30 ft. high. The construction is shown in such detail by the drawings that little explanation is needed. It will be noticed that a V-shaped section was cut out of the invert at each end; the purpose of this was to make the unit load on the foundation-bed as nearly uniform as possible.

Arch Culvert, Kalamazoo, Mich.—In diverting a small stream known as Arcadia Creek which runs through the city of Kalamazoo, Mich., use was made of a concrete-steel arch culvert having the section shown by Fig. 168. This culvert is 1,080 ft. long, with a clear width of 9 ft. 10 ins., and a clear height of 6 ft. Its grade varies from 0.4

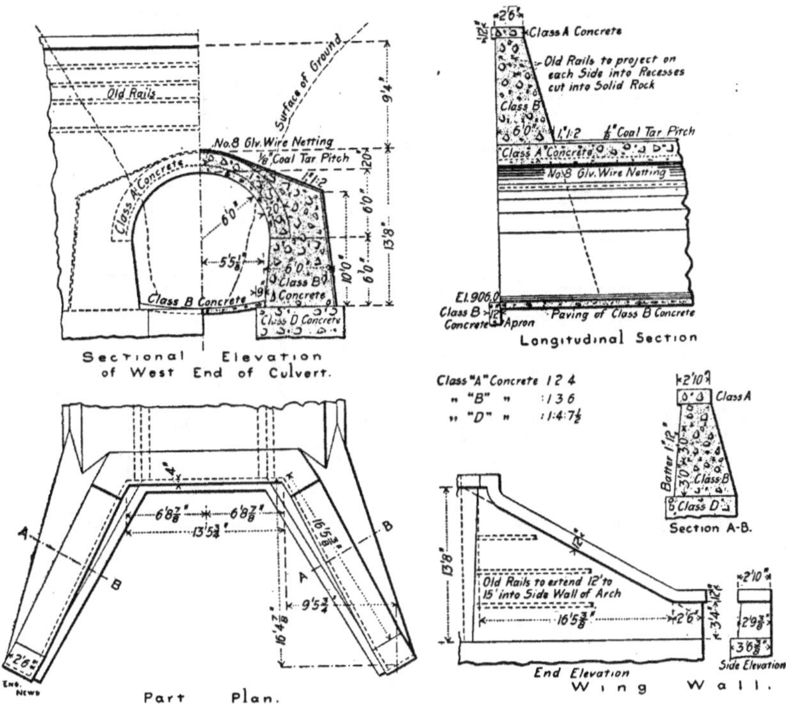

FIG. 166.—Arch Culvert with Expanded Metal Reinforcement, N. Y. C. & H. R. R.R.

to 0.5 per cent. The masonry is entirely of concrete composed of sand, gravel, and Portland cement. The proportions are generally about one part of Portland cement to six parts of sand and gravel. The upper arch, however, was made of a somewhat richer concrete than this, and the lower arch with a somewhat poorer concrete. These proportions were not rigidly adhered to, the composition being generally made richer under street crossings where there is a light covering of earth, and also where particularly treacherous soil underlaid the foundations.

The sect'on is reinforced by a woven steel-wire fabric, as shown in the cross-section. The members of this fabric which extend around

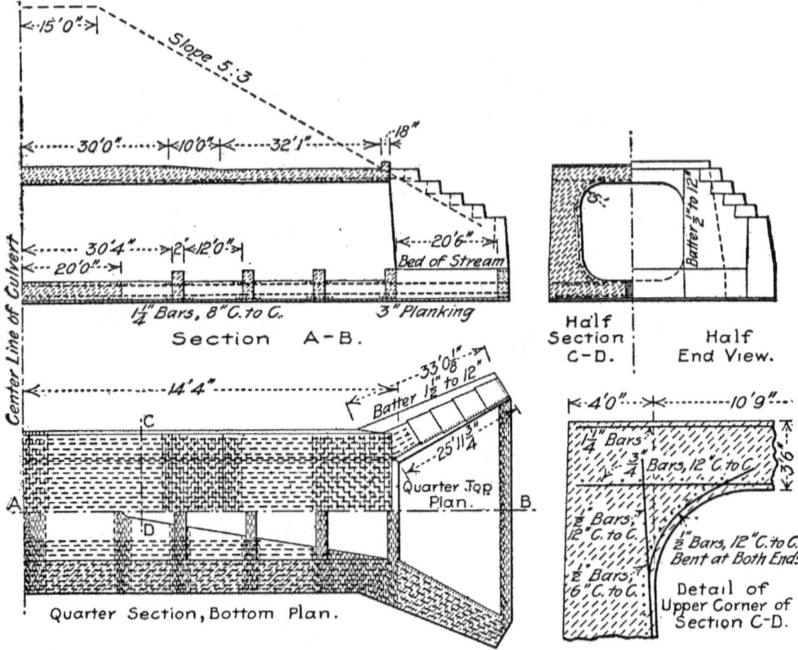

FIG. 167.—Box Culvert Reinforced by Corrugated Bars, C., B. & Q. R.R.

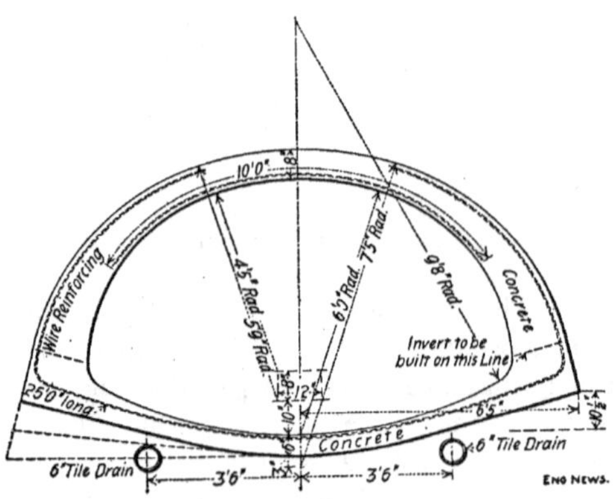

FIG. 168.—Arch Culvert at Kalamazoo, Mich.

the culvert are of No. 11 steel wire, and two layers of the fabric are used in all cases, thus making a total length of wire surrounding the culvert of 175 ft. per lin. ft. There is an average of about five wires per lineal foot, enclosing the conduit, except where the outer and inner reinforcements overlap; in this case there is double this number.

The bearing portion of the concrete in the inverted arch was changed in form, as shown by the dotted lines, Fig. 168, according to the character of the earth on which the concrete was to be laid. In some parts of the work the conduit rests on rather soft material. For a considerable portion of the way it rests upon quicksand. In ground of this character two lines of tile drains are laid under the invert, and these by removing the excessive water from the quicksand, made it firm and a good foundation. Before backfilling about the arch its exterior surface was well brushed with neat cement grout. The interior of the conduit was also finished in the same way.

Spandrel Construction.

The most common spandrel construction for reinforced concrete arches, as for masonry arches, is a solid filling between spandrel walls. For arches of considerable span and rise a hollow construction is sometimes employed to reduce the load on the main arch, and for æsthetic reasons. The most common form of hollow construction consists of two or more spandrel arches over the haunches of the main arch. Such arches are usually worked out so as to constitute a prominent architectural feature of the bridge. Where this is not purposely sought the arches are replaced by a simple arrangement of columns carrying a roadway platform and enclosed by thin curtain-walls at the sides of the bridge. This is usually a less expensive construction than spandrel arches.

Solid-filled Spandrels.—Solid-filled spandrels usually consist of an embankment of earth, sand, cinders, or other cheap and durable filling enclosed between solid retaining-walls or spandrel walls. Sometimes a filling of meagre concrete is employed in place of the earth or other loose material, and in these cases the spandrel walls become a corporate part of the filling. Where the spandrel walls retain a loose filling they have the usual trapezoidal retaining-wall section. Usually they are not reinforced, nor are they bonded into arch ring otherwise than by the natural cohesion of the concrete. Many of the illustrations of arch bridges in the preceding section of this chapter show the details of the spandrel walls employed. Fig. 140, illustrating the spandrel walls for the Zanesville, Ohio, Y Bridge, shows a method of bonding the walls into the arch ring by means of dowels. A reinforced-spandrels wall construction is shown by Fig. 149. This wall was used in the Seeley Street Bridge, Brooklyn, described on p. 226. It will be observed that the reinforcement is designed to strengthen the wall as a retaining-wall, and it will appear at once that any of the retaining-wall reinforcements described on p. 307 can be used for bridge spandrel walls. The filling between spandrel walls may be sand, gravel, earth, cinders, or any other suitable loose material. This filling is deposited in thin layers and thoroughly compacted by ramming or rolling or by flooding it in with water. An example of concrete filling is furnished by the 120-ft. space Melan arch built over the Yellowstone River in the Yellowstone National Park in 1903. The filling concrete was composed of 1 part Portland cement, 4 parts sand and 9 parts unwashed gravel, and in it there were embedded boulders amounting to about 20 per cent. of the total volume. The spandrel walls were composed of a 1–2–4 broken-stone concrete and were a corporate part of the filling.

Hollow Spandrels with Curtain-walls.—For spans of great length and rise a number of recent designs have called for a cellular spandrel construction enclosed by curtain-walls. Notable examples of these will be found in the long-span arches described on page 231. In all of these the roadway platform is carried on a staging of braced columns and beams which is enclosed by thin curtain-walls at the sides of the bridge. A study of the drawings will make clear all the essential details of these designs.

Spandrel Arches.—Spandrel arches combine the advantage resulting from a reduction of the load on the main arch ring and the foundations with the opportunity to introduce architectural effects of great beauty when well handled. Another advantage which is claimed for the spandrel-arch onstruction is that in doing away with long and heavy solid spandrel walls much of the trouble with temperature strains is obviated. The saving in weight by using spandrel arches instead of solid-earth fill and spandrel walls is excellently illustrated by the figures given by Mr. H. W. Parkhurst, M. Am. Soc. C. E., for the Big Muddy River Bridge on the Illinois Central R.R. Comparative calculations of weight for a 140-ft. span were made for spandrel-arch construction and solid-fill construction, with the following results:

Solid-filled Spandrel.

	Tons.
Main arch, 1396 cu. yds. at 2 tons.	2,792
Parapet and coping, 404 cu. yds. at 2 tons.	808
Pier complete, including old masonry, 1,825 cu. yds. at 2 tons.	3,650
Filling and ballast.	1,600
Total.	8,850

Spandrel Arches.

Main arch, 1,400 cu. yds. at 2 tons.	2,800
Spandrel arches and copings, 600 cu. yds. at 2 tons.	1,200
Pier complete, 1,555 cu. yds. at 2 tons.	3,110
Ballast.	525
Surplus weight of steel.	25
Total.	7,660

The spandrel-arch construction for this bridge is described by Mr. Parkhurst as follows:

The skeleton of the spandrel arches is a complete self-supporting and internally braced structure. It consists of eleven lines of longitudinal horizontal steel rails, spaced 3 ft. center to center, tied and also braced at this fixed distance by rods of 1 in. diameter, spaced about 18 ins. center to center, each having two nuts at each end which are screwed up tightly against the webs of the rails. These webs are punched with 1⅛-in. holes at intervals of 3 ins. for their whole length, and any rod will fit and can be secured in any hole. The rails are joined end to end by the ordinary splice-bars. This set of rails is held at proper elevation by vertical posts spaced also 3 ft. apart transversely, and at proper intervals for the spans of the arches. Each post is connected to the next by similar 1-in. rods, and is also connected to the longitudinal rail above it by adjustable diagonals which lie within the haunches of the arches. The vertical posts rest upon and are bolted to transverse horizontal rails set in recesses in the upper surface of the main rib of the concrete work. All rails have 1⅛-in. holes punched through their webs. This was done with a threefold purpose: (1) To save accurate templet work in laying out location of holes for connections; (2) to make it possible to add other connecting ties or struts or to make changes; and (3), in the writer's opinion, of considerable importance, to provide for bond between the steel and the concrete, the mortar of the latter entering and extending through these holes and so firmly binding both together.

The steel structure was erected, adjusted, and lightly bolted together by ordinary labor, and required no riveting and no expert assistance. Careful records show that large portions of it were completed at slightly less than ¼ ct. per lb. The self-supporting character of this reinforcing material was advantageous in several ways: (1) A level floor could easily be laid on the top, furnishing a very satisfactory working platform for making concrete; (2) Narrow runways were laid on it at trifling expense and used with confidence by the workmen; (3) The most important advantage was that the molds for the concrete could be wired to the steelwork, permitting cheaper construction in a part of the work where the quantity of concrete was small in proportion to amount of lumber needed for molds.

The more common construction is to reinforce the rings only of the spandrel arches. In the Laibach Bridge described on p. 215 the three spandrel arches over each haunch have solid concrete piers and reinforced arch rings; the reinforcement consists of 4-in. I beams curved and spaced 3.3 ft. apart center to center.

A modification of spandrel-arch construction consists in replacing the ring by rows of columns and the intervening arches by a rectilinear platform. The Chatellerault Bridge described on p. 208 is a notable example of this type of spandrel construction.

Expansion-joints.—Expansion-joints to provide for the action of temperature strains are usually constructed in the spandrels where they meet the abutment and at one or more points between abutments and crown. These joints cut the spandrels vertically from parapet-wall coping to the arch ring, and they are sometimes mere planes of weakness in the concrete and sometimes actual joints with a compressible filling. Joints of the first kind are formed by ending one section of wall against a vertical lagging to which a timber-strip is attached so as to form a groove, usually a V or U in shape, in the finished face. The concrete having hardened, the lagging is removed and the succeeding section of wall is constructed by concreting against the grooved face of the preceding section without the usual grout washing or other provisions for bonding old and new concrete work. This construction leaves a plane of weakness at which it is assumed any movement or rupture produced by temperature strains will occur. When actual separation of the concrete surfaces is desired one or more layers of felt, corrugated paper, or other partially elastic material may be pasted against the face of the first section and the concrete of the second section of wall deposited against this cushion.

Fig. 169 shows the method of providing for expansion adopted in building a Melan arch bridge at Washington, D. C. Each spandrel wall was provided with a vertical expansion-joint over each springing-line and at a point about 10 ft. each side of the crown. These joints were tongue-and-groove joints as indicated. In addition the sections of the wall between expansion-joints were reinforced against temperature strains and given lateral stiffness by means of the coupled bar-stiffeners illustrated.

In constructing the Big Muddy River bridge for the Illinois Central R.R., Mr. H. W. Parkhurst, M. Am. Soc. C. E., introduced an unusually elaborate system of expansion-joints which are described by him as follows:

The piers and abutments were carried up above the springing-line of the arches, and the voussoirs numbered. Nos. 1 and 1 were built as a portion of the pier or abutment. Then (the centering and molds being completed) the several voussoirs were built in the order in which they are numbered on the plan, voussoirs numbered 2 being located

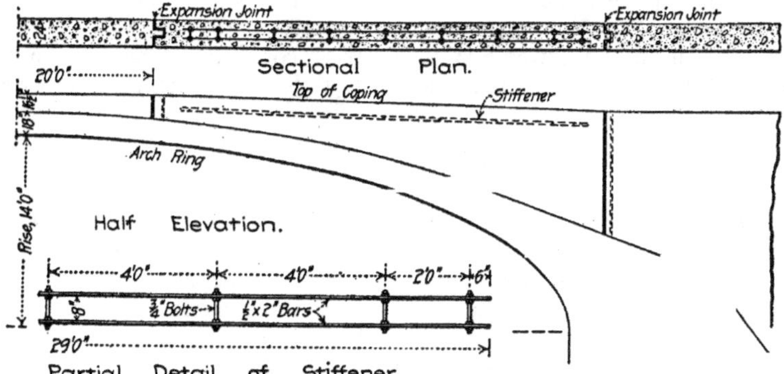

FIG. 169.—Expansion-joint and Spandrel-wall Stiffener, Rock Creek Bridge, Washington, D. C.

fourth in order from No. 1. Those numbered 3 being separated from those numbered 2 by three other sections; those numbered 4 being half-way between 1 and 2, and so on, as shown in Fig. 170. The succession was so arranged as to keep a symmetrical and as nearly as

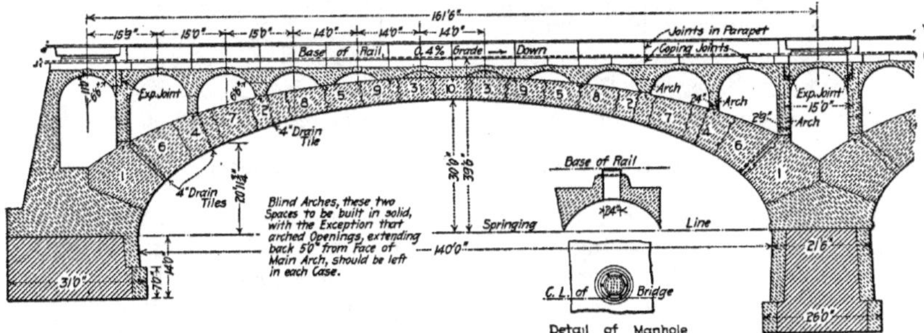

FIG. 170.—Section of Big Muddy Bridge, showing Segments of Arch Ring.

possible uniformly distributed load on the centering, and the number of voussoirs was planned so that at a certain stage of the work the arch would have alternate blocks completed, the keystone being one of the unfinished blocks.

The remaining blocks were then filled in, working toward the center, and the keystone being the last built. Each voussoir keys into the next by two projections on each side made by recesses built into the block first completed. This was done by securing planks of proper size on the face of the form dividing the mold into voussoirs. The blocks numbered 4 were held in place to prevent slipping until the blocks numbered 6 were built. No other blocks had any tendency toward movement. The face-joints between voussoirs were made by securing strips of wood of triangular section to the molds, a sharp right-angled edge adjacent to a thin joint being avoided on account of the possibility of flaking or cracking off, should there be a very heavy strain on the extreme outer layer of concrete. The real joints are about 8 ft. apart, and intermediate false joints of similar style were made on the exterior faces of each voussoir. All the radial joints extend across the molding with which the top of the main arched rib is finished, and corresponding joints extend across the soffit from face to face of arch.

Additional provisions for expansion are made in the construction of the transverse spandrel arches, in the connections of the longitudinal rails by splices having more motion, but specially by the sliding arches and expansion-joints at each pier and abutment.

Assuming that the piers and abutments are fixed and that the arches must expand, a joint is made at each end of each arch, at its juncture with the adjacent immovable portion of the structure. There are six expansion-joints (Fig. 171), essentially alike. They are intended

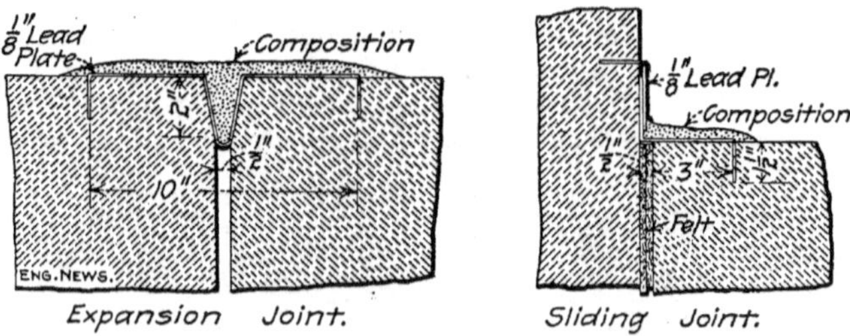

FIG. 171.—Expansion Joint, Big Muddy Bridge.

to provide expansion or freedom of motion for that portion of the structure on which the moving load brings its immediate effect. It will be noted that the character of this portion of the structure differs materially from that of the lower portion of the bridge. In the latter everything is massive and as nearly immovable as possible; in the former, by the combination of steel and concrete, an amount of elasticity is imparted to the structure which is designed to insure its durability under any and all conditions of exposure and service. The steel and concrete are proportioned to allow for the maximum range of expansion due to changes of temperature, also to absorb possible shocks due to derailments or to the sudden setting of brakes on swiftly moving trains, and other similar conditions.

The section of concrete in the crown of the small transverse arches is purposely made a minimum, to provide for elastic expansion at this point. In addition to this, however, at each of the expansion-joints referred to above, sufficient provision is made to cover any possible range of contraction and elongation due to changes in temperature for each arch. This expansion-joint extends from the haunch of the main rib to the level of the track, and is concealed and protected by the projection of the pier or abutment, where the spandrel arches recess into the same. There are two surfaces to this expansion-joint, one surface being in a longitudinal plane with the extension of the face of the arch, and the other at right angles to it.

The longitudinal surfaces both of pier and arch work are covered with heavy asbestos board, thus securing a sliding surface. The transverse joint is filled with several thicknesses of corrugated asbestos board, so that under pressure it could close, and when the pressure is relaxed could expand again, preventing entrance of foreign material into the joint. To assist in protecting this joint, the top is covered with a lead plate, which is folded into the joint, and filled with pitch or asphaltic cement, as shown in the plans.

These expansion-joints are constructed one at each abutment, and two at each pier; the last spandrel arch at each end of each main arch can move slightly in a longitudinal direction at the piers or abutments. This movement is further provided for in the construction of the sliding blind arches built into each pier and abutment, which are fitted with sliding bearings, and with an expansion-joint similarly packed with corrugated asbestos board, which joint extends across the whole width of the bridge floor, from coping to coping. This makes a section of floor approximating 15 ft. long and the width of both tracks, which is separated by a longitudinal sliding joint, lined with asbestos board, between the fixed portions of pier and abutment copings, etc., and the central section of floor. The skeleton steelwork in the floor is, however, made continuous across these sliding joints, and at the same time allowed to move longitudially by incasing the transverse tie-rods with pipe which is buried in the concrete of the floor, thus permitting sufficient longitudinal movement for expansion or contraction. The skeleton ironwork thus ties together the ends of piers, and holds them at the proper distance apart.

Having provided expansion-joints not only on this bridge, but on several other structures, it may be interesting to state that careful observations up to date indicate but very slight movement. On this bridge, brass expansion-gages, each made in two parts, and graduated to read by means of verniers to thousandths of a foot, were set in the copings across each expansion-joint, and careful readings of these gages were taken up to the completion of the work. The longest series covers four months from January 20 to May 23, 1903, and was for the west side of the north span, this having been completed, and the opening laid first. The results of these readings show an extreme movement of 0.012 ft., of which 0.007 ft. was observed at the north end and 0.005 ft. at the south end of span.

The maximum temperature of each day was noted, and the gage read simultaneously. It is evident that no accurate method of arriving at the average temperature of the whole mass of the structure could be devised, and that only general deductions might be made from such incomplete data. The record of gage observations does not indicate that concrete is extremely sensitive to heat and cold. While some irregularities of movement were noted, the gage readings showed contractions in cold and expansion in warm weather up to the limits, as above stated, of 0.012 ft. for 140 ft. span from temperature in January to that of May.

The procedure here described of providing planes of weakness in the arch ring is unusual in reinforced arch rings. The almost universal practice is to construct the arch ring as a single homogeneous monolith so far as is practicable, every effort being made to prevent planes of weakness in any portion.

In a number of instances the expedient has been adopted of omitting expansion-joints and resorting to a strong reinforcement to resist temperature strains and prevent cracks at the critical points. A notable example of such practice is furnished by the 120-ft. Melan arch built in the Yellowstone National Park in 1903. This bridge had a spandrel filling of meagre concrete built up in one piece with the spandrel walls, and it was decided to make the structure act as one solid mass without any provision for expansion or contraction. As the greatest strain tending to crack the concrete would come over the springing-lines where the arch proper joins the approaches, it was decided to embed steel rods in the concrete about 2 ft. below the top and extending parallel to the bridge axis from the ends of the abutments to near the crown of the bridge. It is stated by the engineer that, very slight cracks have appeared over the spring-line on both ends of the bridge, showing that the precautions taken to prevent them were not entirely sufficient. The only possible ill effect that can come from these cracks will be due to the infiltration of water, and care will be taken to prevent this.

Waterproofing.—To facilitate drainage and to exclude accumulated water from penetrating the arch ring the top of the arch and the lower portions of the spandrel walls are waterproofed. Frequently this waterproofing consists of nothing more than a plaster coat of cement and sand mortar to smooth the surfaces, but in many cases a special waterproof material is used in addition to or in place of the plastering. The most used waterproofing has been a layer of asphalt made by mopping hot asphalt into the concrete until a coat from $\frac{1}{4}$ to $\frac{1}{2}$ in. thick is formed. The objection urged against this material is that asphalt exposed to the action of water has but a limited life at the best, and when located under a spandrel filling and a pavement which must be removed to make repairs no material can be considered as satisfactory which has a life of not more than ten years under ordinary conditions. The waterproofing which has been used on a number of recent concrete-steel bridges designed by the Osborn Engineering

Company of Cleveland, Ohio, consists of a 1-in. coating of 1 Portland cement to 2 sand mortar applied to the back of the arch after the work is completed. The opinion of the users is that this is the most satisfactory waterproofing so far used by them. The practice of Mr. Edwin Thacher, M. Am. Soc. C. E., is to plaster the inside of the spandrel walls with a 1 to 2½ sand mortar. The back of the arch and the lower 6 ins. in height of the spandrel walls are then coated with a coat of semi-liquid mortar consisting of 1 part Portland cement, ½ part thoroughly slacked lime, and 3 parts sand spread to leave a smooth finish. After this plaster has set it is given a heavy coat of pure cement grout. Several of the methods for waterproofing tanks described on p. 293 are capable of being employed for arch bridges, but they have not been frequently used for this purpose.

Drainage.—The drainage of concrete-steel arch bridges of more than one span is accomplished exactly as in masonry arches. The usual construction is to build into the concrete a wrought-iron pipe reaching diagonally from the center of the space over the pier to the soffits of the arch through which it projects about an inch. The surface of the pier top is dished toward the upper end of the pipe, which is covered with wire netting to exclude coarse material, and a bed of broken stone is laid over all to receive the earth-fill.

Parapet Walls.—Parapet walls when built of concrete are sometimes molded in place and sometimes built of blocks previously cast, which are laid up exactly as cut stone would be laid. When molded in place the wall is commonly reinforced. This reinforcement is generally very simple, consisting of a channel or other shape lengthwise of the coping and verticals, reaching downward from it through the wall. In most cases, however, no reinforcement is employed, the wall being molded in sections with expansion-joints between them. These joints are placed as frequently as 10 ft. by some engineers. The drawings of Fig. 140 show a novel form of parapet-wall construction in which the cornice is built up of cast blocks and the parapet wall was molded in place. The method of anchoring the several parts together is particularly noticeable.

CHAPTER IX.—EXAMPLES OF REINFORCED-CONCRETE CONDUIT CONSTRUCTION.

Reinforced concrete is employed in conduit construction in two forms. The most common construction is concrete molded in place on suitable centers, and this is practically the only construction employed in the United States. In Europe reinforced concrete pipe molded in sections, which are jointed and laid like vitrified sewer-pipe, is largely employed for conduits of medium and small diameter. Several examples of this cast-pipe construction are given in the following paragraphs, but the greater space is devoted to examples of conduits molded in place. Some of these are notable in both size and the importance of the service to which they are put, and afford a striking indication of the confidence which engineers are coming to have in the strength and durability of concrete-steel under what are ordinarily considered to be rather trying conditions.

Aqueducts.

Simplon Tunnel-works, Switzerland.—To carry water from the River Rhone to the power plant of the Simplon tunnel-works at Brieg, Switzerland, a conduit consisting for part of its length of a concrete-steel flume and for the remainder of its length of a cylindrical steel penstock was employed. The concrete-steel flume was rectangular and 1.9×1.9 m. (6.23×6.23 ft.) in section and was 2 kilometers (1.86 miles) long, with a grade of 1.2 meters per kilometer. Fig. 172 is a

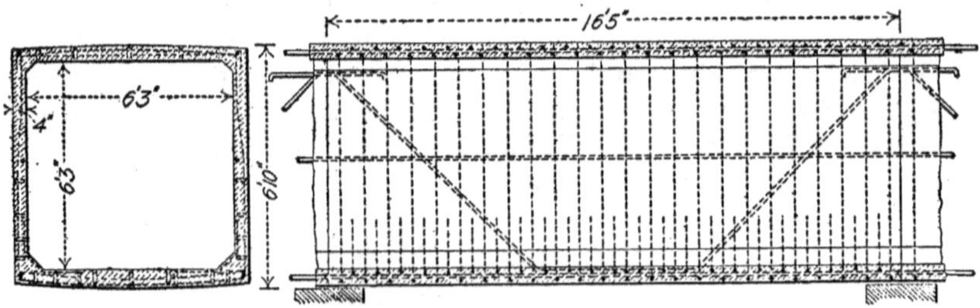

Fig. 172.—Rectangular Conduit for Simplon Tunnel Works.

transverse and part longitudinal section of the flume, and shows clearly the arrangement of the reinforcement and of the concrete walls. The roof of the conduit was designed to carry a superimposed load of 800 kgs. per square meter (164 lbs. per sq. ft.) and an internal upward pressure of 300 kgs. per square meter (61.5 lbs. per sq. ft.). The flume is supported on piers of masonry or bents of concrete-steel spaced 5 m. (16.4 ft.) apart. To provide for expansion and contraction during construction open joints were left over the piers; these were filled before turning the water into the flume. With the water in the flume there is very little expansion or contraction of the structure. The few leaks from percolation were soon closed by the lime in the water. The cost of the flume was 100 francs per lineal meter, or only about 10 per cent. more than the estimated cost of a wooden flume of the same dimensions.

Pecos Irrigation Co.'s Aqueduct.—An open aqueduct or flume built for irrigation purposes in New Mexico by the Pecos Irrigation Company is shown in cross-section by Fig. 173. This aqueduct is carried

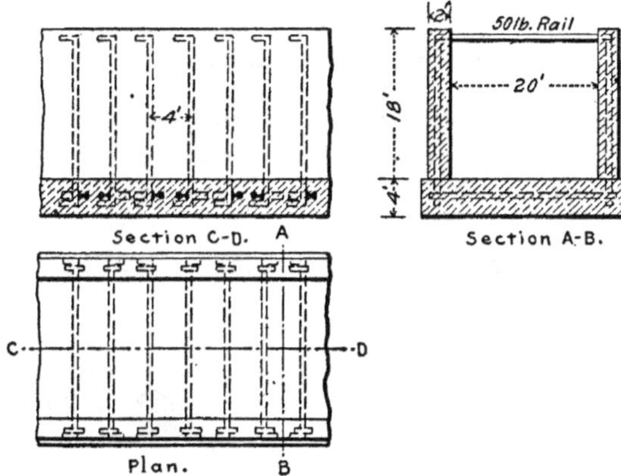

FIG. 173.—Rectangular Aqueduct, Pecos Irrigation Works, Texas.

on four concrete arches of 100 ft. span and 25 ft. rise. The concrete used was composed of one barrel of Portland cement to one cubic yard of mixed sand and gravel. The reinforcement consists of railway rails weighing 50 lbs. per yard spaced 4 ft. apart in the side walls and bottom, with their ends bent as illustrated, and of tie-rails between the opposite side walls at the top. The cost of the aqueduct complete, including the supporting arches, was $45,000, and it is interesting to note that the cost of the centers and forms amounted to $1 per cubic yard of concrete.

Weston Aqueduct, Boston, Mass.—In constructing the large Weston aqueduct for the water-supply of Boston, Mass., use was made of concrete-steel construction in carrying the structure under the tracks of the New York, New Haven & Hartford R.R. The normal polycentric section 9½ ft. high and 10 ft. wide was flattened as shown by Fig. 174, and throughout this flattened section a roof reinforcement of curved I beams was used. These I beams were spaced 36 ins. apart. To

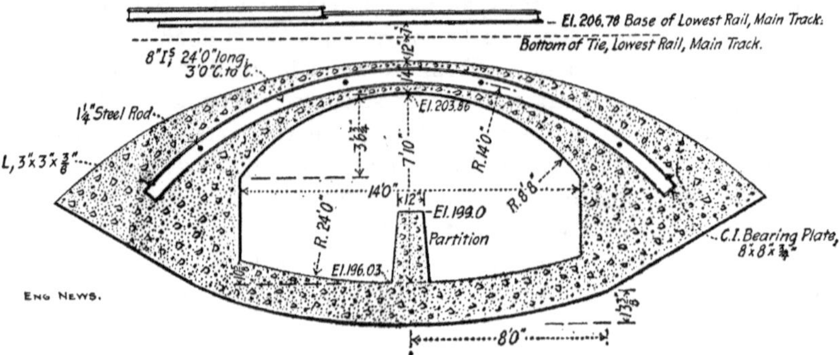

FIG. 174.—Section of Weston Aqueduct, Boston Water Supply.

prevent leakage through the cracks sometimes occurring in the joint between two days' work, use was made throughout this aqueduct of the lead water-stop shown by Fig. 175. The expectation is that the hinge in the lead will withstand any slight movement which may accompany cracking at joints and still prevent the passage of water through the crack.

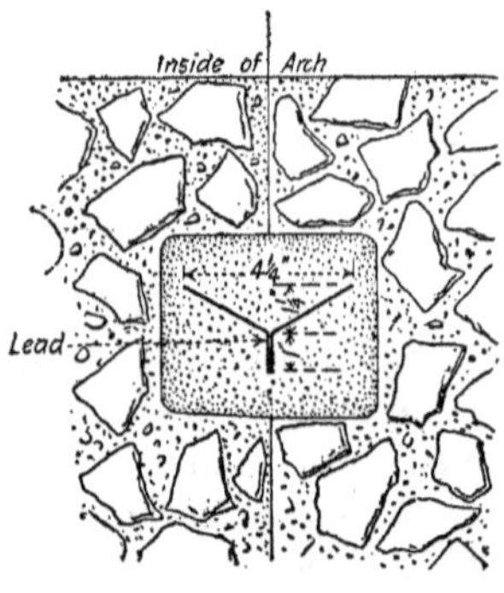

FIG. 175.—Detail of Water-stop at Junction of two Sections of Weston Aqueduct.

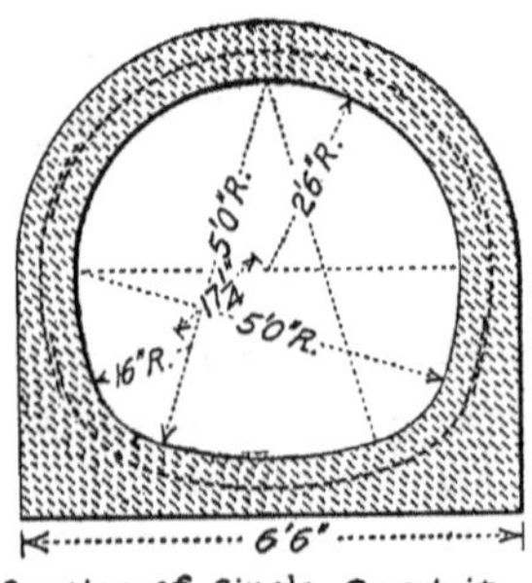

FIG. 176.—Single Conduit, Cedar Grove Reservoir, Newark, N. J.

Cedar Grove Reservoir Conduit, Newark, N. J.—About 7,000 ft. of 5-ft. reinforced-concrete conduit was employed in constructing the Cedar Grove reservoir for the Newark, N. J., water-works. One line, a single conduit 4,000 ft. long, extended from a point near the regulating inlet gate chamber at the north end of the reservoir to the inlet stand-pipe at the south end; another line, a double conduit 1,500 ft. long, extended from the outlet gate chamber to the outlet channel. Fig. 176 is a cross-section of the single conduit and Fig. 177 is a cross-section of the double conduit. The reinforcement consists of a circumferential ring of expanded metal of No. 10 gage and 3 in. mesh with lapped joints. The concrete is composed of 1 part cement, 2 parts sand, and 5 parts 1½-in. broken stone.

These conduits were constructed of concrete-steel after carefully comparing its cost and advantages with those of cast-iron pipe. In respect to cost the comparison worked out as follows: For the double

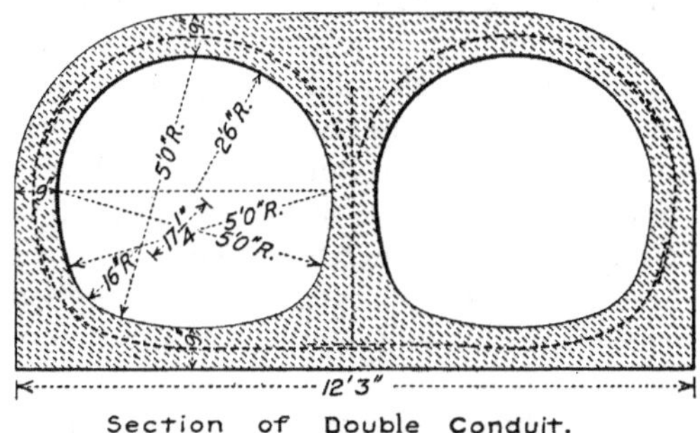

FIG. 177.—Double Conduit, Cedar Grove Reservoir, Newark, N. J.

conduit there were required 1.5 cu. yds. of concrete and 48 sq. ft. of expanded metal per lineal foot, and for single conduit 0.75 cu. yd. of concrete and 26 sq. ft. of expanded metal per lineal foot. With concrete costing $6.20 per cubic yard, and expanded metal costing 5 cents per square foot, these quantities gave the cost of double conduit as $11.75 and the cost of single conduit as $5.95 per lineal foot. For single conduit the cost given is about one-half the cost of 5-ft. cast-iron pipe, when the cost of anchoring the latter to prevent it from floating when empty is added. The concrete-steel pipe had enough weight to make anchoring unnecessary. As the conduit was within the reservoir, it was considered that any leakage which the concrete would allow would not be objectionable. In this connection it may be noted that the conduits will be under 50 ft. head of water, part of the time on the inside and part of the time on the outside. The inside pressure, however, will be applied only for short intervals, at times when the reservoir may be empty.

Philadelphia Filter-conduits.—A cross-section of the supply-conduit which conducts the filtered water to the clear-water basin at the Torresdale filter plant of the Philadelphia, Pa., water-works is shown by Fig. 306. This conduit is 2,200 ft. long and is progressively 7.5 ft.,

9 ft., and 10 ft. in height. The discharge-conduit which conveys the water from the clear-water basin to the tunnel conduit leading to the pumping station is 850 ft. long and 10 ft. high. A by-pass conduit 8 ft. high and 800 ft. long connects the supply- and discharge-conduits. All these conduits are of reinforced concrete of the general form shown by Fig. 306, and all will carry water under a pressure of 20 ft. head. The reinforcement of each conduit consists of expanded metal of sufficient section to carry all tensile strains in the shell due to water pressure. It was found possible to meet this requirement by using either two layers of No. 4 gage 6-in. mesh or one layer of No. 4 cut double width with 6-in. mesh. Preference was given to the one layer of metal, and this size was used for the discharge-conduit. It was found necessary, however, to use two layers of the standard No. 4 gage 6-in. mesh in the supply-conduit, the two layers being wired together and put in as one. The metal was located as is shown by Fig. 306 with 6-in. laps. The concrete was composed of 1 part cement, 3 parts sand, and 5 parts ¾-in. broken stone for the body of the shell and of 1 part cement, 1 part and 1 part granolithic grit for 1 in. deep on the inside.

Water-works Conduit, Bone, Algeria.—An example of the use of reinforced-concrete conduit molded in sections and subsequently laid is furnished by the aqueduct built in 1893 to supply the city of Bone, Algeria, with water. This aqueduct is about 22 miles long and about 2 ft. in diameter, and is of the so-called sidero-cement construction invented by the French engineer Bordenave. Fig. 178 shows the form of rein-

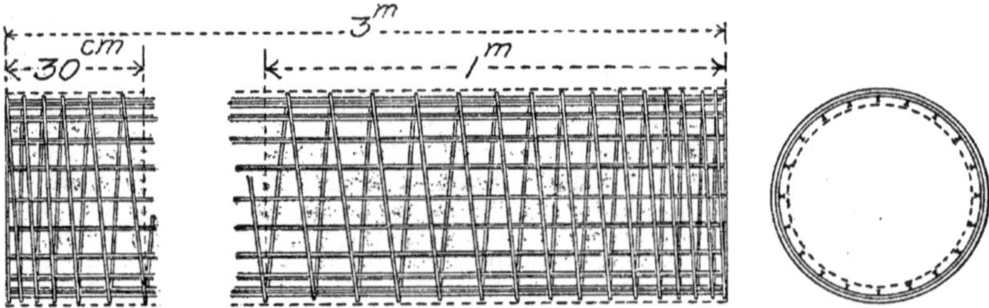

FIG. 178.—Reinforcement for Bordenave Cast Pipe.

forcement employed by this engineer. It consists of T bars wound in helical form around longitudinal parallel bars of the same section, the two sets of bars being connected at intersections by wire ties. This reinforcement is embedded in a Portland-cement and sand mortar. The practice is to manufacture this conduit in convenient lengths which are set end to end in constructing the aqueduct or sewer. The couplings between adjacent lengths of pipe are formed by means of collars, of exactly the same construction as the pipe, which are slipped over the adjoining ends and fastened by filling the annular space between collar and pipe with mortar. At Bone the pipe used had an interior diameter of 0.6 m. (1.96 ft.) and was designed for heads of from 8.5 m. (27.9 ft.) to 24 m. (78.7 ft.). The reinforcing-bars were 12×5 mm. (0.47×0.2 in.)×1.4 mm. (0.05 in.) thick. For pressures of 15 m. (49 ft.) head the circumferential bars were spaced 81.6 mm. (3.2 ins.) apart and the pipe was 40 mm. (1.58 ins.) thick, and for a head of 25 m. (82 ft.) the circumferential bars were spaced 48.6 mm. (1.91 ins.) apart and the pipe was 45 mm. (1.77 ins.) thick. The longitudinal bars were spaced 83.5 mm. (3.29 ft.) apart circumferentially of the pipe.

Water-works Conduit, Jersey City, N. J.—The most notable example of reinforced-concrete aqueduct work in the United States is doubtless the 8½-ft. aqueduct built for the new water-supply for Jersey City, N. J., in 1903. There are between three and four miles of this conduit, of the form and construction shown by Fig. 179. As will be observed, there are heavy and light sections for stiff earth and rock, a soft-earth section and a section for embankment. The reinforcement for all of these sections consists of twisted square steel bars; circumferential bars ⅜ in. square spaced 12 ins. apart, and longitudinal bars ¼ in. square spaced 2 ft. apart. All bars are wired together at intersections and are spliced by 1-ft. laps wound with wire. Only about 826 ft. of conduit of the heavy rock section were built where the cover was about 15 ft., and there were only about 420 ft. of the embankment section The section for soft earth was used in lengths of a few feet in a number of places where the bottom of the trench was soft earth. The concrete used was composed of 1 part cement and 7 parts sand and broken stone of 2 in. maximum size. The stone was broken trap-rock and the run of the crusher was employed.

Sewers.

Drainage Canals, New Orleans, La.—In constructing the drainage works for New Orleans, La., extensive use was made of reinforced concrete for roofing the canals. These canals vary in size, but the general construction of all is the same and consists of a concrete base and invert, brickwork side walls, and a flat-plate concrete roof reinforced with corrugated bars. Fig. 180 is a transverse section of one of the larger canals and shows the general construction clearly. The roof or cover construction of concrete steel is the feature of prin-

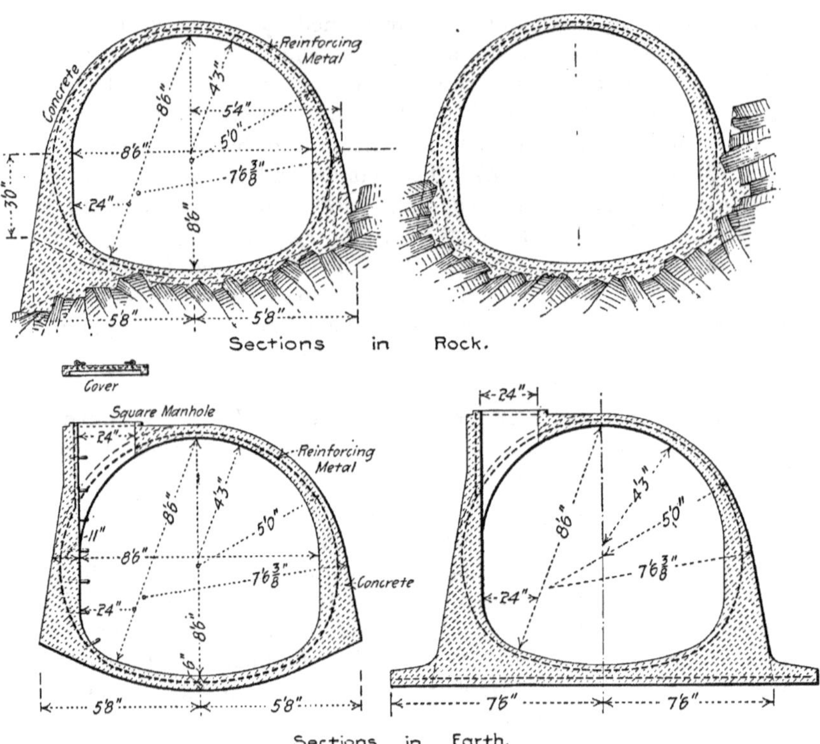

FIG. 179.—Sections of Water-supply Conduit for Jersey City, N. J.

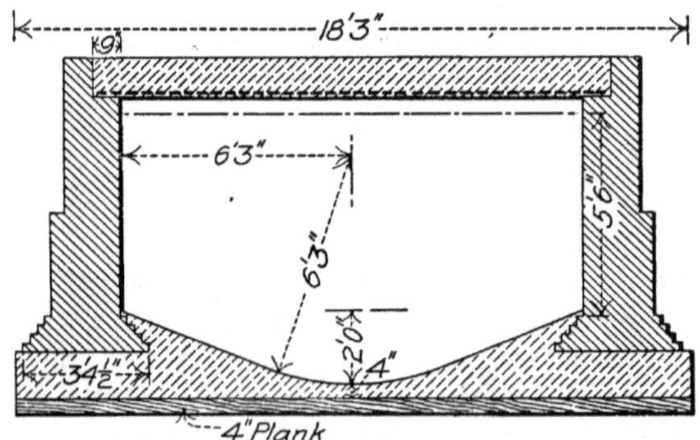

FIG. 180.—Section of Drainage Canal at New Orleans, La.

cipal interest in the present place. These roofs are flat plates of concrete composed of 1 part cement and 3 or 5 parts gravel, and reinforced by ½-in. corrugated bars set 1 in. from the bottom and spaced from 4 ins. apart for the 13-ft. spans to 10 ins. apart for the 5-ft. spans. The following statement shows the thickness of plate used for the various spans:

Span, Feet.	Plate, Inches.	Span, Feet.	Plate, Inches.
5	5.85	9	8.59
6	6.58	10	9.23
7	7.26	12	10.61
8	7.92	13	11.23

These covers withstood the following tests after twenty-eight days: Weights of [10,000+(span in feet ×1,200)] pounds placed along the longitudinal axis on 6×6-in. supports spaced 6 ft. apart.

Intercepting-sewer, Harrisburg, Pa.—Fig. 181 is a transverse section of an intercepting-sewer constructed at Harrisburg, Pa., in 1902.

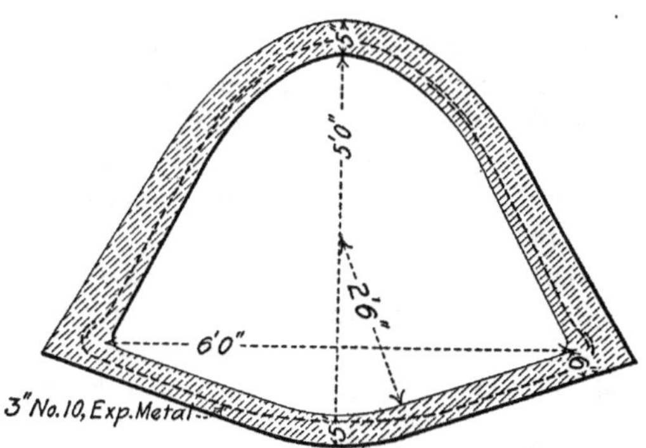

Fig. 181.—Section of Intercepting Sewer at Harrisburg, Pa.

There were 6,780 ft. of this 5-ft. sewer and 7,635 ft. of 4-ft. sewer of exactly the same form of section. This section has a parabolic roof and an invert composed of a circular curve and two tangents. The dimensions of the invert and roof for the 5-ft. sewer are shown by the illustrations. The 4-ft. sewer has the same thickness of roof and invert, but the main dimensions are of course smaller. In the invert the thickness of the concrete varies from 5 ins. at the center to 6 ins. at the sides, and in the roof it varies from 5 ins. at the crown to 9 ins. at the base. This thickness of the walls being sufficient to support the vertical loads coming upon the sewer, it was not deemed essential to locate the reinforcement along the lines of greatest tension. The reinforcement used was expanded metal and was arranged as is clearly shown by the drawing. The concrete was composed of 1 part cement, 2½ parts sand, and 4½ parts gravel or broken stone of 1¼-in. size.

Sewer System, Lancaster, Pa.—The drawings of Figs. 182-*a*-*b*-*c*, show the cross-sections of the reinforced-concrete sewers designed for the city of Lancaster, Pa., by Mr. Samuel M. Gray, M. Am. Soc.

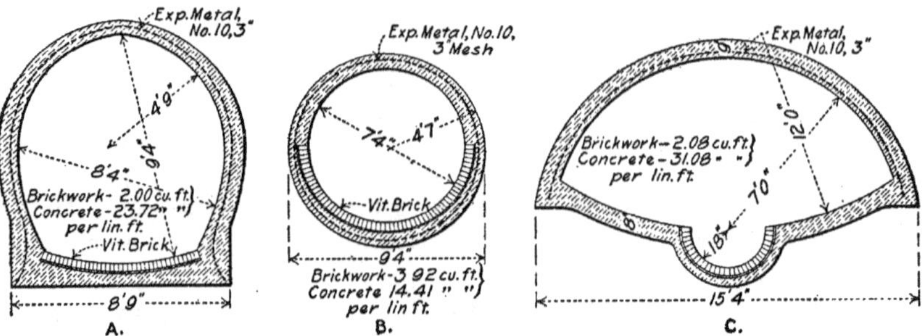

FIG. 182.—Sections of Various Forms of Sewers at Lancaster, Pa.

C. E. Comparative designs were made for brick sewers and for concrete-steel sewers, with the result that the latter proved to be the cheaper. It was also found that the smaller friction on the concrete surface as compared with the brick enabled them to be designed 2 ins. smaller in diameter than brick sewers of the same capacity. The drawings are largely self-explanatory. Fig. 182-*a* shows a horseshoe sewer of large size; a brick sewer of the same form and capacity had a sectional area of 65.24 square feet, and called for 27.6 cubic feet of brickwork and 22.08 cubic feet of concrete per lineal foot. A smaller horseshoe section showed the following comparison between brick and concrete:

Item.	Brick.	Concrete.
Sectional area, square feet.	50.4	49.06
Brickwork, cu. ft. per ft...	24.68	2.0
Concrete " " " " ...	18.52	23.1

Fig. 182-*c* shows a section of semicircular concrete-steel sewer; a birck sewer of the same form and capacity had a sectional area of 70.84 square feet, with 33.12 cubic feet of brickwork and 13.08 cubic feet of concrete per lineal foot. Fig. 182-*b* shows a circular concrete-steel sewer 7 ft. 4 ins. inside diameter; a brick sewer of the same form and capacity had a diameter of 7 ft. 6 ins., a sectional area of 31.17

square feet, with 15.23 cubic feet of brickwork and 14.08 cubic feet of concrete per lineal foot of sewer.

Parc Agricole d'Acheres, Paris, France.—In 1892–3 about 20 miles of concrete-steel pipe subway and circular sewers were constructed to distribute sewage to the Parc Agricole d'Acheres at Paris. This is perhaps the most extensive single application of concrete-steel sewer construction ever made. For the present purpose this conduit will be divided into two items: first, an elliptical pipe gallery and a 3-meter circular sewer of Monier construction, and, second, a 1.8-meter circular sewer and numerous distributing-pipes of smaller diameter of Bonna construction.

Pipe Gallery.—Fig. 183 is a transverse section of the pipe gallery,

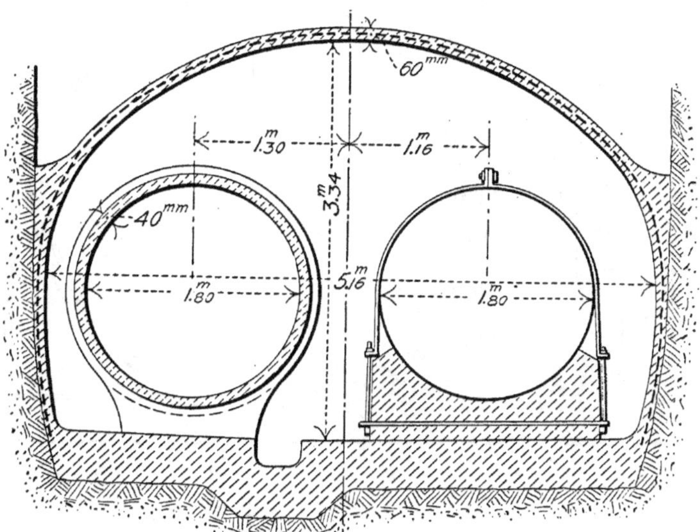

FIG. 183.—Section of Pipe Gallery, Parc Agricole d'Acheres, Paris.

which is 2,351 m. (7,713 ft.) long and carries two lines of sewers, one of which is the 1.8-m. (5.9-ft.) sewer of Bonna construction. As will be seen from the drawing, the bottom of the gallery is a thick layer of unreinforced concrete, and the sides and top are of concrete-steel. The reinforcement consists of a Monier network of round bars with a mesh of 0.11 m. (4.33 ins.). The curved members of the reinforcement are each a single bar without splices, the feet of which rest on channels. The longitudinal bars are placed inside the curved bars from their bottoms to 1 m. (3.28 ft.) above the springing-lines, and outside of them for the remainder of the arc. The aggregate section of the curved bars at the crown is 2 per cent. of the total section. The concrete of the arch is composed of Portland cement and sand in the proportion of 450 kilogrammes of cement to 1 cubic meter of sand; the arch plate is 8 cm. (3.15 ins.) thick and is lined with a 1-cm. (0.4-in.) coat of Vassy-cement mortar on the crown and of Portland-cement mortar on the sides. The interior dimensions of the gallery are 5.16 ×3.31 m. (15.98×10.85 ft.) and it cost 372 francs per meter, or about $22 per lineal foot.

Fig. 184 shows the construction of the 3-m.-(9.84-ft.) diameter circular sewer of Monier construct'on. This sewer rests upon a footing

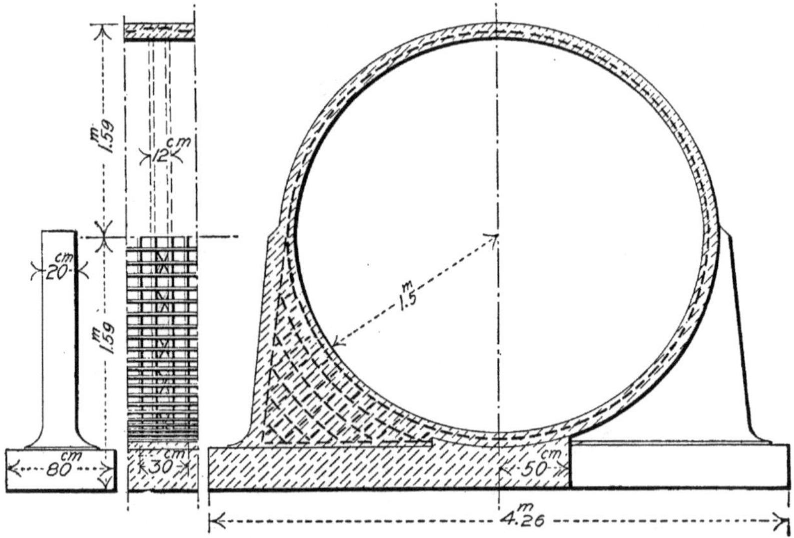

Fig. 184.—Circular Sewer, Parc Agricole d'Achères, Paris.

of unreinforced concrete against which it is braced by brackets of reinforced concrete spaced 4.2 m. (13.78 ft.) apart and similar brackets of unreinforced concrete spaced 6.8 m. (22.3 ft.) apart. When the sewer crosses depressions the brackets are carried down to the ground like piers and arches of 3.4 m. (11.15 ft.) span and 1 m. (3.28 ft.) wide are sprung between them to carry the footing. The reinforcement is clearly indicated by the drawing, and the concrete is of the same composition as that of the pipe gallery described in the preceding paragraph. In fact the gallery and the sewer are of the same general construction. This 9.84-ft. sewer is 561 m. (1,840 ft.) long and cost 214 francs per meter, or about $13 per lineal foot.

Cast-pipe Sewer.—The pipe gallery and 3-m. sewer previousl described were constructed in place; for the 1.8-m. (5.9-ft.) sewer inside the gallery and for the distributing-sewers cast-pipe construction of the Bonna type was adopted. Of the 1.8-m. (5.9-ft.) sewer only that portion subjected to a pressure of less than 22 m. (72 ft.) head was built of concrete-steel. The pipe was all made in special plants in lengths of 2.5 m. (8.2 ft.) and was laid by setting the lengths end to end and joining them by special collar-joints. For pressures greater than 13.6 m. (44.6 ft.) head each length was lined with a riveted steel pipe made of sheets 3.5 mm. (0.1 in.) thick for pressures up to 15.35 m. (50 ft.) head, and 4.5 mm. (0.2 in.) thick for pressures greater than this. The reinforcement consisted of rings and longitudinals of bars of cruciform section spaced from 9.5 to 20.4 cm. (3.75 to 8.16 ins.) apart, according to the interior pressure. Fig. 185 shows

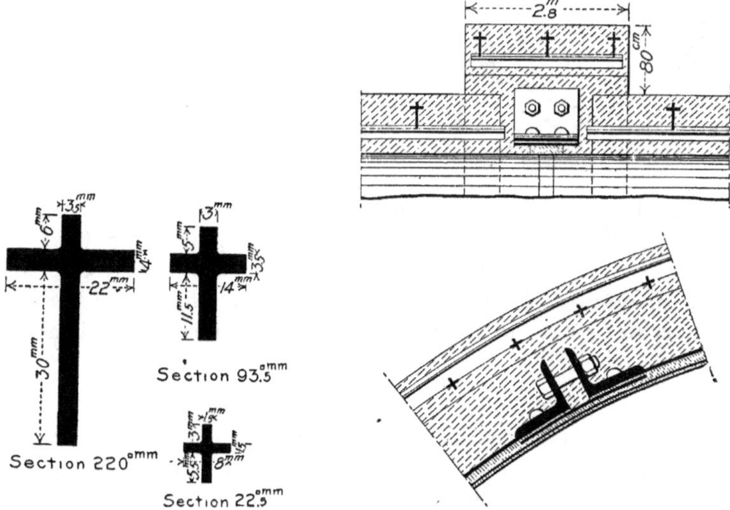

FIG. 185.—Reinforcing-bars for Bonna Cast-pipe Sewer.

FIG. 186.—Detail of Coupling for Bonna Cast-pipe Sewer.

the dimensions of the cruciform bars used; the smallest of the three were introduced as a supplementary reinforcement between the main rings and longitudinals in the pipe having the wider spacing stated above. Fig. 186 is a detail of the coupling between lengths of pipe; the coupling between lengths of unlined pipe consisted of the outside collar alone. The total length of this 1.8-m. conduit is 1,461 m. (4,793 ft.), of which 281 m. (922 ft.) were lined. The cost of the lined conduit was about 300 francs per meter, or say $18 per lineal foot, and the cost of the unlined conduit was about 200 francs per meter, or say $12 per lineal foot.

The distributing-sewers of Bonna construction varied from 0.3 m. (11.8 ins.) in diameter to 1.1 m. (3.6 ft.) in diameter and were designed to withstand a pressure of 40 m. (131.2 ft.) head. Like the 1.8-m. sewer described above they were composed of separate lengths of pipe coupled together end to end. Each length contained an embedded steel tube. Fig. 187 shows the construction of the tubing, and Fig. 188

Fig. 187.—Section of Lining for Concrete Pipe under Pressure.

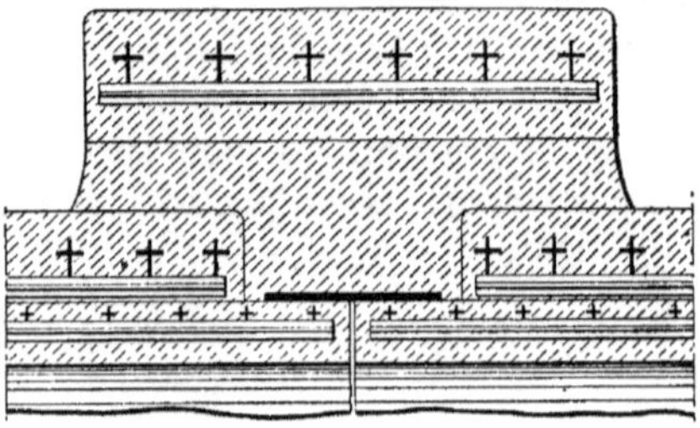

Fig. 188.—Coupling for Cast Pipe of Small Section.

is a detail of one of the couplings. The double reinforcement shown by Fig. 188 was employed for all diameters, except the 0.3-m. (11.82-in.), for which only the inside reinforcement was used. The cost of these distributing-sewers varied from 75 to 90 francs per meter for the 1.1-m. diameter to from 11 to 14 francs per meter for the 0.3-m. diameter.

Flues.

Reinforced concrete has been employed in a number of instances in constructing flues for conveying smoke and gases in metallurgical works. In Germany there are a number of these structures of Monier construction. They are rectangular and from 6 m. to 12 m. (19.7 to 29.4 ft.) square in section and have their side walls braced on the outside by triangular buttresses placed at intervals. The Anhalt Lead and Silver Works, at Alexisbad, Germany, have a flue of this construction 500 m. (1,640 ft.) long. Figs. 189 and 190 show the construction of a dust-flue built in 1899 to carry smoke and gases from the roasting-furnaces of the Arkansas Valley Smelting Company to the chimney and to collect the valuable dust deposited during their journey. Fig. 189 is a cross-section showing the main arch members of the metal skeleton. These are simply channel-bars bent to the proper curve and having their ends set in a concrete base-wall. Between these walls is a concrete floor resting directly on the ground. The channel-iron ribs are connected longitudinally by flat iron members attached to the

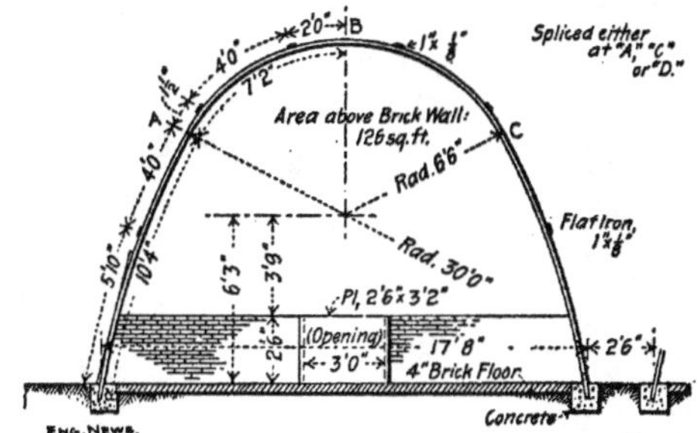

FIG. 189.—Section of Dust Flue for Arkansas Valley Smelting Works.

ribs by clinched staples. Fig. 190 shows this connection and also shows the arrangement of the expanded metal which was fastened

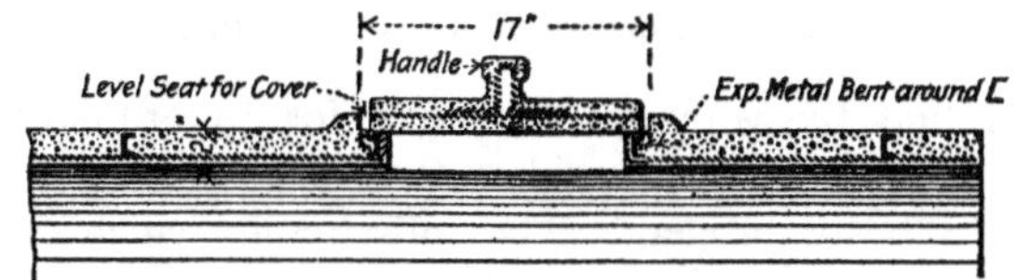

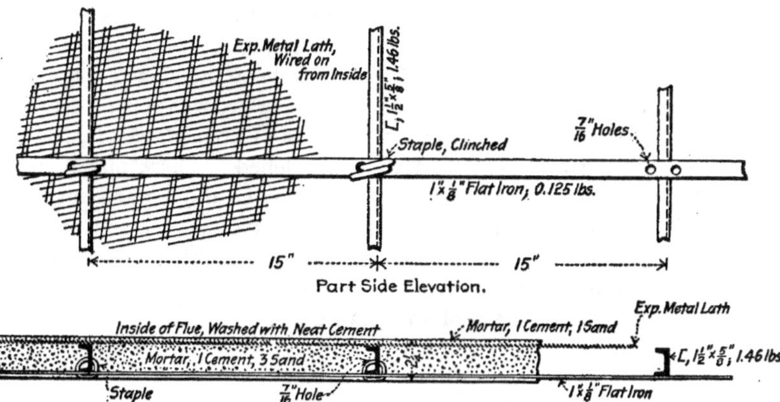

FIG. 190.—Details of Dust Flue for Arkansas Valley Smelting Works.

to the inside of the main skeleton. The reinforcement was embedded in a mortar composed of 1 part cement and 3 parts sand for the sides and in a slag concrete for the crown portion. The total thickness of the concrete wall was 2¼ ins.

Subway for Pipes and Wires.

A description of a pipe subway for the Paris sewage farms has been given on a preceding page. Fig. 191 shows a small subway for pipes and electric wires built in 1901 at Bloomington, Ill. It will

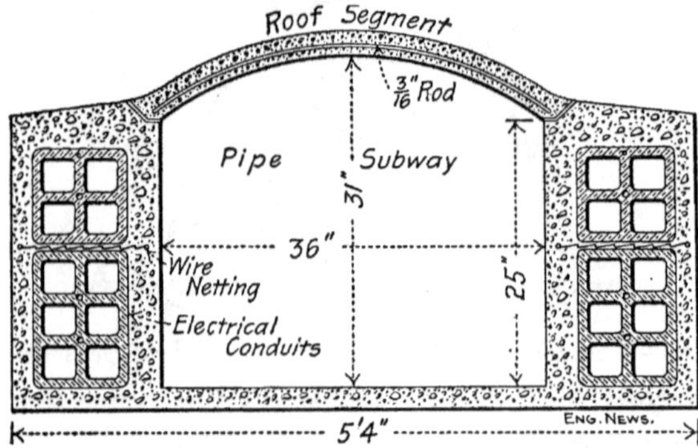

Fig. 191.—Section of Pipe and Wire Subway, Bloomington, Ill.

be observed that the wires are provided for by ordinary clay conduit embedded in the concrete side walls, while the pipes are carried through the subway proper. The only reinforcement of the side walls consists of a sheet of wire netting to strengthen the thin sheet of concrete between the two tiers of clay conduit and to tie the opposite sides of the wall together. The concrete for the side walls and the bottom of the subway was composed of a 1–3–5 mixture. No allowance was made for the adhesion of the concrete to the glazed clay conduit, but it was found that the bond between the two was of considerable strength. The roof of the conduit was made of arch-plates of Monier concrete cast separately and laid when hardened. These plates were 2 ins. thick and were composed of 1 part cement and 3 parts torpedo sand. The use of separately cast roof-plates avoided the use of centers and hastened the work of construction and backfilling materially. The total length of subway built was about 5,000 ft., and the total volume of concrete of the roof was 3.25 cu. ft.

CHAPTER X.—EXAMPLES OF REINFORCED TANK AND RESERVOIR CONSTRUCTION.

THE first structural application of reinforced concrete by Joseph Monier is said to have been in the construction of tubs for holding trees and shrubs. As early as 1868, however, the inventor had employed the new material in constructing water-tanks of large capacity, and the success of these was so great that in a few years Monier concrete-steel was being extensively adopted in Germany and Austria for tank and reservoir construction for a variety of purposes. These structures are now numbered by the score and include surface and subsurface reservoirs, elevated water-tanks, and vats for manufacturing purposes of both cylindrical and rectangular form. The position of the tank, whether partly or wholly beneath the surface or elevated, introduces some differences in construction, as does also the shape, whether cylindrical or rectangular, but the general system of reinforcement which is employed is substantially the same in all cases. It consists of a rectangular mesh network of round bars and rods, the size of the mesh and of the members which compose it varying with the loads to be carried. For the purpose of more detailed description the bottom, side wall, and roof can best be considered separately.

For tanks partly or wholly buried in the ground the bottom construction consists of a flat plate where the depth of water carried is small and of a spherical shell when the load to be carried is greater. When spherical bottoms are employed they are placed with the convex surface downward and are practically always reinforced. Flat-plate bottoms are, however, often built of unreinforced concrete. The bottoms of elevated tanks are practically always spherical and convex upward. The side-wall reinforcement of cylindrical tanks consists, first, of a series of horizontal rings placed close together at the bottom and gradually increasing in distance apart toward the top, and, second, of a series of vertical rods spaced uniformly around the tank and connected by wire ties to the horizontal rings. In rectangular tanks the side-wall reinforcement consists of horizontal and vertical rods. At the corners formed by the adjoining sides the horizontal rods are usually bent around so as to make the reinforcement continuous and thus reduce the chances of cracks developing. The junction of the side walls and the tank bottom is formed in a similar manner. Roofs for tanks of Monier construction are either flat plates or spherical arches, the latter being employed when any considerable load is carried by

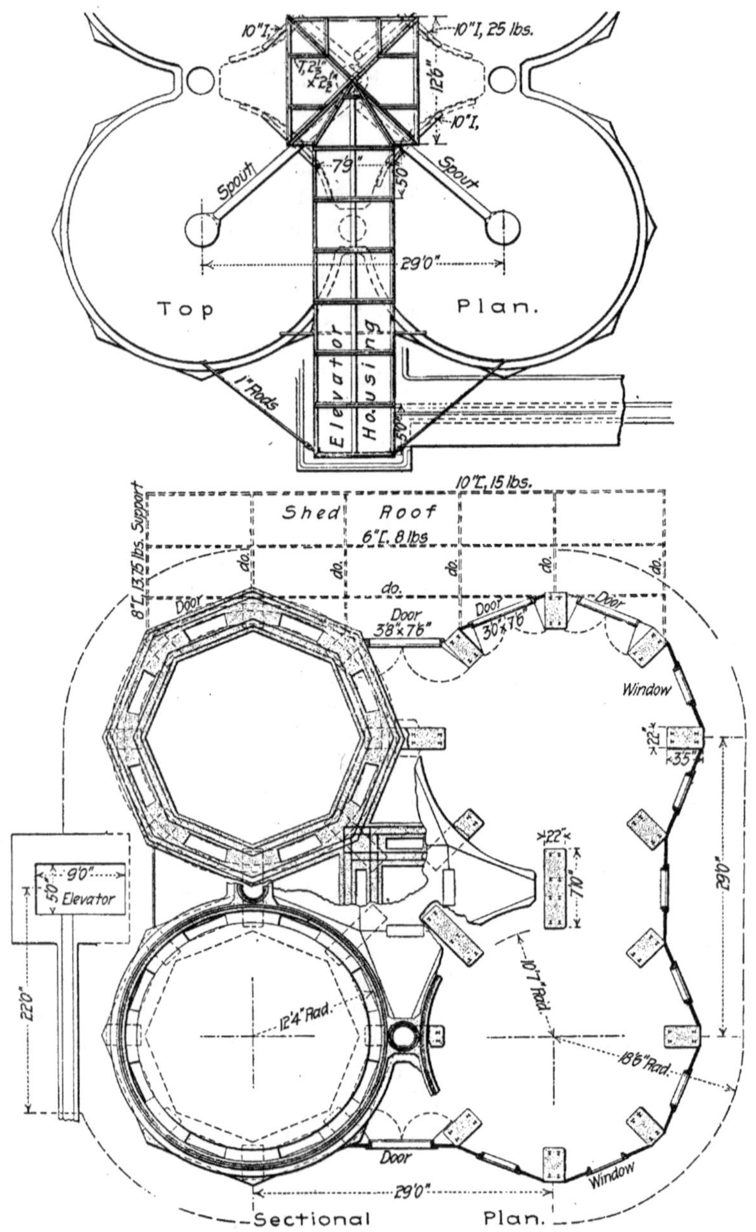

FIG. 193 *a*.—Cement Storage Bins of Monier Concrete Steel, Illinois Steel Company, Chicago, Ill.

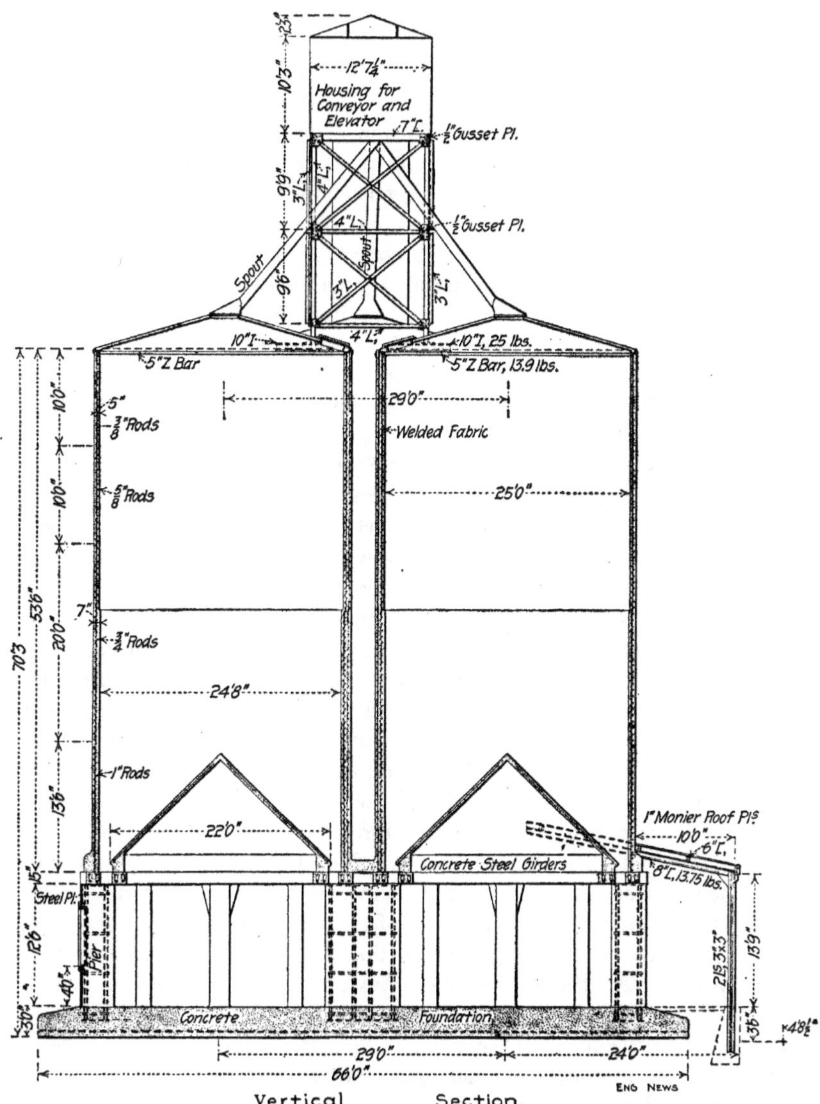

FIG. 193 *b*.—Cement Storage Bins of Monier Concrete-Steel, Illinois Steel Company, Chicago, Ill.

the roof. The reinforcement of spherical arches in tank roofs and bottoms consists of concentric rings and of radial members fastened together by wire ties. The mixture in which the reinforcement is embedded in all Monier construction is a Portland cement and sand mortar mixed, usually in the proportion of one to three or four.

The general scheme of reinforcement adopted by Monier is followed in nearly all other forms of tank construction with bar reinforcements. The walls are reinforced by vertical and horizontal bars, the latter spaced wider apart as the distance above the bottom increases. With mesh reinforcement, such as expanded metal or any of the wire fabrics, a double layer near the bottom and a single layer above secures the same object, or the same strength of reinforcement may be used from top to bottom. The following examples will explain current practice; they include both rectangular and cylindrical tanks and bins for holding dry materials.

Cement-storage Tanks, Illinois **Steel Company, Chicago, Ill.**— Figs. 193 and 194 illustrate the construction of Monier concrete-steel adopted for a group of elevated cylindrical tanks built to store cement at the plant of the Illinois Steel Company at South Chicago, Ill., in 1902. The group consists of four circular tanks set at the corners of a square and with the irregular space between them forming a fifth tank or bin. The total storage capacity of the five bins is 25,000 barrels of cement.

The tanks are 25 ft. diameter inside, and 29 ft. center to center, the space between each pair of adjacent walls being closed by a cylindrical shaft 30 ins. diameter, and the entire structure being monolithic. The foundation is a continuous floor or bed of concrete 3 ft. thick, 66×66 ft. with corners of 18 ft. 6 ins. radius. About 5-in. above the base of this is embedded a netting of 9-in. mesh formed by $\frac{5}{8}$ ins. round steel rods, tied together at their intersections by No. 18 wire. Upon this concrete bed is a series of piers 12 ft. 6 ins. high and 1 ft. 10 ins. thick. Those near the outer circumference of the tanks are mainly 3 ft. 5 ins. long, but others near the central portion are from 6 ft. 8 ins. to 7 ft. 10 ins. long. All the smaller piers have four steel rails embedded in the concrete, the rails being connected by spacing-bars riveted to them, and resting on $\frac{1}{2}$-in. steel plates embedded in the concrete floor about 15 ins. below the surface. The piers are capped with similar plates. The larger piers have six or eight rails. Upon the piers rest concrete-steel girders 15 ins. deep and 4 ft. wide, with vertical openings at intervals for the discharge-spouts. Through each girder run four horizontal lines of steel rods near the top, and four other lines bent to form truss-rods, with sheets of wire netting on each side of each pair of rods.

The cylindrical tanks, 53 ft. 6 ins. high, rest upon this system of girders, the base being 13 ft. 9 ins. above the level of the concrete floor. The walls are 7 ins. thick in the lower part and 5 ins. in the upper part, the thickness being increased where they unite with the circular walls which close the spaces between the tanks. Within the concrete walls is embedded a continuous sheet of netting of No. 9 wires electrically welded at their intersections and forming rectangular

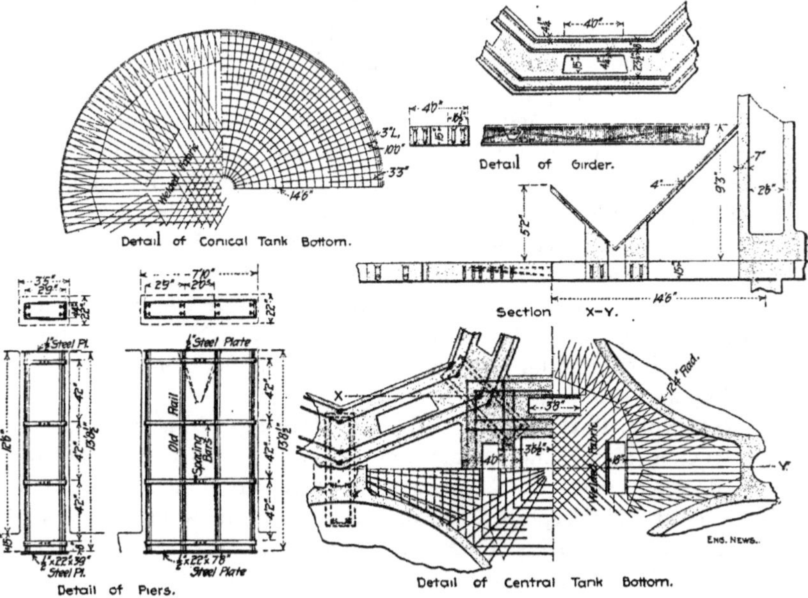

FIG. 194.—Details of Concrete-steel Bins for Illinois Steel Company.

meshes 1×4 ins. Around this (and alternately inside and outside of it) are horizontal rings of rods 4 ins. apart, tied to the netting by wire. These rods vary from 1 in. diameter near the base to $\frac{3}{8}$-in. near the top, while the top is finished with a ring formed by a 5-in. Z bar. The roof is conical, 2 ins. thick, with a manhole at the edge and an opening at the top for the spout. It was thought that with a hopper bottom and central discharge, material such as cement would give trouble by bridging, and the bottom is therefore made conical, with eight discharge-openings in the annular space between the base of the cone and the side of the tank. These openings are about 15×48 ins. and each serves two spouts leading to conveyors which carry the cement to the packing and shipping department. The conical bottom is 4 ins. thick, reinforced with rods and netting, and its diameter at the base is 22 ft. The concrete in the foundation floor and piers is composed of 1 part Portland cement, and 3 parts coarse sand, and 4 parts stone. That of the tanks is composed of 1 part Portland cement to 3½ parts sand, no stone being used. It was mixed moderately wet and lightly rammed in wooden forms. These forms were constructed of 3-in. planks 3 ft. long, kept together by means of three horizontal angle-irons supported by vertical-angle irons clamped together above the wall. The concrete was poured in and tamped inside the forms, which were raised, in 45° sections, 28 ins. high, every twenty-four hours. The work was carried on by day and night so as to prevent any setting between the layers of concrete.

The cement is brought from the mill by a horizontal screw conveyor and delivered into the boot of a vertical bucket elevator, which again delivers it to a horizontal screw conveyor above the tanks, by which it is carried to the spout of one or other of the tanks. The elevator runs in a rectangular shaft 12×14 ft., and the distributing conveyor in a chamber or housing 10 ft. 3 ins. ×12 ft. 4 ins., supported by a steel bent. The shaft and housing are built of a framing of steel angles, sheathed with Monier concrete-steel plates 2×5 ft., ⅝ in. thick, made of cement mortar and wire netting. The floor of the conveyor housing is of similar plates 1 in. thick.

Reservoir, Port Deposit, Md.—The drawings of Fig. 195 show the construction of a reinforced-concrete reservoir 16 ft. in diameter and 4 ft. deep, covered with a domed roof, which was built for the Tome Institute at Port Deposit, Md., in 1900. The side walls and floor of this reservoir are 6 ins. thick, the outer 5 ins. of thickness being a 1–3–6 Portland-cement concrete and the inside 1 in. of thickness being a 1–1 cement mortar laid simultaneously with the concrete for the sidewalls and plastered onto the concrete for the bottom. The roof is a dome with a span of 16 ft. and a rise of 2 ft., its thickness being 4 ins. at the crown and 5 ins. at the edge. The reinforcement was expanded metal in 5×8-ft. sheets. The wall reinforcing sheets were bent over at top and bottom so as to bond into the floor and roof. For reinforcing the roof the sheets were cut into triangles of 2½ ft. base and 8 ft. altitude, each sheet making two triangles. The concrete used consisted of Portland cement, sharp sand, and fine broken granite containing the screenings.

In constructing the small reservoir tank at Port Deposit, Md., the floor was built first without the finishing coat of mortar. The sides were then built, the forms being made in sections 6 ft. long and 2 ft. high for convenience in tamping the concrete and placing the pipes through the walls. For building the dome a center formed of earth covered with sand and papered with heavy roofing-paper was employed. The earth center was carried on a patform level with the top of the side walls. This platform framing consisted of two cross-stringers 3 ft. apart at the center and of shorter stringers running at right angles to the main stringers and extending to the perimeter of the tank. This framing was supported on posts on the bottom of the tank and carried

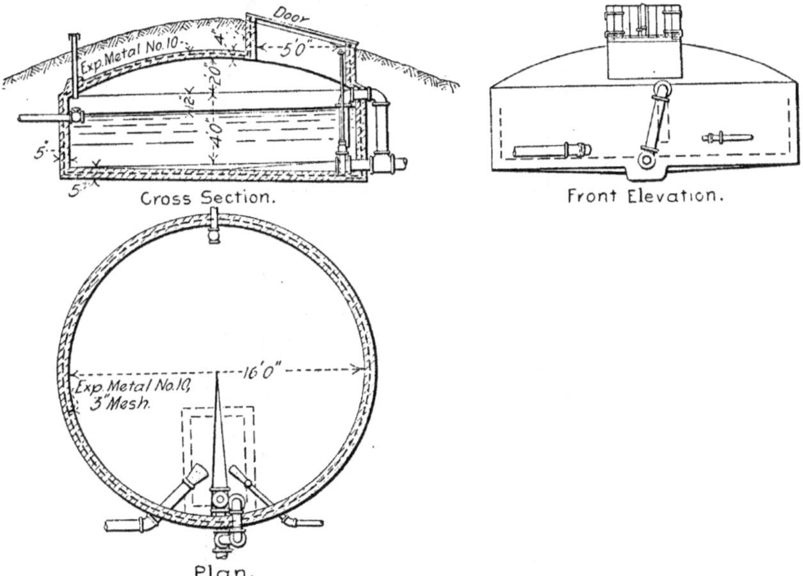

FIG. 195.—Reinforced Concrete Reservoir, Port Deposit, Mich.

a flooring of 2-in. plank. The earth for the center was kept damp so as to pack well, and it was molded to a wooden template. After completion, a wash of neat cement was applied to the inside of the reservoir by means of brooms.

Water-tower, Fort Revere, Mass.—One of the most interesting examples of deep-tank construction existing in the United States is furnished by the water-tower erected at Fort Revere, Boston Harbor, in 1903. A vertical section of this tower is shown by Fig. 196. The construction consists of a base of reinforced concrete carrying eight wall piers of reinforced concrete filled between with curtain-walls of brick. Surmounting the piers is a reinforced concrete floor projecting beyond the walls to form a water-table with a deep fascia. Above this floor is an observatory with a timber roof. From the ground surface to the top of the finial the height is about 93 ft. The base-wall

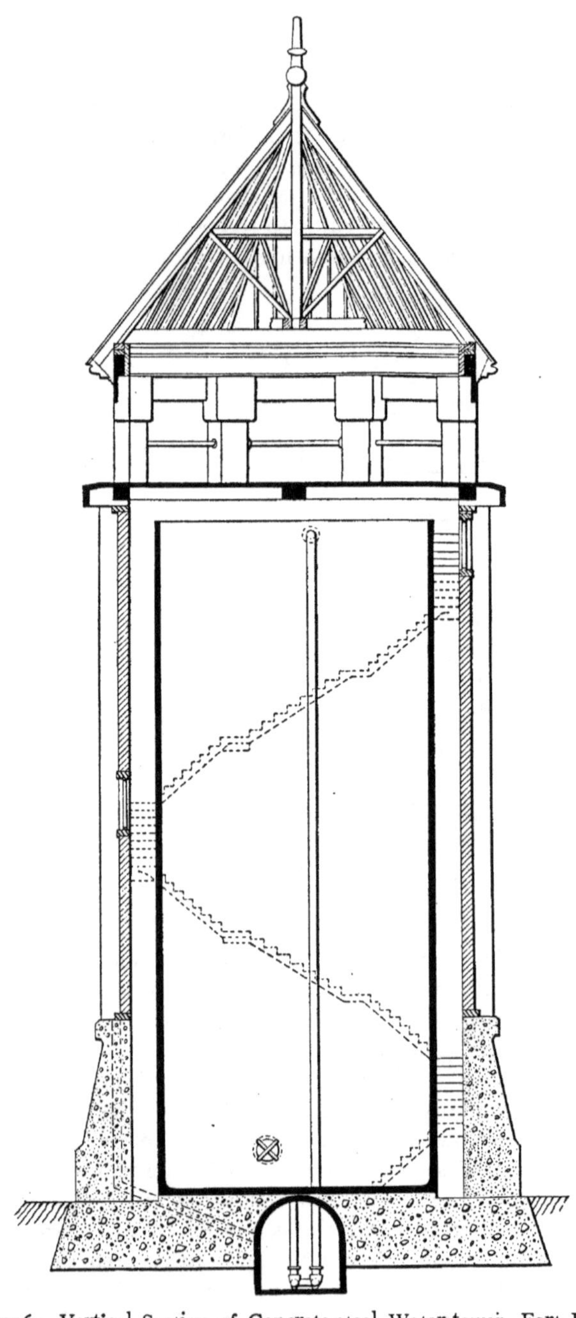

Fig. 196.—Vertical Section of Concrete-steel Water-tower, Fort Revere, Massachusetts.

and pier construction are clearly shown by the horizontal and vertical sections, Figs. 197 and 198.

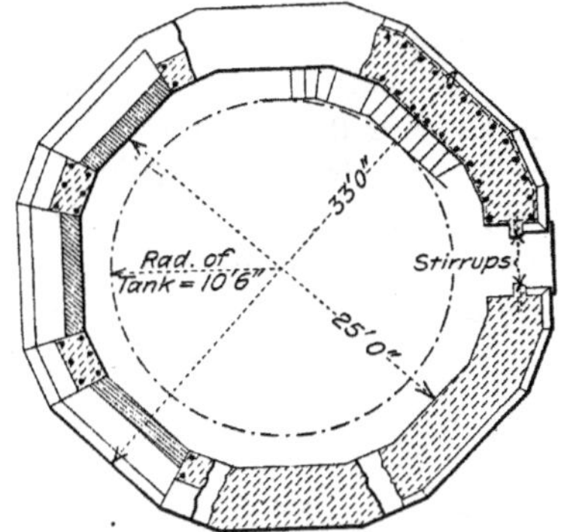

FIG. 197.—Horizontal Section of Base of Fort Revere Water-tower

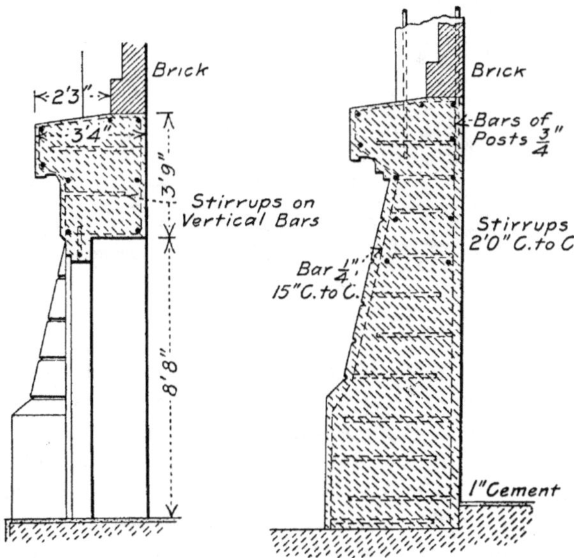

FIG. 198.—Vertical Sections of Base of Fort Revere Water-tower.

The tank or stand-pipe inside the tower has a diameter of 20 ft., leaving an annular space between the tank and inside wall which is occupied by a spiral stairway leading to the observatory. The tank is 50 ft. high with a shell $7\frac{1}{2}$ ins. thick at the bottom and $4\frac{1}{2}$ ins. thick at the top. The bottom of the tank is 4 ins. thick. The method of reinforcing the shell and bottom is shown by Figs. 199 and 200. Considering the shell first, the reinforcement will be seen to consist of two sets of vertical rods set 2 ins. apart transversely of the wall and of two sets of horizontal hoops, one encircling each set of verticals. The bars in each system of verticals are spaced 16 ins. apart, but the bars of one set are staggered with those of the other set so that the shell has a vertical every 8 ins. circumferentially. The horiontal hoops have welded joints throughout the first two-thirds of the height and wire-wound lap-joints throughout the top one-third of the height. The welded hoops are all ½-in. rods and the others are ⅜-in. rods. The vertical spacing of the hoops increases as the height of the shell increases.

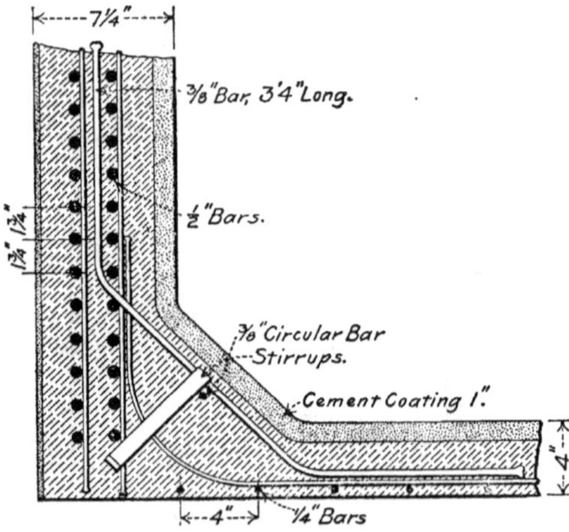

Fig. 199.—Section of Tank Side and Bottom, Fort Revere Water-tower.

For the ½-in. hoops there are 23 spaces of 1¾ ins.; 41 spaces of 2 ins.; 34 spaces of 2½ ins.; 22 spaces of 3 ins.; 13 spaces of 3½ ins., and 23 spaces of 3¾ ins. For the ⅜-in. hoops there are nine 3-in. spaces, six 3½-in. spaces, and six 3¾-in. spaces, the inner and outer hoops at each level up to this elevation being in the same horizontal plane. For the remaining 16 ft. the two sets of hoops are staggered and the spacing varies from 2 ins. to 7½ ins. One of the notable features of the construction is the bonding together by the reinforcement of the bottom and the shell of the tank. This is clearly shown by Figs. 199 and 200. The tank is lined throughout with 1 in. of 1-to-1 mortar to prevent leakage. This tank was designed and built by the American agents of the Hennebique system of construction.

Rectangular Tanks.—The drawings of Fig. 201 show the construction of a set of two rectangular tanks constructed at the factory of Zinsser & Company of New York City. These tanks were designed to hold a solution of chloride of lime and were reinforced with electrically

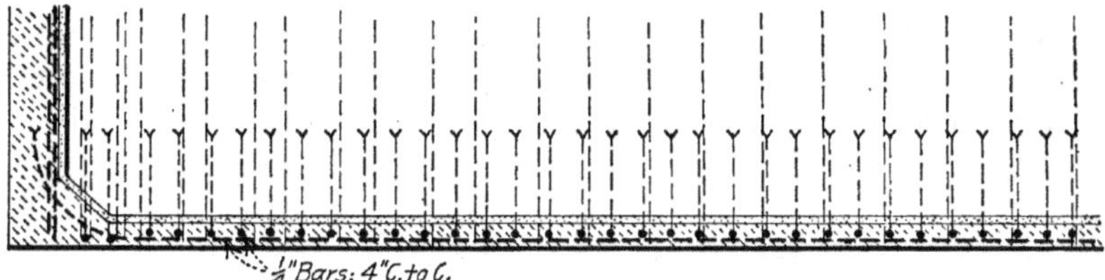

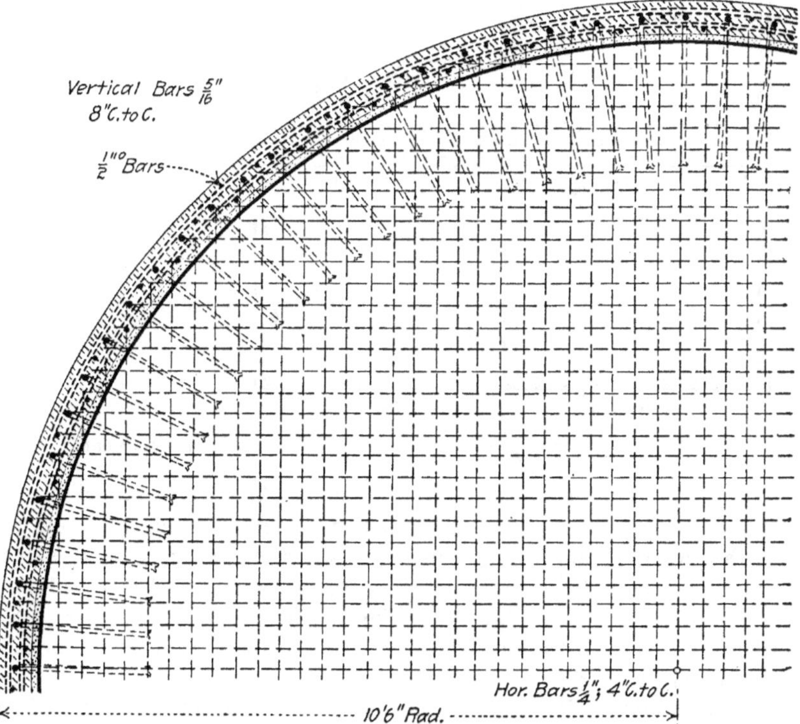

FIG. 200.—Reinforcement of Tank Bottom, Fort Revere Water-tower.

welded wire fabric having a 3×8-in. mesh and made of 0.3-in.-diameter wire. The concrete was a 1–2–4 mixture of trap-rock broken to $\frac{3}{4}$-in. maximum size. The construction is very simple and shows clearly the adaptability of reinforced concrete to small-tank construction.

Grain-elevator Bins.—Reinforced concrete has been employed in Europe for constructing grain-elevator bins for about twenty years, and some of the European grain-elevators of this material are of considerable size and importance even when judged by American standards. In the great majority of these European elevators, the bins have been made rectangular and are supported on columns of concrete rising

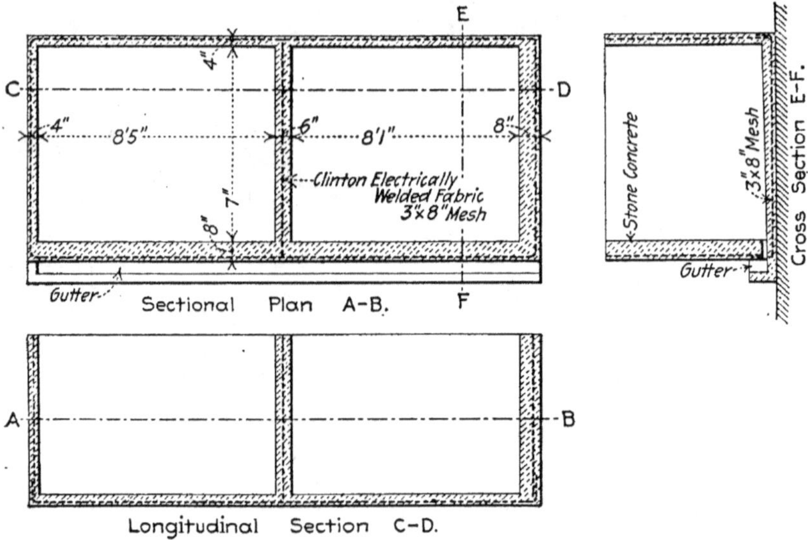

FIG 201.—Rectangular Tanks with Wire Fabric Reinforcement.

only to the bottoms of the bins. The few American grain-elevators which have been designed of reinforced concrete include examples of both cylindrical and rectangular bin construction, and in some cases the bins have been built resting directly on the ground without column supports of any sort. The following examples are fairly representative of the ideas of American designers so far as they have been developed in actual practice.

Canadian Pacific Elevator, Port Arthur, Ont.—The Canadian Pacific elevator at Port Arthur, Ont., consists of nine circular bins 30 ft. in diameter and 90 ft. high, arranged in three rows of three bins each, with the adjacent sides in contact. The quadrilateral spaces between the circular bins were made to form a second set of four bins. The shells of the main tanks extend below the surface of the ground and rest on spread footings. Above ground the shells are 9 ins. thick except where the adjacent tanks touch, and here the acute angles between the convex surfaces are filled in for a width of 7 ft. and a maximum thickness of 2½ ft. The shell reinforcement consists of circular hoops in pairs, a hoop near each surface. These pairs of bars are spaced 12 ins. apart vertically of the tank, but the size of bars used decreases from the bottom upward; for 15 ft. they have a cross-section of 1 sq. in.; for 35 ft. a cross-section of 0.88 sq. in.; for 20 ft. a cross-section of 0.75 sq. in., and for 10 ft. a cross-section of ½ sq. in. In addition to the hoops there are 27 vertical bars spaced equidistant around the shell; these bars are all ½ in. in diameter. Where adjacent tanks touch and have the walls thickened as described above, the two shells are clamped together across the thickening by two 1¼ ×15/16-in. straps every foot in height, whose ends hook over the hoop bars of the two shells. There is also in this thickened portion a horizontal sheet of wire netting at every foot in height. The hopper bottoms of the tanks are not reinforced, and they rest directly on a prepared earth-fill.

The concrete walls of the bins were made in movable cylindrical forms (Fig. 202) 4 feet high. The curved surfaces of the forms were

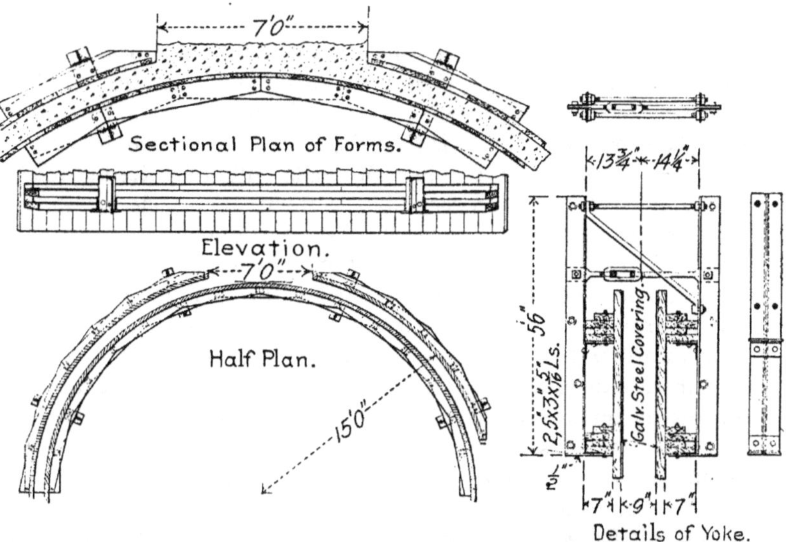

Fig. 202.—Forms for Circular Bin Construction, Grain Elevator, Port Arthur, Ont.

made of 2-in. vertical planks spiked to inside and outside circular horizontal chords. The chords were made like ordinary arch centers with four thicknesses of 2×8-in. scarf-planks bolted together to break joints and make complete circles inside the tanks and circular segments of 270° or less on the outsides of the tanks. The molds were faced on the inside with No. 28 galvanized steel, and were maintained in concentric positions with a fixed distance between them by means of eight U-shaped steel yokes in radial planes. Each yoke consisted of an inside and outside vertical post with a radial web and flanges engaging the inner and outer faces of the circular chords. The posts projected about 2 ft. above the tops of the molds and were rigidly connected there by means of heavy braces and an adjustable tension-rod. These yokes were bolted to the inner and outer chords of the molds and virtually united them in a single structure. The lower ends of the vertical yoke-posts were seated on jack-screws which were supported on falsework built up inside the tanks as the walls progressed. The inside surface of the mold was a complete cylinder, but the outer surface was made in two, three, or four sections to allow for the connections between the concrete walls at the points of tangency of the different tanks.

Grain Elevator, Montreal, P. Q.—The drawings of Fig. 203 show the bin construction designed by Mr. J. A. Jamieson for elevator construction at Montreal Harbor. As will be seen, the bins are rectangular and are carried on concrete-filled steel columns which extend through their full height and continue above as a portion of the frame of the house. In the design shown the outside bin walls and the bin bottoms are reinforced concrete, but the interior bin walls are a steel trough-plate construction. It is possible, however, if desired, to replace the steel construction of the interior walls with reinforced-concrete construction. The concrete-steel bin bottoms are so clearly shown by the drawings that they need no further description.

Bins, Narragansett Bay Coal Depot.—An example of the use of reinforced concrete in large bin construction is shown by Fig. 204. This is a half-transverse section of one of the large coal-storage bins erected at Narrangansett Bay for the U. S. Navy Department. There were two of these structures, each 725 ft. long and divided into five compartments by transverse partitions. As will be seen, only the bottom and sides of the bins proper and the dividing partitions are of concrete, the remainder of the structure being a framework of structural steel. The reinforcement was electrically welded wire fabric.

Reservoir Roofs.—In many cases the use of reinforced concrete in reservoir construction has been limited to the roof or covering. For this purpose practically any of the systems employed for arch and floor construction are available. The following examples from current practice serve to illustrate this particular application of concrete steel.

Water-works, Louisville, Ky.—Fig. 205 shows the concrete-steel

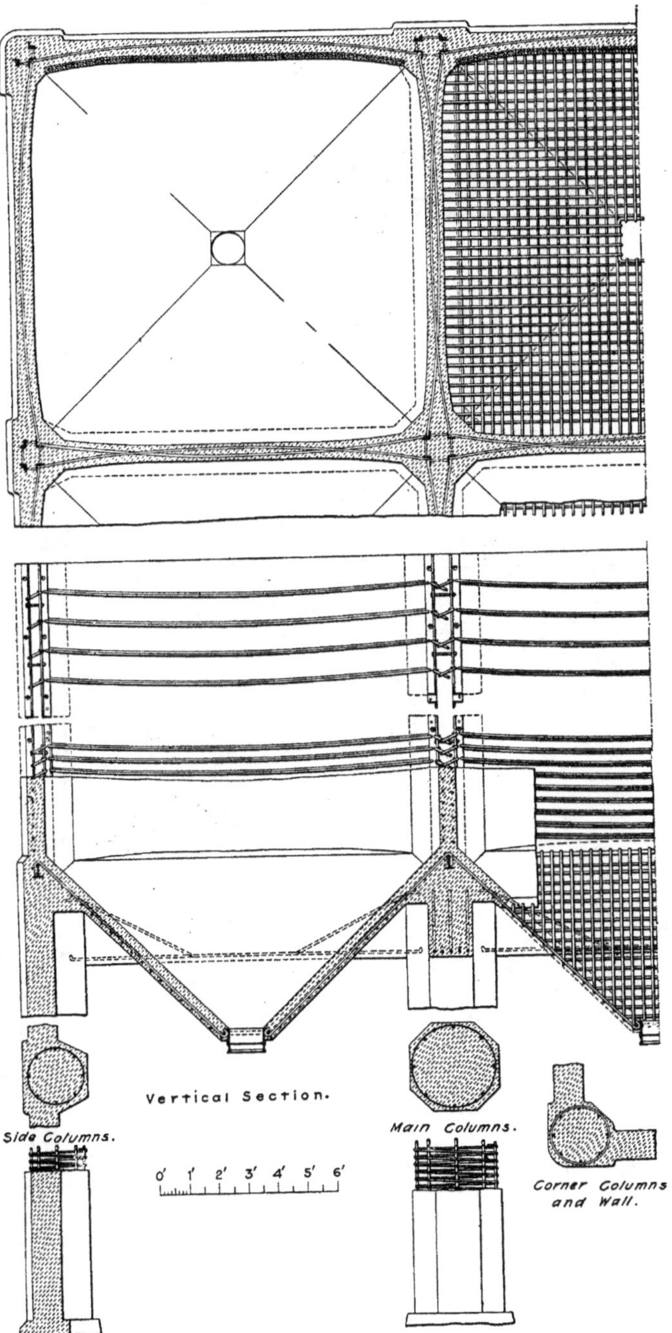

FIG. 203.—Grain Elevator Bins for Montreal Harbor, Canada.

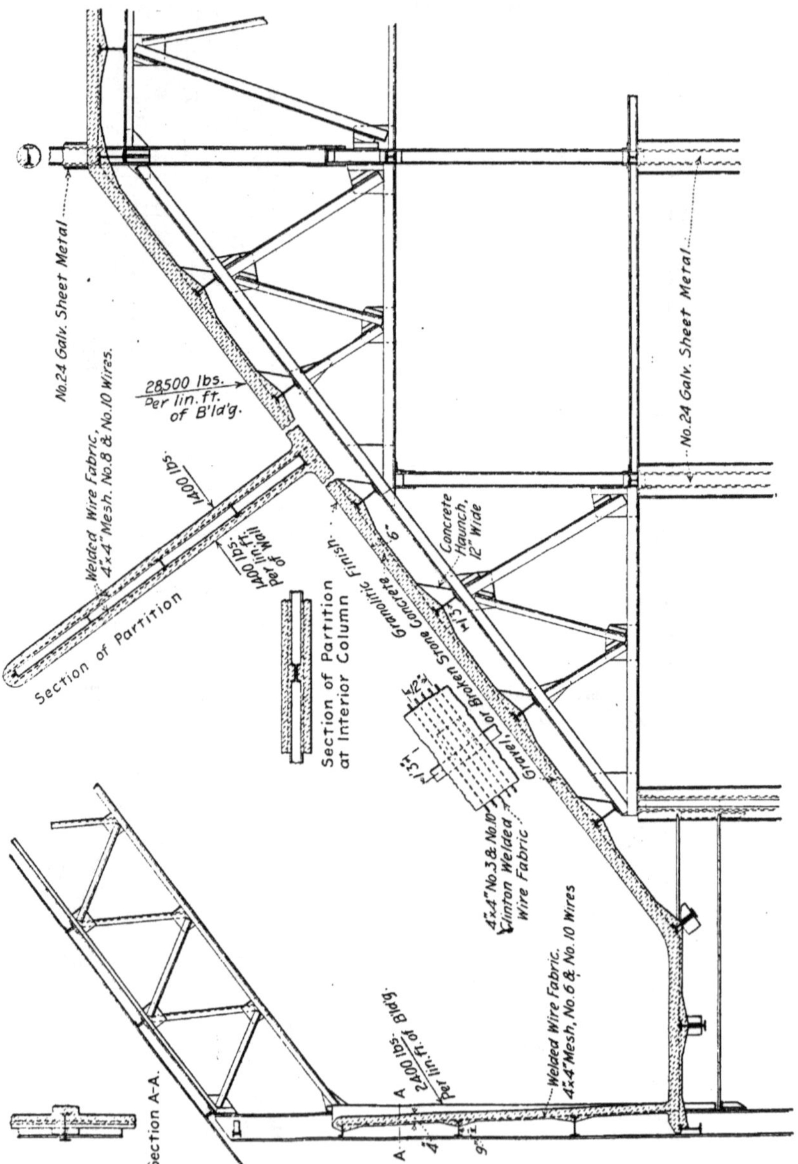

FIG. 204.—Coal Pocket of Reinforced Concrete Carried by Steel Framework.

groined-arch roof construction employed in building a 25,000,000-gallon covered clear-water reservoir at Louisville, Ky. The reservoir is nearly rectangular in shape, 392 and 394×460 ft., and is divided into four nearly equal compartments by three division-walls. The exterior walls, the division-walls, and the columns supporting the groined roof are constructed of unreinforced concrete. The columns are spaced

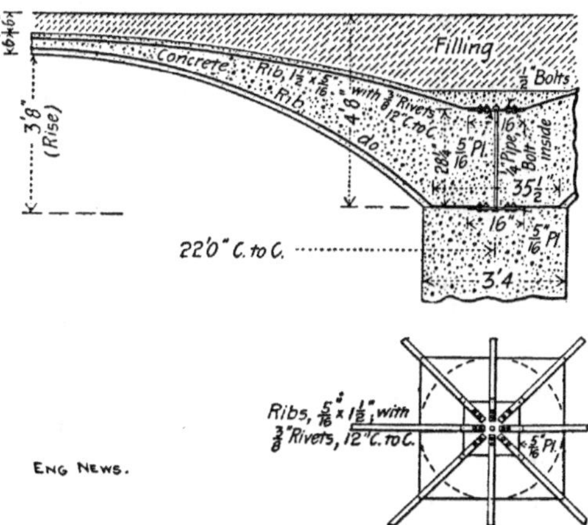

FIG. 205—Groined-arch Roof of Reinforced Concrete for Clear Water Reservoir, Louisville, Ky.

22 ft. apart center to center, and are about 3.4 ft. in diameter and 21.11 ft. high.

The arches are groined concrete-steel arches of approximately 19 ft. span and 3.8 ft. rise. The intrados and extrados in these arches are arcs of circles of 13.775 ft. and 32.256 ft. radius respectively, the concrete being 6 ins. thick at the crown and 36 ins. at the spring-line in radial lines, and are constructed of Portland-cement 1–2–4 concrete. These arches have built and embedded within them concrete-steel ribs 1½×5/16 in. in section, placed in eight half-pairs, resting upon each pier, four being placed at the groins of the arch and four midway between the apices of the groins. These ribs meet upon and are riveted to steel plates 16 ins. square and 5/16 in. thick (two plates in each pier, and in the side division-walls where the ribs terminate) placed 28¼ ins. apart vertically, and stayed in the middle by one ½-in. rivet passing through a short section of ½-in. iron pipe to stay the plates on. The two pairs of steel ribs intersecting each other at the apex of each groined arch are fastened together in a similar manner by a ¼-in. rivet and a piece of a ¼-in. wrought-iron pipe as a stay. This arrangement results in connecting the steel ribs of every groined arch with those of every other through the 16-in. square plates built in the top of the piers. In placing the steel ribs and laying the concrete, care was taken to embed every portion of each rib solidly in the concrete, as also the plates to which the ribs were attached. At intervals of about 12 ins. in the length of each rib there is a $\frac{3}{8}$-in. rivet through the middle of the rib, with head on one side and burr on the other, as a means of securing an immovable connection between the concrete and the steel ribs.

In constructing the groined-arch roof of the clear-water reservoir for the Louisville, Ky., filter plant, the first work was the erection of the centers. After the concrete in the pillars had thoroughly set four vertical wooden posts were set close to each pillar and connected by a horizontal rectangular frame cap at their tops. The caps carried the ends of the centers. The center for each groined arch was made up of four parts each triangular in plan. The bases of the triangles formed the rectangle connecting the pillars, and their apices met at the center of the arch and were supported then on a 6×6-in. vertical post resting on the floor of the reservoir. This construction of the centering is shown by Fig. 205. Generally the centers were erected for a group of six to nine arches. The next step was to end the reinforcing-ribs for these arches, and they were placed and the connections riveted and were blocked up by wooden spacers. The concrete was wheeled to the work in barrows and tamped in place working from the springing-lines toward the crowns. Concreting was carried on without intermission until each group or set of arches was completed. The blocking and spacers were removed as the concreting advanced. Water-pipes laid along the work and provided with pet-cocks kept the concrete covered with a spray of water until it was thoroughly set.

Rockford, Ill.—Fig. 206 is a transverse section of a covered reservoir with a concrete-steel roof which was constructed at Rockford, Ill., in 1894. This reservoir is rectangular and 66.26 ft.×156.56 ft. in plan and was constructed of concrete throughout except for a brick facing for the upper portion of the sides. Only the roof was of reinforced concrete. This roof is a ribbed arch which spans the reservoir transversely. The ribs are spaced 7 ft. apart and increase in depth from crown to haunch; each is reinforced by an intradosal channel laid face downward. The arch ring proper is of uniform thickness, 2 ins., and is reinforced by expanded metal placed near its soffit. Fig. 206 gives details showing the arrangement of the rib and arch-ring reinforcement, and the other structural details. The roof was erected on centers arranged as shown by the dotted lines in Fig. 206. The concrete used was composed of 1 part Portland cement, 2 parts sand, and 5 parts fine gravel. The soffits of the ribs and arch ring were plastered with a 1 Portland cement $2\frac{1}{2}$ sand mortar. The total

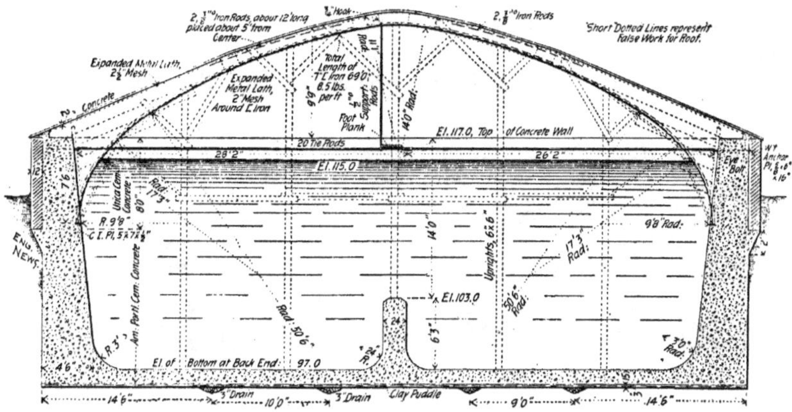

FIG. 206.—Arch Roof for Reservoir at Rockford, Ill.

cost of the reservoir was $18,506, and the cost of the roof alone was $2,000.

Newton, Mass.—In constructing a 127×165×50-ft. covered reservoir at Newton, Mass., in 1902, a roof of concrete-steel of the construction shown by Fig. 207 was employed. The walls of this reservoir are

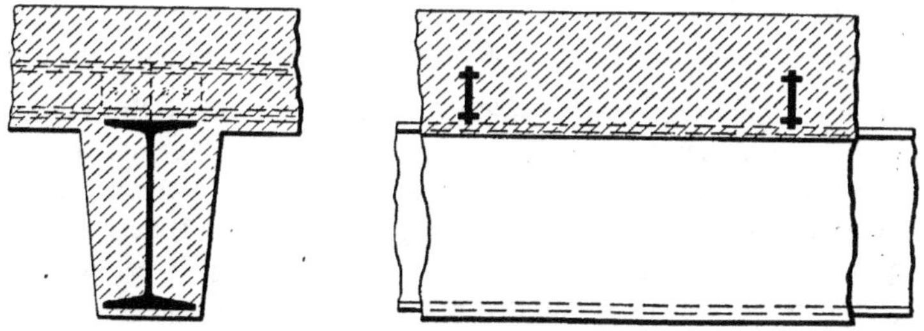

FIG. 207.—Reservoir Roof at Newton, Mass.

of rubble masonry and the floor is of 4 ins. of concrete. Brickwork pillars were arranged in rows 11 ft. $10\frac{7}{8}$ ins. apart one way and 11 ft. 8 ins. the other way. Steel I beams 12 ins. deep and weighing $31\frac{1}{2}$ lbs. per foot are carried by the pillars. These beams run in the direction of the 11 ft. 8 in. span, and each is long enough to cover two spans. As shown by the drawing they were haunched around by concrete and carry a concrete plate reinforced by Columbian bars spliced at adjoining ends. The concrete used was a 1:2:5 Portland-cement, broken-stone concrete. The contractor's price for the covering was 29 cents per sq. yd.

Liverpool, England.—In constructing the reservoir for the new Hoylake and West Kirby water-works near Liverpool, England, the engineers adopted the reinforced-concrete roof construction shown by Fig. 208. This reservoir is trapezoidal in plan, about 155 ft. long,

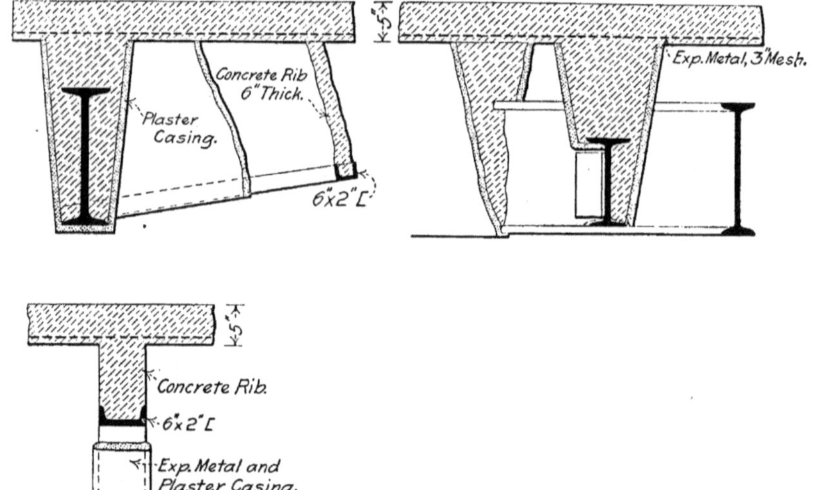

FIG. 208.—Details of Concrete-steel Reservoir Roof, Liverpool, England.

27 ft. deep, and 116 ft. wide midway of the two ends. Steel columns embedded in concrete were set on a concrete floor 12 ins. thick and spaced 16½ ft. apart longitudinally and 15 ft. apart transversely to carry the roof. These columns carry longitudinal I beams 18 ins. deep, dividing the roof into transverse bays of 16½ ft. span. These long transverse bays are subdivided into 7½-ft. panels by secondary girders carried at the ends by the 18-in. I beams. About one-quarter of these girders are 12-in. I beams and the remaining three-quarters are arched ribs of concrete. These main and secondary joists carry a concrete roof-slab reinforced with expanded metal. The drawing shows the method of embedding the main and secondary I-beam girders. Their lower flanges were wrapped with expanded metal, after which they were haunched around with concrete by means of suitable forms. A plastering of concrete was then applied to the whole surface of the ribs. The construction of the concrete secondary girders

is the familiar Golding type of ribbed-plate construction. The channels were bent to a curve having a rise of 1 in. per foot and had their ends carefully fitted to the 18-in. I-beam girders. The concrete for these ribs was composed of 1 part of cement to 3 parts of aggregates. The roof-slab is continuous and uniformly 5 ins. thick, reinforced at the bottom with sheets of expanded metal of No. 10 gage and 3-in. mesh. The concrete was composed of 1 part cement and 4 parts aggregates. The aggregates were shore sand and $\frac{3}{4}$-in. broken stone. The concrete was mixed twice dry and twice wet, with not too much water. The roof was waterproofed with a bitumen sheeting and covered with 18 ins. of earth. It was designed to carry a uniform load of 336 lbs. per square foot with a factor of safety of four. In a paper describing this work the engineer, Mr. Alfred J. Jenkins, Assoc. M. Inst. C. E., gives the following formula for working load:

$$W=6\frac{t^2}{s^2}.$$

In this formula s = span in feet, t = thickness of slab in ins., and W = safe uniform load with a factor of safety of 4. The section of expanded metal required is given by the following rule: The sectional area of the strands should equal one-two-hundredth the sectional area of the concrete slab.

Waterproofing.—One of the requisites of tanks and vats for liquids is that they shall not leak. Leakage may be prevented either by using an impervious concrete or by waterproofing the walls and bottom. The possibilities for securing an impermeable concrete are discussed in Chapter XII. Except in the general respect of using a concrete rich in cement and employing a wet mixture, practice has not developed much experience with impermeable mixtures. The usual resort adopted has been some sort of waterproofing. For water-tanks European practice favors the use of an inside layer or plaster coat of rich cement and sand mortar. An example of this construction is furnished by the water-tower constructed at Fort Revere, Mass., and described in a preceding section of this chapter. An old formula for waterproofing concrete cisterns is to wash the inside with two or more coats of neat cement grout; this procedure seems to give very good results for heads not exceeding 10 ft., but for greater heads it does not suffice. A coating of asphalt is also frequently used for waterproofing tanks. The usual practice is to apply the asphalt hot with a mop until a coat about $\frac{1}{4}$ in. thick has been secured. The neatest solution of the problem of securing water-tight tanks is by all odds the use of an impermeable concrete, and setting aside mixtures involving the use of special chemicals, this is most nearly secured by means of a wet mixture rich in cement. This contruction, with the addition of a surfacing of cement mortar, is the one that is employed universally in Europe where reinforced concrete tanks and vats are in use in large numbers. It would appear to be best in all respects to build up the mortar surfacing and the concrete together after the manner described in Chapter XVII, but where this cannot be done the following mode of procedure may be adopted: Apply the mortar coating to the bottom and sides as soon as possible after the forms are removed and in the following manner: Carefully clean the surfaces with water and then apply a wash of neat cement grout; while the grout is still wet apply the first coat of plaster about ¼ in. thick; apply the succeeding coats at intervals of about one hour until the desired thickness of mortar has been secured.

CHAPTER XI.—EXAMPLES OF REINFORCED-CONCRETE RETAINING-WALLS, DAMS, CHIMNEYS, ETC.

THE examples of reinforced-concrete work grouped together in this chapter include such of the more important engineering works as do not belong to foundations, buildings, bridges, conduits, and tanks and bins. No attempt has been made to present an exhaustive list of special structures which have been built of concrete-steel, as this would consume a large amount of space without giving adequate return in the shape of useful information. The examples chosen are all such as are likely to come up for solution in general engineering practice.

CHIMNEYS.

The use of reinforced concrete in chimney construction is illustrated by a number of important structures in the United States. The design of these structures has followed closely that of brick chimneys, an inner and an outer shell separated by an annular air-space being employed. The following examples show this construction in detail.

Pacific Electric Ry., Los Angeles, Cal.—The chimney built during July and August, 1902, for the new Los Angeles power-house of the Pacific Electric Railway Company is 180 ft. high above its base, which is 15 ft. 6 ins. below the level of the ground. The exterior diameter of the chimney above the shoulder, where it first assumes a circular form, is 15 ft. 2 ins., the inner diameter being 11 ft. The shoulder is 51 ft. above the base, and is immediately above the two flues which enter the chimney from opposite sides.

The accompanying plans (Fig. 209) show clearly the construction of the stack. It consists of two concentric walls, independent of each other from base to top, and separated by an air-space of from 11 to 16 ins., increasing in width toward the top. The outer shell, above the shoulder, is 9 ins., 6 ins., and 5 ins. thick, respectively, up to the cap, in sections of about equal height, while the inner shell is 5 ins., 4½ ins., and 4 ins. thick, respectively, from bottom to top, in corresponding sections.

The inner shell ends 4 ft. below the cap and is free to elongate by heat independently of the outer shell. By reference to the sections it will be seen that at intervals of 30 ins., measured around the chimney, the air-space is contracted for a length of 6 ins., and reduced to a width of 2¼ ins. At every 5 ft. in height this is again reduced to ¾ in. by the introduction of a concrete brick in the wall. In this manner the oscillation of either shell independently of the other is checked, and the outer shell may sway in the wind ¾ in. without bringing pressure upon the inner shell.

The chimney was built with a reinforcement of square, cold-twisted steel bars, placed vertically and horizontally in each shell. The horizontal reinforcement consists of rings of ¼-in. bars, placed at intervals averaging 18 ins. in the inner shell and 24 ins. in the outer. Vertical bars of ¾ in. were placed 1 ft. apart in the lower third of the stack above the flues, 2 ft. apart in the middle section, and 4 ft. apart in the top section of the outer shell. In the inner shell ¼-in. bars were used, spaced about 3 ft. apart in the circumference of the work.

The upper 4 feet above the ornamental cap consists of a shell of concrete but 2 ins. thick, reinforced by a sheet of expanded metal. The decorative cap, 7 ft. high, consists of 28 blocks molded on the ground and hoisted to position. As shown by the drawings, these blocks are hollow and are formed by a shell of 2 ins. of concrete, with stiffening cross-partitions, all reinforced by ¼-in. steel rods, embedded in the concrete. Each block of the cap weighs 1,250 lbs., and the entire cap was hoisted and set in three days.

The original design of the chimney contemplated a straight shaft, 15 ft. in diameter throughout from the base up. Subsequent alterations in the plans of the boiler-house changing the relative positions of boilers and chimney, necessitated the provision of two flue-openings on opposite sides, instead of one. The extra opening thus provided would have materially reduced the section area of the chimney at this point, and it was deemed necessary to increase the diameter to 18 ft. for a height of 45 ft. above the base.

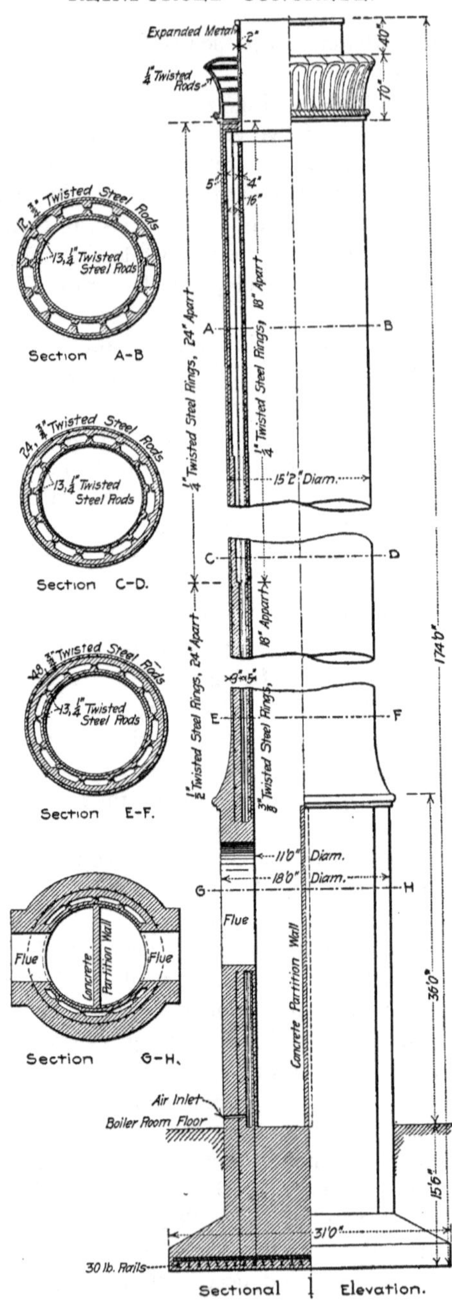

FIG. 209.—Reinforced Concrete Chimney for Electric Railway Power-house, Los Angeles, Cal.

The concrete for the outer shell consisted of 1 part Portland cement, 2 parts sand, and 6 parts crushed granite, and that for the inner shell consisted of 1 part Portland cement, 2 parts sand, and 4 parts broken sandstone. The chimney contains approximately 20,000 cu. ft. of concrete, and 850 barrels of cement were used in its construction. The steel embedded in the concrete consists of 10,000 lbs. of twisted bars, and 4,000 lbs. of old rails, which were placed 12 ins. apart, in two layers in the base. The weight of the chimney is approximately 1,430 tons, and the distributed load on its base is less than 2 tons per square foot. The foundation is laid in a heavy river-gravel stratum.

Central Lard Co., Jersey City, N. J.—In 1901 a concrete-steel chimney 108 ft. high was built at the works of the Central Lard Company in Jersey City, N. J. The general construction of this chimney was substantially the same as that of the Pacific Electric Railway Company, and only its main dimensions and a few of the special details will be given. The chimney is 108 ft. high above the top of the foundation and has a uniform diameter of 11 ft. 4 ins. The flue is 8 ft. in diameter. The shell is double from a point 5 ft. above the foundation, the inner shell having a uniform thickness of 4 ins. and the outer shell varying in thickness from 7 ins. at the bottom to 4 ins. at the top, the stepping-off in thickness being done on the inside. The outside of the inner shell and the inside of the outer shell have opposite vertical ribs or buttresses like the Los Angeles chimney. The reinforcement of the shells consisted of circumferential rings and vertical bars of twisted square steel. The foundation construction is clearly shown by Fig. 210 and differs from that for the Los Angeles chimney in using twisted rods for the reinforcement instead of steel rails. The maximum stress in the concrete was 350 lbs. per square inch, compression. The total amount of concrete used in the chimney was 4,974 cu. ft., of which 1,460 cu. ft. were in the foundation and 3,514 cu. ft. were in the stack proper. From these figures the total weight of the chimney, at 144 lbs. per cubic foot for the concrete, is found to be 362 tons. The total cost of the chimney was $3,500. The plans for the chimney were made by the Ransome Concrete Company of New York City.

Pacific Borax Co., Constable Hook, N. J.—The chimney of Ransome concrete-steel built for the Pacific Borax Company of Constable Hook, N. J., in 1898 is chiefly interesting because of the firebrick lining used for the hottest zone and the fact that the coping is in one piece with the outer shell, while the two shells are rigidly connected by a solid ring at the top of the chimney.

La Clede Fire-brick Co., St. Louis, Mo.—The form of reinforcement designed by Mr. Carl Weber of St. Louis, Mo., consists of

T bars connected at intersections by means of special sheet-metal clamps, and one of its most important applications has been in the

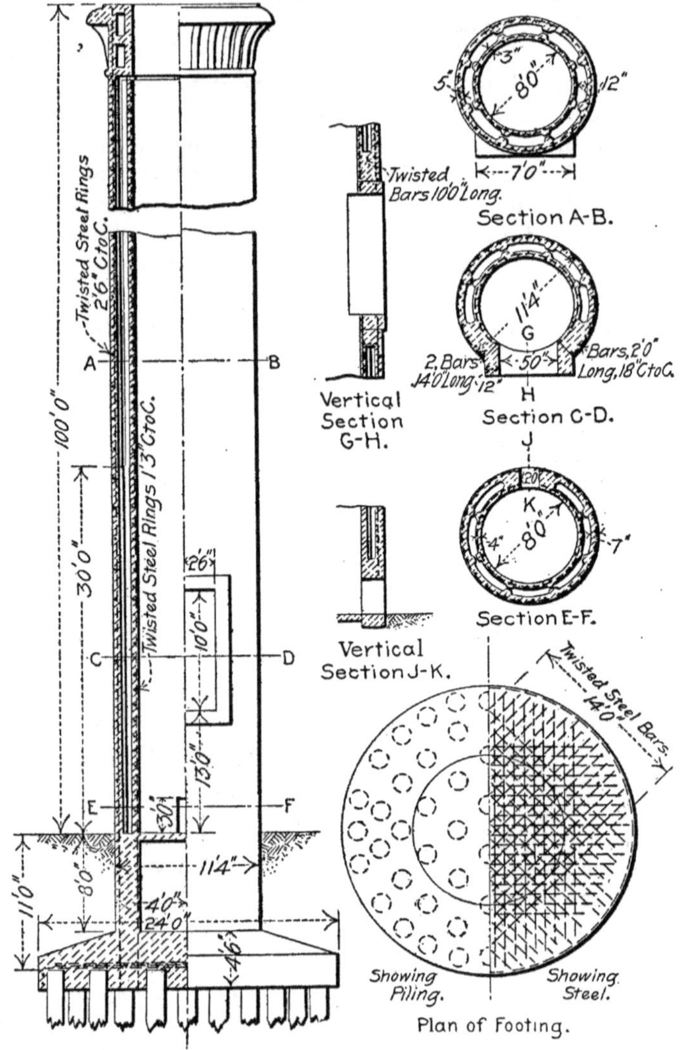

FIG. 210.—Reinforced Concrete Chimney for Central Lard Company, Jersey City, N. J.

construction of a concrete-steel chimney built for the La Clede Fire-brick Company of St. Louis, Mo. This chimney is 130 ft. high and has an inside diameter of 5 ft. The materials used in its construction are river sand and Portland cement reinforced with steel T bars $1\frac{1}{4} \times 1\frac{1}{4} \times \frac{1}{2}$ in. Up to a height of 65 ft. from the base the stack consists of two shells, the outside shell being 6 ins. thick and the inside shell 4 ins. thick, with an intervening air-space of 3 ins. In the concrete mass of the outer shell are 20 vertical steel T bars, 2 ft. apart, running from the foot of foundation to the top of the chimney, while 10 steel bars 1 ft.$\times$1 in.$\times\frac{1}{8}$ in. are in the same manner used for strengthening the inner shell. Every $2\frac{1}{2}$ ft. a horizontal ring of the same material encircles the vertical bars, being connected to them by steel clamps. The concrete-steel base, on which the stack rests, 20 ft. below ground-level, is 5 ft. deep and 16 ft. square, and is built on solid rock. Above the height of 65 ft. there is only a single shell, the thickness of which tapers off in proper intervals to 5 ins., 4 ins., and, finally, 3 ins. The air-space, directly above the grade, is connected by four square openings with the outside atmosphere, allowing the air to enter, which at the upper terminal point of the inner shell will orce itself through specially provided inclined pipes into the shaft of the chimney proper.

The forms used in building this stack were made of wood, forming rings of $2\frac{1}{2}$ ft. in height and divided into six sections which were held together by iron hooks. In operation two such rings were used for the outside and the inside of the chimney, while properly curved 3-in. molds provided for forming the intervening air space.

The method of procedure was as follows: After one form was completely filled with concrete, properly tamped around the vertical bars, the second form was placed on top of the first, and likewise filled in with concrete. The hooks connecting the several sections of the lower forms were then opened, the single parts, which were previously secured with ropes, pulled up and again placed on top of the last form, which remained adhering, and so on, till the top of the chimney was reached, two rings a day being easily completed. A very light frame staging was provided on the inside of the chimney for attaching the ladders and supporting a pulley-beam, used to hoist material by hand. The weight of the whole chimney was about 120 tons without footing. Its outside surface was coated with a cement wash, to secure a uniform color.

Railway Ties.

The use of reinforced concrete cross-ties has been limited to short sections of railway track where they have been placed for experimental purposes. As yet the experience gained has been too limited to warrant the consideration of concrete-steel ties as an established improvement, but the success with them has, on the other hand, been sufficient to hold out much promise to railway engineers that they may ultimately develop into a satisfactiory substitute for wood and steel. The ties illustrated are among those which have given the most satisfactory results.

Kimball Cross-tie.—The tie illustrated by Fig. 211 is one designed

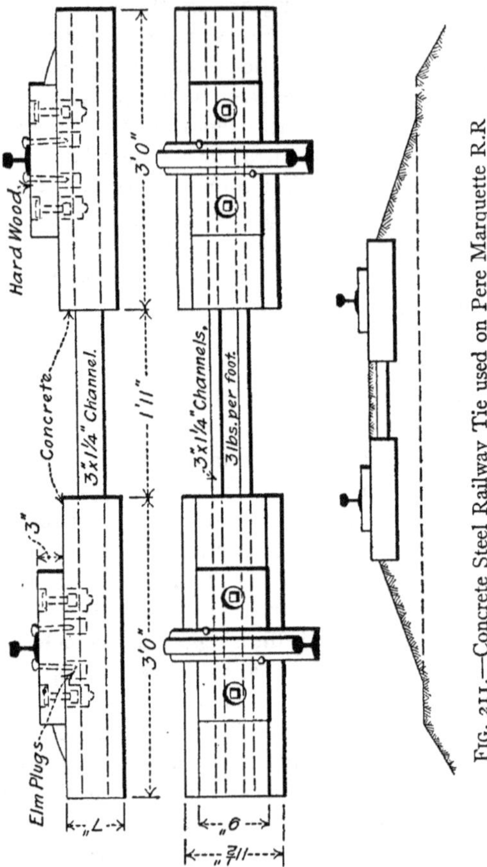

Fig. 211.—Concrete Steel Railway Tie used on Pere Marquette R.R

by Mr. A. H. Kimball, M. Am. Soc. C. E., and now being tested on a section of the Pere Marquette Ry. It consists of two rectangular blocks of concrete, each 3 ft. long and 7×9 ins. in cross-section, rigidly connected by two steel channels which extend through both blocks from end to end and bridge the 1 ft. 8 in. space between them. These channels are placed back to back and 2 in. apart. Hard-wood blocks 3 ins. thick by 9 ins. wide and 18 ins. long, designed to cushion shocks, distribute pressure, and serve as spiking-blocks, are secured to the top of the concrete blocks and centered with regard to the rail. Cast-iron sockets, that also serve to space the channels, are molded in place in the concrete and serve as anchors for holding down the wood blocks. These sockets receive suitable bolts, head down, which extend up through holes in the blocks and receive nuts which are screwed down into concrete bores in the top of the block and sealed so as to be water-tight. Elm plugs set on end of grain are embedded in the concrete to receive the ends of the spikes, and the blocks are bored for the same purpose. The exposed portions of the channels between blocks are painted with a neat cement grout. The concrete is made either with gravel or with broken stone, both of 1½ in. maximum size; the proportions of the mixture are as follows:

	Broken Stone.	Gravel.
Portland cement, parts....	6	6
Sand, parts...............	3	..
Aggregate, parts.........	11	14

The estimated cost and weight of each tie is as follows:

Item.	Cost.	Weight, Lbs.
Iron and steel...........	$0.58	52
Concrete................	0.50	374
Wood blocks............	0.12	10
Royalty.................	0.10	
Total..............	$1.30	436

Burbank Cross-tie.—The drawings of Fig. 212 illustrate a concrete-steel tie designed by Mr. Burbank and used on the Hecla Belt Line Railway in Bay City, Mich., since May, 1903. In this tie the concrete is molded around two metallic members. These are flat bars, the uppermost of which is twisted. Wood blocks are partly embedded in the concrete underneath the rails. The ends of the top plate are carriel over these blocks and contain punched spike-holes. The end of the lower plate have four pins which serve to lock the plate into the concrete. The other details are clear from the drawings.

Adriatic Railway, Italy.—The drawing of Fig. 213 illustrates the reinforced concrete cross-tie which has been tested in a section of track near Ancona since 1900. The section is triangular except at the rail bearings, where it is rectangular. These ties weigh about 286 lbs. and cost from $2.16 to $2.40 apiece. The total section of the reinforcing-rods is about 3 sq. ins. The other details are shown by the drawings.

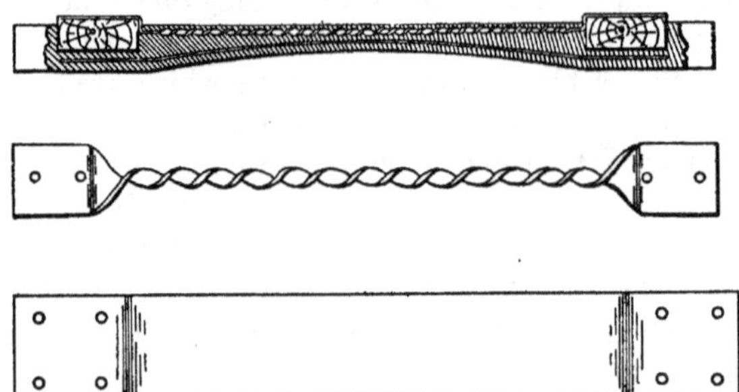

FIG. 212.—Concrete Steel Railway Tie, Hecla Belt Line Railway, Michigan.

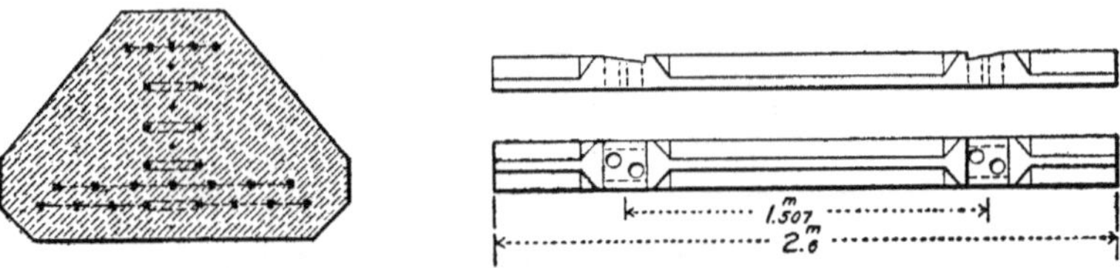

FIG. 213.—Reinforced Concrete Cross-tie, Adriatic Railway, Italy.

TUNNEL LININGS.

The use of concrete-steel for tunnel lining has been much less frequent than the use of unreinforced concrete. In most instances also where concrete-steel tunnel lining has been employed the reinforcement has been confined to the roof-arch. The Aspen Tunnel described in a succeeding paragraph is the only example in America of a concrete tunnel lining reinforced in both the side walls and the roof-arch, and here the primary idea of reinforcing all sides of the lining was due to the advantages which the heavy steel ribs afforded in other respects than as a reinforcing skeleton pure and simple. In Europe there has been practically no use of concrete-steel for lining tunnels of large section, although it has been highly developed for subway and conduit work. This is possibly due to the intricacy of the forms of reinforcement which have gained the greatest vogue in Europe; a wire network or rod and stirrup skeleton is hardly to be thought of under such conditions as, for example, confronted the builders of the Aspen Tunnel. In fact any form of reinforcement to be applicable for tunnel lining must have comparatively few members of rigid section which can be erected in large units. All things considered, the opportunities for securing any advantages over unreinforced concrete by using a concrete-steel lining are probably comparatively few in number.

Open-cut Street Tunnels.—In constructing parts of both the Boston Subway and the New York Rapid Transit Railway concrete tunnel sec-

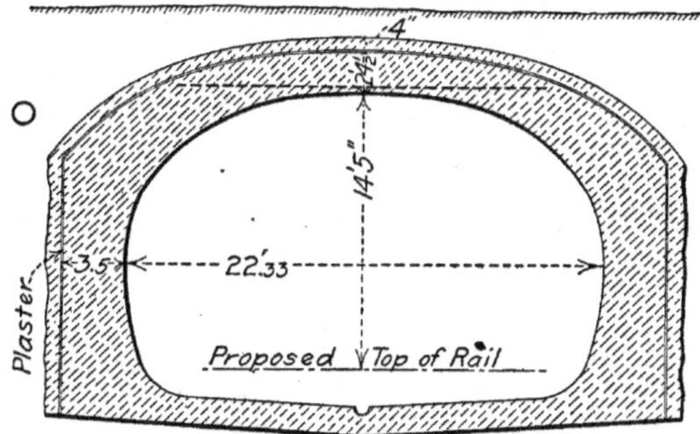

FIG. 214.—Section of Boston Subway with Rod Reinforcement in Roof.

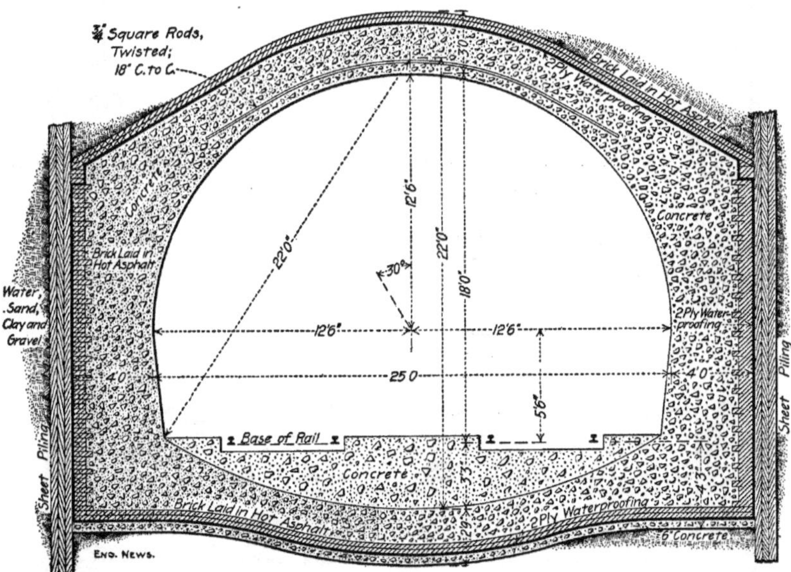

FIG. 215.—Section of Harlem River Tunnel, New York City, with Bar Reinforcement in Roof.

tions with roof reinforcements were employed. These sections were constructed in open cut. Fig. 214, showing a section of the Boston Subway crossing under Summer Street at a shallow depth requiring a flat roof, illustrates the method of roof reinforcement employed. It

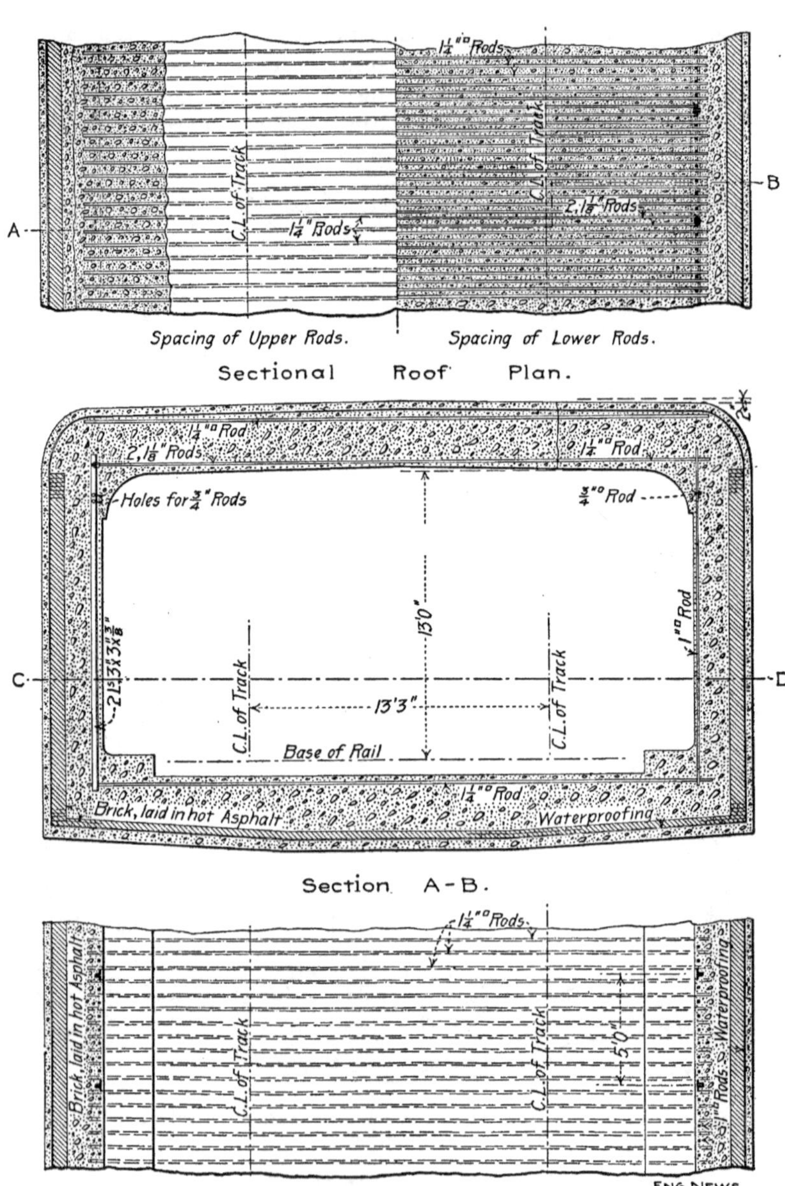

FIG. 216.—Section of Battery Park Loop of New York Rapid Transit Railway.

consists simply of horizontal bars or rods carried across the roof at intervals of from 4 ins. to 12 ins. The roof reinforcement adopted in the New York Rapid Transit Railway was somewhat different than that used in the similar work at Boston and is sufficiently illustrated by Fig. 215, showing a section of the land portions of the Harlem River Tunnel. From this section it will be observed that the reinforcement consists of rods bent to the curve of the arch and spaced 18 ins. apart and parallel to each other. A more typical example of reinforced concrete tunnel lining is shown by Fig. 216. This illustrates a section of the Battery Park loop of the New York Rapid Transit Railway.

Aspen Tunnel, Union Pacific R. R.—In constructing the Le Roy-Bear River cut-off on the Union Pacific R.R. in 1901 a tunnel 5,900 ft. long was carried through the Aspen Ridge near Aspen, Wyoming. This tunnel was excavated through rock and was designed to be lined with timber. For a portion of the tunnel, however, the rock was very unstable, and to withstand the great pressure developed it was necessary to line some 713 ft. with a stronger construction. For this construction a concrete-steel lining of the form and dimensions shown by Fig. 217 was adopted. The steel reinforcement consists of parallel T-beam ribs spaced from 12 ins. to 24 ins. apart according to the pressure. The T beams of which the ribs are made are 12 ins. deep and weigh 35 lbs. per foot. Each rib is composed of three segments connected by riveted fish-plates, and is footed on cast-iron shoes. The ribs are embedded in concrete which reached from 4 ins. to 6¾ ins. inside of the ribs and extends backward to the wall of the excavation, a distance varying from 2 ft. to 3 ft. The concrete foundation extends across the tunnel in a thick mass and has embedded in it a transverse reinforcement of old railway rails. All concrete was composed of 1 part Portland cement, 3 parts sand, and 6 parts broken

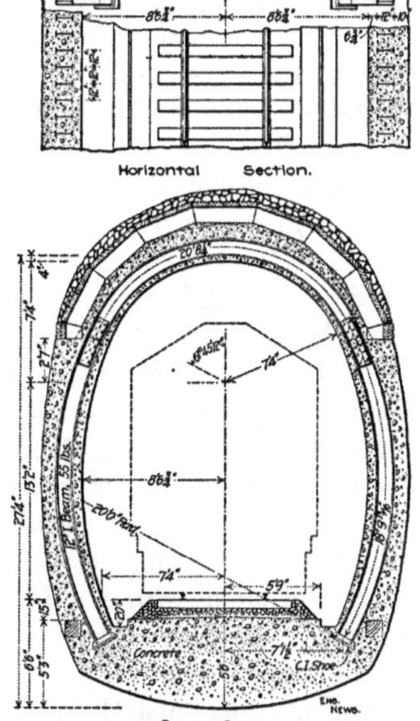

FIG. 217.—Section of Lining for Aspen Tunnel, Union Pacific R.R.

stone. The amount of concrete averaged about 8.75 cu. yds. per lineal foot of tunnel.

Tunnel Revetment.—In Europe concrete-steel has been employed in a number of instances in the revetment of tunnel linings and the waterproofing of aqueducts. Two forms of construction have been practiced in this work; in one the revetment is attached directly to the masonry to be protected, and in the other it is separated from the old structure by an air-space. The first construction is considered to be the most economical, but it has the disadvantage of suffering any distortion which occurs in the original masonry. By using the separate construction this danger of distortion and cracking is avoided, and any seepage through the old masonry is collected in the air-space and can easily be drained away. The cost of construction is, however, somewhat greater.

Belt Railway Tunnel, Vienna, Austria.—In 1901 the so-called Steudeltunnel on the Vienna Belt Railway was revetted with an interior lining of Monier concrete-steel. The original lining of the tunnel was of brick masonry which had become disintegrated in places and was otherwise in a damaged and leaky condition. The first work was to remove the disintergated brickwork and to fill these and all other voids with mortar and grout. A concrete arch reinforced by a Monier network was then built inside the old masonry. This arch was made 8 cm. (3.15 ins.) thick at the crown and 1.5 dcm. (5.9 ins.) thick at the springing-lines. The reinforcement consisted of 10 mm. and 7 mm. (0.4 in. and 0.3 in.) rods, and was fastened to the old brick lining by hooks. The reinforcement was embedded in a 1-to-3 Portland-cement mortar rammed close against the old masonry. The tunnel being double-tracked, only one-half of the arch was built from the springing-line to the crown at one time, thus leaving one track always open for traffic. The revetment had a superficial area of 309 sq. in. and cost 12,600 frs.

Acheres Sewer, Paris, France.—The Acheres sewer of Paris at one part of its length crosses a valley on masonry arches and is itself constructed of stone with a circular section. Owing to its exposed position the expansion and construction due to changes in temperature caused fissures and leakage, and to remedy the leakage a revetment of Bonna concrete-steel was applied. The interior of the masonry sewer was first rendered smooth and of true profile by a plaster coat of cement mortar. A tube of sheet lead was then laid down against this plastering and secured in place by a cylinder of rectangular network of cruciform bars—the standard Bonna pipe reinforcement. This reinforcement was then embedded in cement mortar, forming a shell 3 cm. (1.18 ins.) thick; the idea of this construction was that the reinforced mortar shell would provide a tight lining and that the layer of lead between the old masonry and the revetment would allow the former to expand and contract without causing any movement in the latter. The construction has, it is stated, proven perfectly successful for its purpose.

Borsang Tunnel, France.—Fig. 218 shows the method of reveting a roadway tunnel adopted by Mr. Hennebique in 1898. The tunnel lining was of stone masonry and had given trouble by leaking. At intervals of 1.5 m. (4. 92 ft.) arch-ribs, two bricks wide and 3.4 dcm. (13.3 ins.) deep, were built against the masonry lining, and to these ribs the concrete-steel revetment was attached. This revetment was

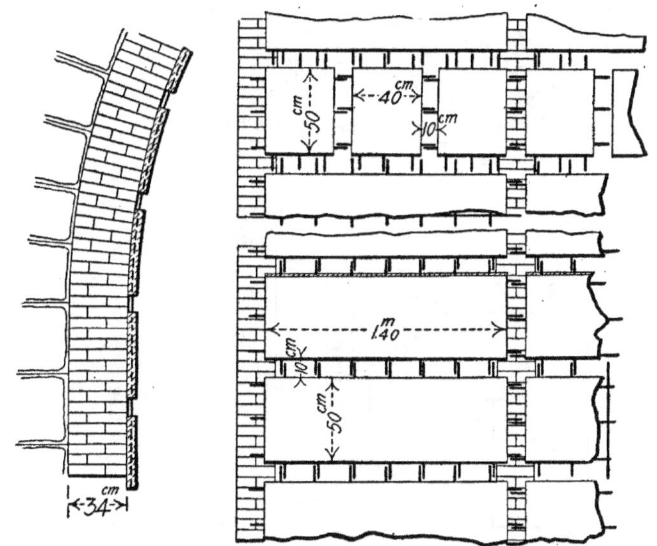

FIG. 218.—Revetment of Masonry-lined Tunnel with Reinforced Concrete.

made up of thin plates reinforced with a rectangular network of bars and cast separately ready for erection. The reinforcing-bars were of such length that they projected from the edges of the plates, which were laid as shown by the illustration with open joints and so that the projecting bars lapped by each other. After the plates were laid these open joints were filled with cement mortar. The plates were laid up by means of rib centers spaced 1.5 m. (4.92 ft.) apart. The construction described gave a cellular lining, from the cells of which the water leaking through the trussed lining was collected by drains.

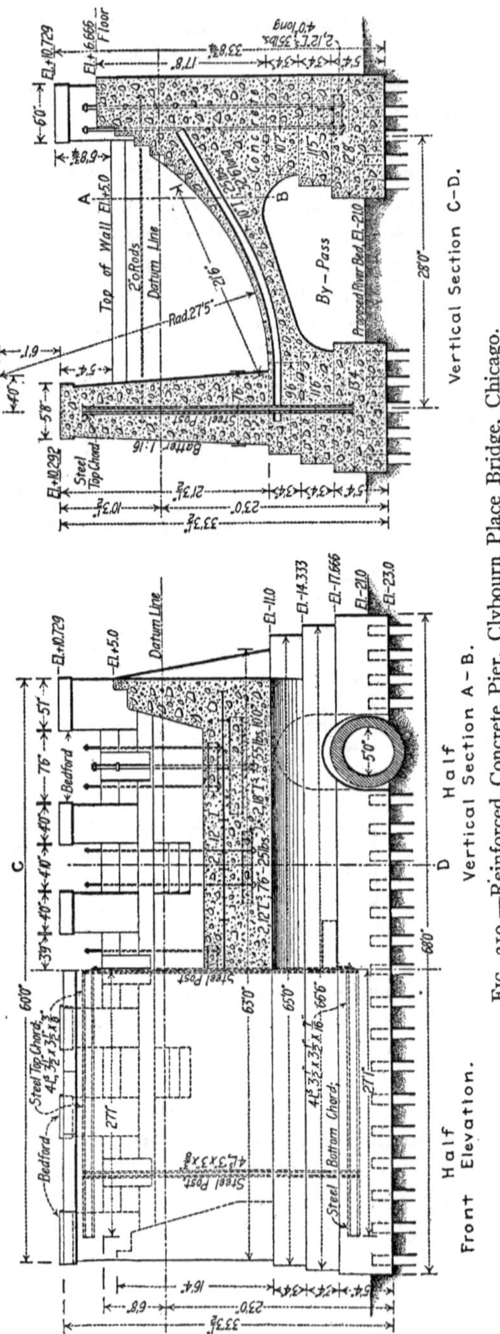

FIG. 219.—Reinforced Concrete Pier, Clybourn Place Bridge, Chicago.

Bridge Piers.

Clybourn Place Bridge, Chicago, Ill.—In constructing the series of new bascule bridges across the Chicago River at Chicago, Ill., it was planned to make extensive use of reinforced concrete in the pier and abutment construction. One of the most important of these bridges was that at Clybourn Place, and the pier and abutment construction of this structure is shown by Fig. 219. It will be observed that the pier and abutment on each side were a single structure, and that a by-pass was carried under each to provided additional waterway. These facts and the fact that the short arm of the bascule requires a suitable tail-pit for its movement account for the peculiar form of structure adopted.

The reinforcement proper is confined to the pier and the bottom sheet of the tail-pit. The pier reinforcement consists of a rectangular framework consisting of two chords and three vertical posts set vertical and lengthwise of the pier. To reinforce the bottom sheet of the tail-pit 13 I beams 10 ins. deep bent to curve are set parallel and 4 ft. apart except the outermost pair on each side, between which the space is 3 ft. The outermost I beam on each side is 15 ins. from the face-wall. These parallel I beams are connected by tie-rods 1 in. in diameter spaced 4 ft. apart. Fig. 220 shows the details of the reinforcement and of the various anchors built into the abutment and pier.

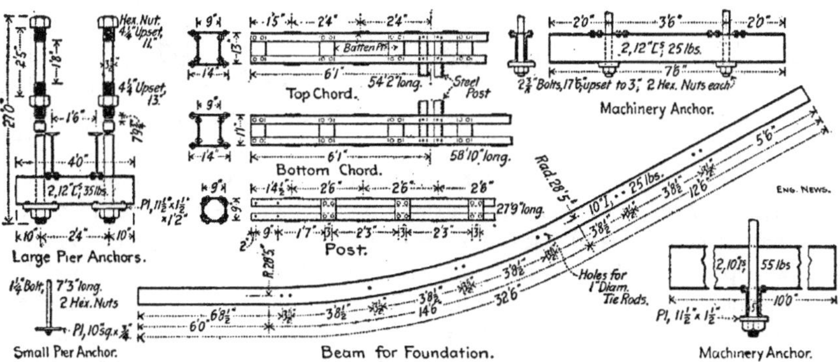

Fig. 220.—Details of Reinforcement for Clybourn Place Bridge.

The concrete used was composed of 1 part Portland cement, 3 parts sand, and 5 parts of 1½-in. brokeen stone.

Perth Amboy, N. J.—Fig. 221 shows the details of the concrete-steel bent construction adopted for a railway trestle built at Perth

Amboy, N. J., in 1902. The trestle runs over coal and ore stock piles, and one of the reasons influencing the selection of solid bents of the construction shown was the facility which they afforded in keeping different lots of ore separate and their resistance to damage by abrasion. As will be readily seen, the concrete bents are simple cross-walls resting on foundation-piles and carried up to the under side of the stringers. Cast-iron boxes are fastened to the top of the wall by anchor-bolts set in the concrete, and the ends of the wooden stringers are notched to fit into these boxes. The trackwork above the stringers is arranged in the ordinary way, with cross-ties, guard-stringers, and a foot-walk.

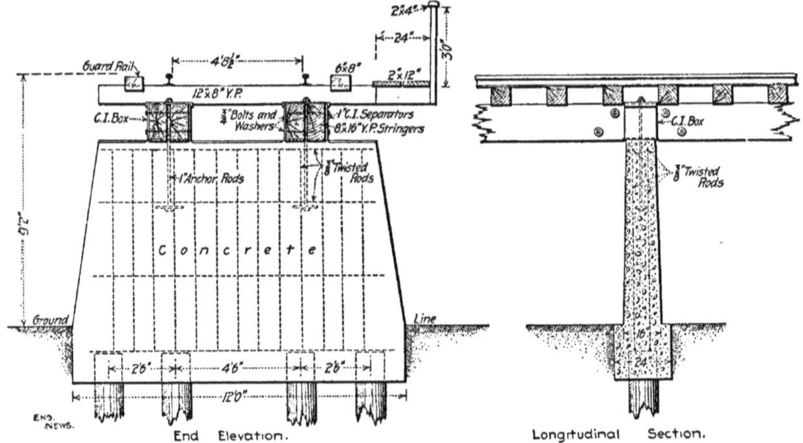

FIG. 221.—Reinforced Concrete Piers for Trestle, Perth Amboy, N. J.

The details of the trestle are given in the drawing. The structure is 1,277 ft. long between the extreme bents, bents being placed 12 ft. apart on centers, and rises at each end on a 3 per cent. grade from a low-approach embankment to a normal height of 9 ft. ins. to top of rail, corresponding to a height of 6 ft. 9½ ins. to top of bents. The stringers are each a double timber of two 8×16-in. yellow-pine sticks spaced 1 in. apart by cast-iron separators. The cast-iron boxes in which the ends of the stringers rest are anchored by 1-in. bolts set in the concrete of the bents. These boxes have the object of protecting the ends of the stringers from the weather. The stringers are anchored at both ends of the trestle, and owing to this fact, to the embedded rods, and to the stability and weight of the concrete bents, no longitudinal bracing between bents was necessary.

The bents themselves are concrete walls 12 ins. thick at the top and 16 ins. at the ground level. The top of the wall is sloped both ways to shed water, except where the boxes rest on it. At the ground level each bent spreads out to form a footing 24 ins. thick, which rests over four 13-in. piles driven about 25 ft. through marshy and made ground to firm oil; the piles were cut off one foot below ground-level, and the concrete footing formed over and around them. Steel reinforcing-rods were embedded in the concrete of the bents in both vertical and horizontal directions; these rods are Ransome cold-twisted rods of $\frac{3}{8}$-in. square steel. As will be seen from the drawing, the vertical rods are arranged in two vertical planes 8 ins. apart in the direction of the thickness of the wall, and are et 18 ins. apart in each row, being staggered in the two rows. The object of the rods is to provide against tension in the concrete due to longitudinal traction of braked trains.

The bents were built in forms made of $\frac{7}{8}$-in. dressed lumber, stiffened by brace-pieces and bolted across at the ends. The inside faces of the forms were well soaped. The forms were allowed to remain in position about forty-eight hours. The concrete used in the bents was composed of 1 part Portland cement, 2 parts sand, and 4 parts broken slag from a lead blast-furnace. It was allowed to set one month before the trestle was put into use, but in the mean time the superstructure had been placed. The ends of the stringers where they rest in the cast-iron boxes were well coated with "sludge," a tarry residue of fuel-oil, and this with the protection afforded by the boxes is thought to insure the arrest of decay.

Retaining-walls.

Arch-bridge Wing-walls, Black Lick, O.—In constructing a small Melan arch bridge near Black Lick, O., in 1902, the wing-walls which confined the heavy embankment approaches were built of concrete reinforced in the manner shown by Fig. 222. This form of reinforcement has been patented by Mr. Frank A. Bone of Lebanon, O. The wing-walls referred to vary from 20 ft. to 60 ft. in length and from 10 ft. to 28 ft. in height. Fig. 222 shows a transverse section of the wall where it was 24 ft. 8 ins. high. The masonry of the wall was of concrete of the proportions of $1\frac{1}{4}$ barrels of cement, $\frac{1}{2}$ cu. yd. of sand, and 1 cu. yd. of crushed limestone, with one-third of the whole volume of large-sized sound freestone embedded in the concrete. The walls have a coat of plaster on each side.

The steel bents shown embedded in the back part of the wall were composed of plates and angles; were placed 4 ft. apart throughout the wall, and were connected together at the bottom with 2×2-in. angle. The bents were figured strong enough to sustain safely the entire tensile strength in the wall at the different points; all the tensile strength of the concrete being allowed to go to increase the margin of safety. Small angle-brackets were riveted to the bents at short intervals to

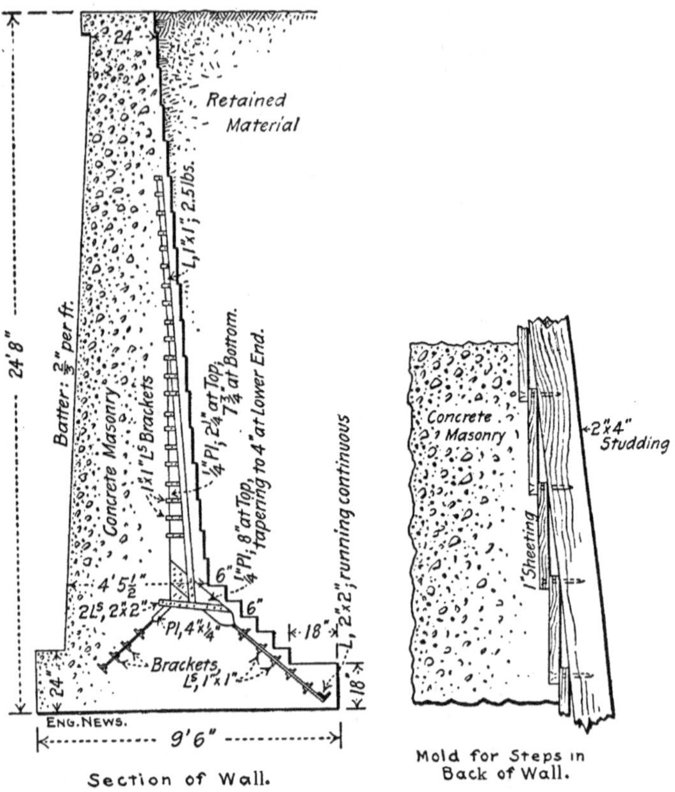

FIG. 222.—Reinforced-concrete Wing-walls for Arch Bridge at Black Lick, Ohio.

prevent any tendency of the bents to slide through the concrete. It will be noticed that the size of the upright member increases as it descends the wall, corresponding to the increase of the strains. The plate in this member also acts to prevent bulging or shearing of the wall.

The ¼×4-in. plate extending diagonally downward toward the toe of the wall is an anchor to prevent any upward or backward movement of the metal at that point resulting from the strains to which the bent is subjected. This anchor-plate is twisted one-fourth way round at its lower end, bringing the edge to the observer. The plate extending down along the heel is twisted one-quarter way round in a like manner.

A novel form of mold or casing was used for the back part of the wall, as shown in Fig. 222, by overlapping the sheeting to form the steps shown. These steps are for the purpose of creating friction between the wall and the material retained. The foundation of the wall is on solid rock.

The principal advantages claimed for this construction is that the weight of the retained material resting on the heel is used as a force to keep the wall in its normally upright position, and that metal is used where and only where the greatest tensile strains come. The front part of any good masonry wall is sufficient to bear the compressive strains. With the use of steel the wall may be made much lighter in the vertical part and the base may be made much wider in proportion, without danger of breaking off. The whole amount of masonry required is from 30 to 40 per cent. less than in an ordinary-shaped retaining-wall, to be amply safe against overturning, bulging, settling, or sliding on the base. With the toe of the wall resting on a firm foundation it is evident that the wall to be overturned by the pressure of the earth from behind must rise upward in the back part together with the material resting on the heel that projects back of the vertical part of the wall. Therefore the weight of the earth on and vertically above the projecting heel multiplied by its distance from the fulcrum at the toe gives the moment of the resistance to the overturning of the wall exerted by the earth resting on said heel. To prove the truth of the above theory the inventor made numbers of experiments with model walls of similar shape with dry sand as the retained material, and found the results of the experiments to agree well with the theory. More recent designs of this wall use plates slightly corrugated for the entire steel bent, thus dispensing with the small brackets on the sides of the bent, and where timber foundations are used the bents are connected to them at the back part of the heel.

Quai Debilly, Paris.—A retaining-wall of unusual magnitude was built at the Paris Exposition of 1900 to support the sides of a sunken street near the Gardens of the Trocadero. For this wall the Hennebique system of concrete-steel construction was employed. The trench for the street is made up of two slopes, with an average gradient of 25 per cent., meeting in a level of about 98 ft. The retaining-walls along this trench are divided into panels averaging 6 m. (19.7 ft.) in length, and each of these panels is made up of a facing from the back of which project three buttresses. The facing and buttresses are connected by two horizontal beams. Another buttress projects from the face of the retaining-wall and is located a certain distance below the street level. This construction is clearly shown by Fig. 223.

By this arrangement of horizontal beams the retaining-wall is assisted in sustaining the earth by the weight of this earth upon the horizontal beams, and does not, as in ordinary retaining walls, depend

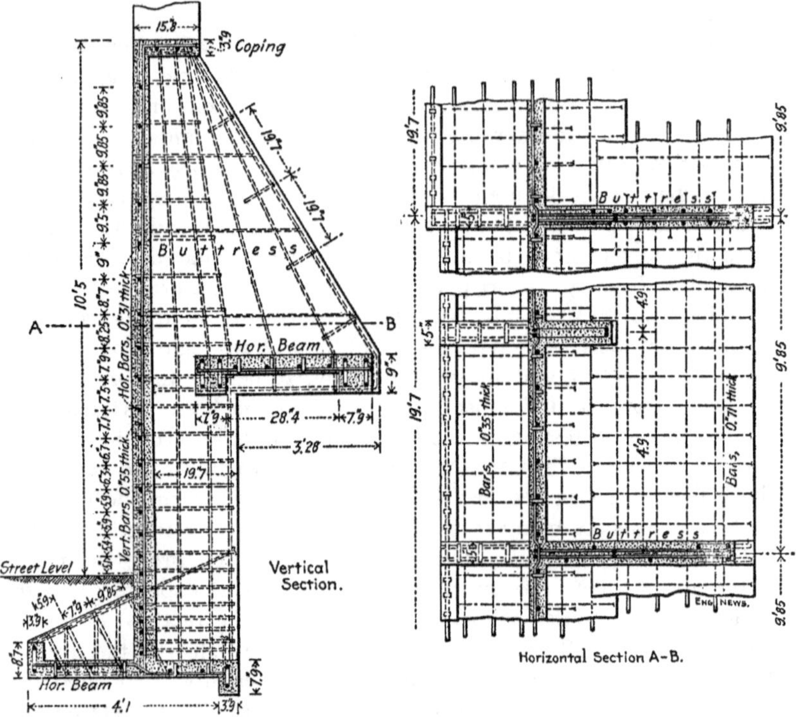

FIG. 223.—Reinforced-concrete Retaining-wall, Quai Debilly, Paris.

upon its weight alone. The employment of the two separate beams at different levels, instead of one of the same total width, results in largely decreasing the thrust of the earth upon the vertical face; and by this arrangement there is also less earth to move in building the wall. The two rear beams, however, are only used in the nine panels on each side of the 98-ft. level, as beyond that the height of the wall is so small that the second beam can be omitted.

In calculating the elements of this wall, an angle of 35 per cent. was assumed for the safe slope of the earth, and a weight of 112 lbs. per cu. ft. was taken for the earth itself. For each lineal meter of the wall it was assumed that a part of the overturning tendency would be counteracted by the weight of the earth upon the rear beam, and that the corresponding buttresses would work in tension; while the other part of this moment would be met by the resistance to compression of the front buttresses and beams. The width of the front horizontal beam was fixed by assuming a safe load of 2,048 lbs. per sq. ft. upon soil of this nature; the width of the back beams was figured with an average factor of safety exceding 2, in calculating the moment of stability of the wall.

The embedded metal-work of the vertical face consists of two series of vertical bars combined with one series of horizontal bars, the spaces between which increase towards the top of the wall. These vertical bars are bent over at right angles at the top to give support for a coping of the same construction as the facing. The illustrations show the construction of the buttresses, which are composed of inclined bars, tied together by straps, and supported by horizontal bars. The horizontal beams are made up of bars in both directions, spaced five bars to the square meter; and at their edges these beams are further strengthened by flanges. The concrete used was made by hand, in the proportion of 300 kilos. of slow-setting cement to 1 cu. m. of gravel and sand.

Ingalls Building, Cincinnati, O.—In constructing the 16-story Ingalls office building at Cincinnati, O., the space under the sidewalks was utilized, concrete-steel retaining-walls being built under the curb, lines of the streets. The wall on the south side and part of the west side is 14 ft. deep, but that on the remainder of the west side is 21 ft. deep. These retaining-walls were built as shown by Fig. 224 and are braced at the bottom by the basement floor and at the tops by girders running back to the wall columns. The other details are clearly shown by the drawings.

Expansion-joints.—Provision against shrinkage and temperature cracks is necessary in retaining-walls of any considerable length. In reinforced walls the necessity is less than in unreinforced walls, since the reinforcement will in a measure at least take up the internal stresses. The wall may in fact be specially reinforced against shrinkage and temperature stresses. The more usual practice, however, is to provide expansion-joints to allow for movement and to localize any cracking that may result from expansion and contraction.

In their specifications for concrete work the American Railway Engineering and Maintenance of Way Association stipulate that in all exposed work of unreinforced concrete theres hall be expansion-joints at intervals of from 30 ft. to 50 ft.; in reinforced concrete these intervals may be materially increased at the option of the engineer. The method of constructing these joints is that adopted by the Illinois Central R.R., whose specifications in this particular are as follows:

Where masonry structures are more than 100 ft. in length, such provision for expansion-joints shall be made as may be specified by the engineer of bridges or his assistants. Generally in the construction of large arches, or of smaller, long concrete arches, the work shall be subdivided into sections of approximately 25 ft. in length, each sec tion being separated from the adjacent one by a vertical joint extending entirely through the bench-walls, arch rings, etc.; but the founda-

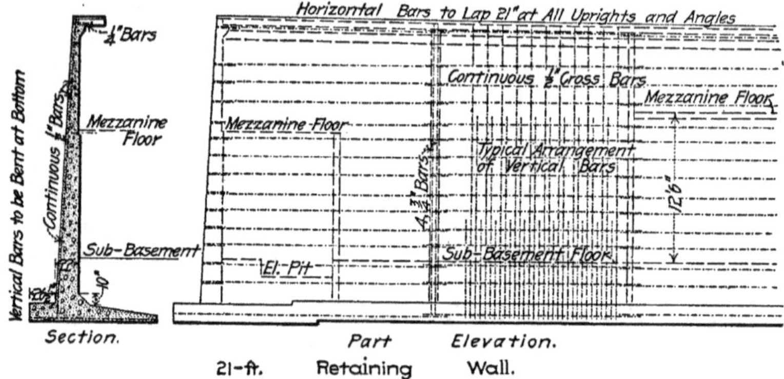

FIG. 224.—Braced Retaining-wall, Ingalls Building, Cincinnati, Ohio.

tion-work shall be stepped as previously explained, and made in one continuous monolithic mass. Temporary vertical partitions shall be put into the molds, against which the concrete shall be thoroughly rammed, where arch culverts are subdivided into short lengths as above specified, these partitions being removed as each section is completed and the next adjacent section being rammed against the concrete already constructed and set. The joints thus made shall not be flushed with mortar, nor shall any attempt be made to make the fresh concrete adhere to the older work, but a small beveled strip of wood shall be set in the angle next to the temporary partition so as to make a "V" groove, defining the joint and leaving a depth of, say, ¾ in. on the finished face of the work, it being the intention that any contraction shall open or that settlement shall effect a sliding action at such vertical joints, rather than to break up the concrete in the separate sections.

Manhole Covers.

The use of reinforced concrete for covers for sewer and subway manholes has been quite extended. These covers are of various forms, but the construction always consists of a reinforcement inserted in the tension side of a concrete slab. The drawings of Fig. 225 show the

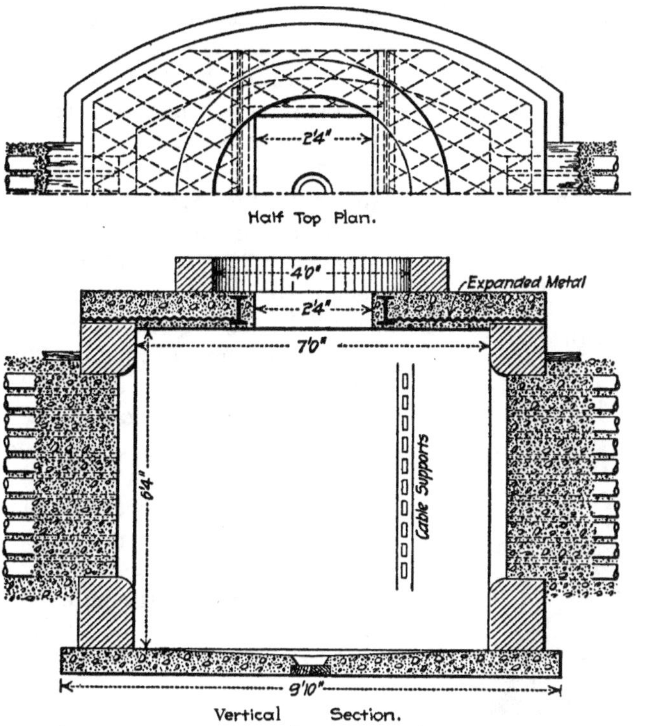

Fig. 225.—Reinforced-concrete Manhole Cover, Union Subway, New York City.

manhole construction employed by the Union Subway Company of New York City.

Lighthouse.

A bold example of reinforced-concrete construction is furnished by the lighthouse constructed by the Russian Government to mark the canal connecting the city of Nikolaief with the Black Sea. This lighthouse is 40.3 m. (132.2 ft.) high and is shown in sectional elevation by Fig. 226. The shaft is hollow cylindrical in cross-section and decreases in diameter from the bottom upward. At its bottom the shaft is attached to the reinforced-concrete foundation shown by Fig. 227 and forms structurally a single piece with it. At its top it carries a cylindrical chamber surmounted by a cylindrical lantern-house capped by a domed roof. These are also connected monolithically to the shaft, so that the structure from foundations to lantern-house is a single piece of reinforced concrete. Inside the shaft there is a spiral stairway also constructed in one piece with the shaft-walls.

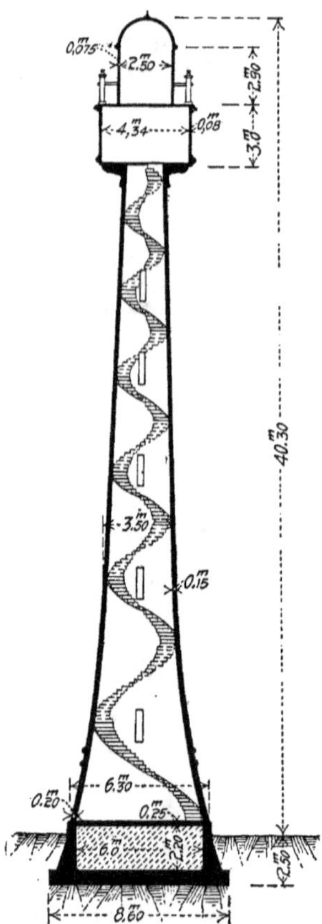

FIG. 226.—Vertical Section of Concrete-steel Lighthouse at Nikolaief, Russia.

The construction of the foundation is shown by Fig. 227, which gives the arrangement of the reinforcing-bars and stirrups. The walls of the shaft are reinforced with longitudinal and circumferential rods, respectively 23 mm. (0.91 in.) and 19 mm. (0.75 in.) in diameter. The arrangement of these rods in the wall is shown by Fig. 228. The total weight of the structure is 348 tons, exclusive of the lantern and fittings. A brick tower of the same dimensions was estimated to weigh 1,365 tons. Under the conditions existing the cost of the reinforced-concrete structure was found to be 40 per cent. less than one of masonry or steel.

DAMS.

The use of reinforced concrete for constructing dams has been exceedingly limited if we except the use of steel plates or of a reinforced revetment simply to render the dam impervious. A dam of reinforced concrete constructed for a small water-power plant at Theresa, N. Y., is shown by Fig. 229. This dam is built of concrete reinforced with Thacher rods and expanded metal; it is 120 ft. long and 11 ft. high and is founded on solid rock. The structure consists of a series of solid concrete buttresses 12 ins. thick and spaced 6 ft. apart center to center, and of a reinforced plate supported on the inclined tops of the buttresses. At the crest the plate is stiffened by a reinforced

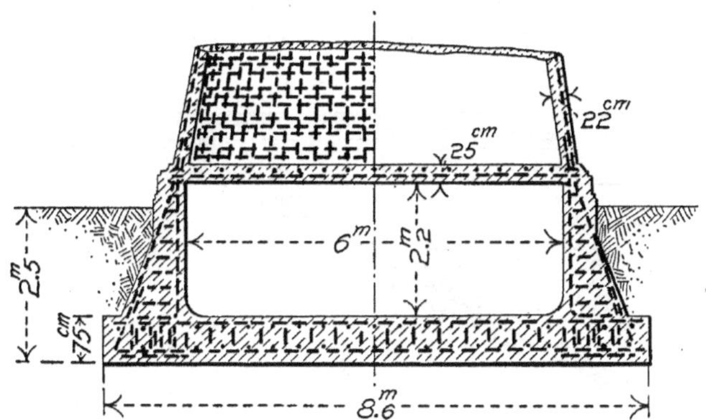

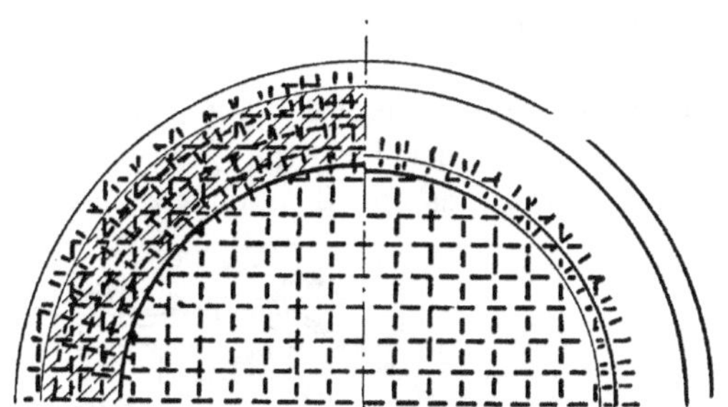

FIG. 227.—Details of Lighthouse Foundation.

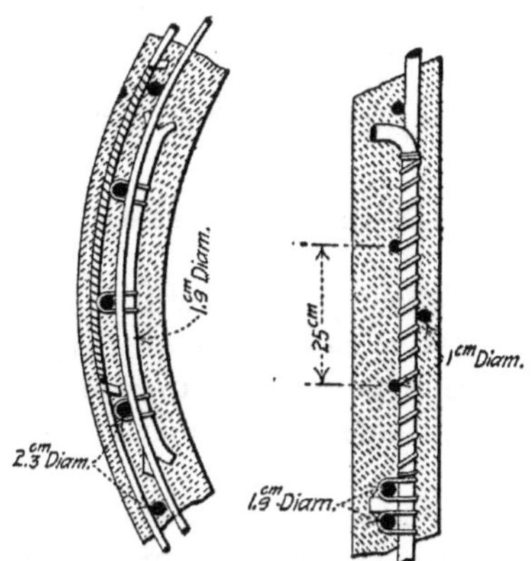

FIG. 228.—Details of Lighthouse Tower Walls.

beam 6×8 ins. in section. The plate was made of concrete composed of 1 part Portland cement, 2 parts sand, and 4 parts broken limestone;

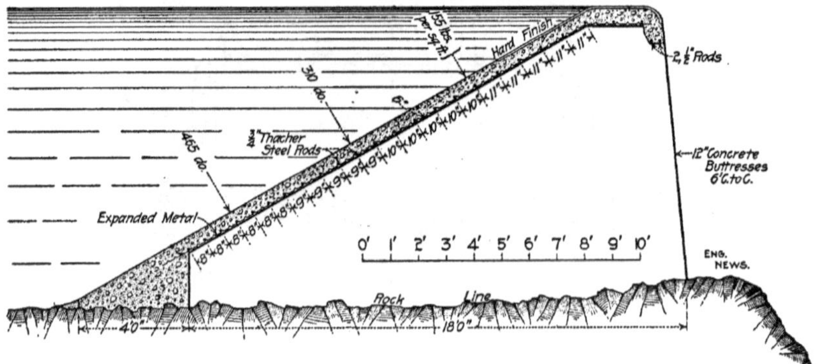

FIG. 229.—Concrete-steel Dam at Theresa, N. Y.

the toe and buttresses were made of a 1–3–6 mixture. The buttresses were anchor-bolted to the rock by 3-ft. 1¼-in. bolts. The drawing shows the spacing of the rods and their dimensions. The dam is so

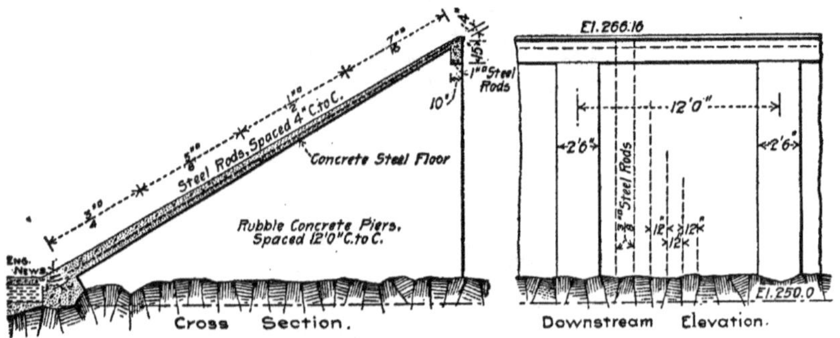

FIG. 230.—Design for Concrete-steel Dam.

constructed that the resultant pressure falls always within the base, and it is therefore a gravity dam under all heads of water. About 125 cu. yds. of concrete were required to construct the dam. This construction has been patented by Ambursen & Sayles of Watertown, N. Y.

The drawings of Figs. 230 and 231 show two forms of concrete-steel dams designed by Mr. H. W. Foster of Standish, Me., the design shown by Fig. 230 being planned for a water-power development at

Standish, but not constructed. In designing the floor of this dam the slope was divided into four nearly equal horizontal sections. The maximum load for each section was calculated and a concrete-steel slab designed for that load. The slab (considered as a whole) was uniformly tapered from 11 ins. in thickness at the bottom to 4 ins. at the top, and reinforced with sufficient steel, varying in size, to furnish a resistance in each section as nearly as possible proportional to the load for that section. The carrying-rods were spaced 4 ins. center to center, and the distributing-rods 12 ins. The piers, which were

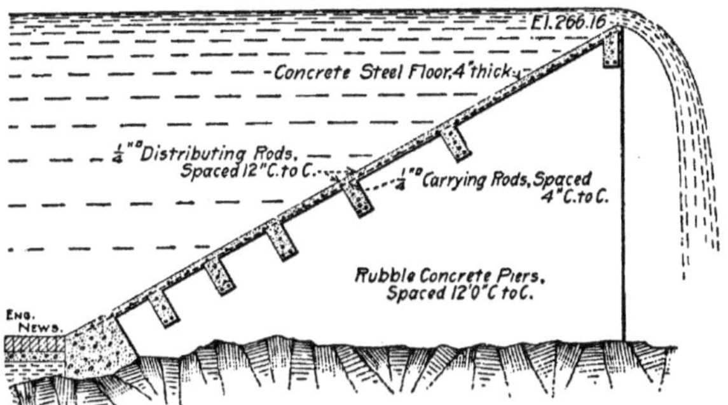

FIG. 231.—Design for Concrete-steel Dam.

designed to be of boulder concrete, were spaced 12 ft. center to center, and were 2.5 ft. in thickness. It will be seen that a reinforced-concrete beam is introduced at the crest of the dam to provide additional strength at this point, in order to resist the shock due to the passage of floating material over the dam. The probable cost of this form of dam is somewhat less than 80 per cent. of that for a solid gravity dam with the same factor of safety. A form of construction somewhat more economical in the use of cement and steel is shown by Fig. 231, in which concrete-steel beams of uniform cross-section but unequal spacing are introduced in connection with a thin concrete-steel floor in place of the single slab, as in the former design. This form of construction is probably not more economical, in this instance, than the first, as the increased cost of the forms would nearly balance the saving in cement and steel.

PART III.

METHODS OF CONSTRUCTION.

CHAPTER XII.—MATERIAL EMPLOYED IN THE FABRICATION OF REINFORCED CONCRETE.

THE component materials of reinforced concrete are iron or steel and Portland cement concrete. The qualities which these materials should possess when employed separately have through long study become familiar to engineers. For use in combination it is necessary that certain properties of concrete should be emphasized and that the metal should have special forms adapted to the special duty that it has to perform.

CONCRETE.

Those properties of concrete a knowledge of which is chiefly essential in calculating and designing constructions in reinforced concrete have been discussed in Chapter I. The qualities which a concrete should have in addition to those there specified will now be discussed briefly from the standpoint of the builder and user rather than from that of the designer.

Quality.—Save in exceptional cases Portland cement alone is employed for reinforced concrete. The aggregates used with the cement are sand and broken stone or gravel. The only exceptions to this practice are the use of cinders and crushed slag with Portland cement for certain forms of fire-resisting floor construction and the use in Europe of a quick-setting natural cement for manufacturing cast pipe reinforced by a metal skeleton. These examples of special practice are considered elsewhere; here only stone and gravel mixtures are discussed. For these only the best qualities of materials are allowable in concrete-steel work, and their proportioning and admixture must be so controlled as to produce a concrete which is sound, homogeneous, and strong to an exceptional degree. When used as a substitute for stone masonry in massive walls, piers, and foundations concrete may suffer local disintegration or have local weaknesses without serious change to the integrity of the structure as a whole. The case is different when it is used as a beam or column in place of steel or timber. Here local weaknesses due to unsoundness or want of homogeneity are elements of positive danger in the structure. To the experienced engineer these facts need no emphasis, but nowadays the widespread use of concrete places its manufacture in the hands of many who are not engineers, and the fact that concrete will endure a large amount of mishandling and still serve fairly well its purpose for ordinary masonry work is likely to give the impression that it can be mishandled safely when used for reinforced beams and arches. This is not the case. Excellence of materials and manufacture is quite as important an element in producing a serviceable and durable concrete-steel structure as is excellence in design. In fact a poor design is more safely allowable for such a structure than are poor materials and workmanship. In receiving this caution against poor materials and workmanship it should not be misinterpreted to mean that impracticable refinements of manufacture are necessary for good results. This is far from the actual fact. Only the ordinary standards of excellence are demanded. These are given in all the leading text-books on cements and masonry work, and the reader is referred to those books for detailed information. An idea of what is considered good practice in this respect may also be obtained from representative specifications. In insisting upon the need of a high-quality mixture for reinforced-concrete work care should be taken not to carry the demand to extremes. To demand for a floor-filling a concrete equal in quality to that in the reinforced plate, or for a bridge abutment a mixture as rich as that in the arch ring, is usually unnecessary for the stability of the structure and is poor engineering. This is a structural principle which has been well worked out in the practice of European engineers of concrete-steel work. As an example, in building the Melan arch at Laibach, Austria, described on pp. 219 and 220, the heavy abutments were begun with a 1-14 concrete and the mixture was gradually increased in richness to a 1–3–5 concrete at the top.

Composition.—The composition of the concrete used in concrete-steel construction varies: In Monier construction a mortar composed of 1 part Portland cement and 3 parts sand is used for thin plates for floors, conduits, tanks, etc., and a mortar composed of 1 part Portland cement to 4 or 4½ parts of sand is used for arch bridges and similar heavy work. Some others of the European construction use a mixture of cement and sand only, but the greater number employ a true concrete of cement, sand, and broken stone or gravel. In the United States concrete is employed almost exclusively. The proportions adopted are 1–2–3; 1–1½–4; 1–2–4, and 1–3–5. The most common mixtures are 1–2–4 and 1–3–5, but there is no uniformity of practice and apparently no settled opinion as to the best mixture for general use or for special classes of construction. In the descriptions of various structures given in the preceding section the composition of the concrete has been stated in each case.

Consistency.—In the United States wet mixtures are now almost universally employed for concrete-steel construction. This choice is dictated largely by practical considerations. With the amount of care which it is ordinarily practicable to secure, a more dense and homogeneous concrete is obtained with a wet than with a dry mixture. It is also practically necessary, in order to insure the thorough embedding of the reinforcement, that the concrete be wet enough to flow readily under moderate ramming into all the interstices of the metal skeleton. This quality is particularly desirable when the reinforcing elements are of small size and are closely spaced. These practical advantages of wet mixtures for reinforced concrete are held by experienced engineers generally to be sufficient to overbalance such objections to them as actually or theoretically exist. In this connection it must be remembered that the fear once anticipated that a serious loss of strength would result from the use of an excess of water has been shown to be largely unwarranted. The most complete and trustworthy experiments to determine the relative strengths of dry and wet mixtures are those made by Mr. Geo. W. Rafter, M. Am. Soc. C. E., in 1896, for the State Engineers Office of New York. Mr. Rafter made 12-in. cubes of dry, plastic, and wet mixtures. In the dry blocks the mortar was only a little more moist than damp earth; in the plastic blocks the mortar was of the ordinary consistency used by masons; in the wet blocks the water was so far in excess that the concrete quaked like liver, under moderate ramming. The resulting average compressive strengths of the three sets of blocks were as follows:

Dry mixture.	156 blocks,	2,470 lbs. per sq. in.
Plastic mixture.	144 "	2,294 " " " "
Wet mixture.	148 "	2,180 " " " "

The blocks tested were all from 18 to 24 months old.

European builders in reinforced concrete are, as a rule, advocates of a dry mixture, but they acknowledge its permissibility only in case the deposition and compacting are performed with exceeding care.

Permeability.—Resistance to the percolation of water is desirable in all concrete work and is an important essential in tanks, reservoirs conduits, subways, and other structures designed to carry liquids or to prevent their infiltration into an enclosed space. A number of experiments have been conducted to determine the practicability of making concrete impermeable to water. In these experiments the object sought has been either the determination of a mixture of cement and aggregates that would resist the percolation of water or the discovery of a substance which by admixture with the concrete materials would render the hardened product impermeable.

The elaborate tests of Mr. R. Feret of the Boulogne Laboratory of the Ponts et Chaussées, extending over five years, led him to the following conclusions:

That in all mortars of granulometric composition the most permeable are those which contain the least quantity of cement.

Of all mortars of the same richness, but of varying granulometric composition, those which contain very few fine grains are much more permeable. They are the more so where, with equal proportions of the fine grains, the coarse grains predominate more in relation to the grains of medium size.

The minimum permeability is found in mortars where the proportion of medium-size grains is small, and the coarse and fine grains are about equal to each other.

Mr. Feret's experiments also showed that the permeability of mortars submitted to a continuous filtration of fresh water or sea water diminished rapidly with time. He also recommends using a too-large rather than a too-small quantity of mixing water.

A series of experiments made by Mr. A. W. Hyde and W. J. Smith in 1889 on the permeability of mortar sand concrete showed that with the same proportions of cement, mortars made with fine sand are less permeable than those made with coarse sand; that with the same quality of sand, permeability decreases as the proportion of cement increases; that mortar made with neat cement is less permeable than any cement and sand mortar.

A third set of tests to determine the permeability of concrete under high-water pressure show results which are in general similar to those already stated. These tests were made by Mr. J. B. McIntyre and A. L. True at the Thayer School of Civil Engineering in 1902, using 1–1, 1–2, and 1–3 Portland cement and sand mortars. The conclusions drawn from these experiments were as follows:

From the tables it will be seen that several of the mixtures were practically impermeable to water at the pressures which we had at our disposal, nearly from 20 to 80 lbs. per square inch.

All the specimens composed of 1:1 mortar in the proportion of 30, 35, 40, and 45 per cent. of the whole mass were impermeable. Some of the specimens composed of 1:2 mortar in the proportions of 40 and 45 per cent. were also impermeable, as well as the mixture of 1:2:4 and 1:2½:4.

All others leaked at the high pressure, and in a general way may be said to have shown a degree of imperviousness in direct proportion to the proportion of mortar in them, with the lower pressures, as well as with 80 lbs. However, as the object was to secure an impermeable.

In a number of instances it has been found that water percolating through concrete absorbed lime in its passage which was deposited on the exterior surface of the concrete in the form of an efflorescence that acted as a very perfect seal against further percolation. These experiences have been taken by a number of engineers as the basis for a theory that slacked lime added to the concrete mixture would serve to make it impermeable. Mr. R. W. Lesley, Assoc. Am. Soc. C. E., advocated this practice in a discussion before the American Society of Civil Engineers in the following words:

The speaker's theory and conclusions, based upon all that has been discovered, and especially upon Feret's remarkable experiments, would seem to indicate that the best addition to cement mortars for the purpose of making them impermeable, according to the theory of an exterior coating by lime carried through the mass during filtration, would be to add to the concrete, at the time of mixing, a reasonable proportion of hydrate of lime, or, in other words, the ordinary slaked lime of commerce. This slaked lime, as has been shown by experiments of the late Professor De Smedt, formerly of the Laboratory of the District of Columbia, does not injure cements or mortars; does not cause expansion; and does not decrease their strength, though retarding their setting slightly. Such an addition would be perfectly safe, as a matter of practice, and would form in the mass a substance which would be carried by filtration and close the pores, form efflorescence or stalactites and stalagmites on the surface, and, in the speaker's judgment, though he knows of no experiments on the subject, would aid largely in making mortars impermeable. Experiments in this line would be most valuable.

Other substances which have been mixed with the concrete materials to render the hardened product impermeable to water are silicate of soda or water-glass and soap and alum. Experiments conducted by Prof. W. *K.* Hatt to determine the effect of these substances on concrete gave the following results:

The effect of silicate of soda is to diminish the strength of both ash and sand mortars more than 50 per cent., and to diminish the absorption of the ash mortars about 50 per cent. The effect of alum and soap mixed in with the mortar at the time it is gaged is to strengthen and harden the ash mortar about 50 per cent. and to diminish its absorption by 50 per cent. A soap solution alone will diminish the absorption (by the action of the alkali in the cement on the soap), but will not increase the strength. The strength of the sand mortar is not greatly affected by the soap and alum, but its absorption is decreased about 50 per cent. The effect on the absorption was measured by comparison of the weight of water taken up by briquettes which were immersed after having been dried out. Check tests were made by measuring the water which percolated from the outside through the walls of hollow cylinders. The speaker believes that this is the first use of a soap and alum solution for waterproofing, in place of the usual gaging water. The method used by the speaker is as follows: A 5 per cent. solution of grodnu alum and water is prepared, and a 7 per cent. solution of soap and water. The alum solution is mixed with the mortar to the amount of one-half the ordinary gaging water. The soap solution is then applied in amount to bring the mortar to the desired plas: ticity. The soap and alum, acting together, cause the precipitation of an insoluble compound in the pores of the mortar.

METAL.

The theoretical considerations governing the quality of the metal to be selected for the reinforcement of concrete are discussed in the opening chapter of this book. Here reference will be made only to the character of the metal used in practice, and to the forms in which this metal is employed. In Europe quite general use is made of wrought iron for reinforcing concrete. The quality of metal used has a tensile strength of from 32 to 36 kilograms per square millimeter (45,500 to 51,200 lbs. per sq. in.) and an elongation of from 8 to 12 per cent. in a length of 20 centimeters. Steel is, however, employed in many European systems of concrete-steel. A soft steel, having a tensile strength of from 42 to 50 kilograms per square milimeter (59,700 to 71,100 lbs. per sq. in.) and an elongation of from 20 to 24 per cent. in a length of 20 centimeters (7.87 ins.), is generally used. In the United States steel is universally used for reinforcing concretes, but the quality of metal chosed by different builders varies from soft structural steel to steel wire and strong high-carbon steel. The metal is also employed in a variety of special forms which are unusual in European practice, which favors the use of rounds, tees, channels, straps, and other commercial shapes. The succeeding paragraphs describe and illustrate the principal forms of metal reinforcement employed in concrete-steel work.

Monier Netting.—Monier reinforcement consists of a netting of rods and wires, with rectangular meshes. Both wrought iron and steel are employed. The two sets of parallel rods are known respectively as carrying-rods and distributing-rods. The carrying-rods are the larger and are depended upon to take most of the tensile strain; the smaller distributing-rods serve to distribute the load evenly over the carrying-rods. The distance apart of the rods is varied with the load to be carried, and it runs in round figures from 2 ins. to 10 ins. Relatively the distributing-bars are spaced about twice as far apart as are the carrying-bars. The two sets of rods are wired together at every third or fourth intersection. This form of netting is built up by hand from commercial bars and wires, and appears when completed as shown by Fig. 232.

Expanded Metal.—Expanded metal is manufactured from structural steel plates by piercing them with parallel rows of slits and then expanding or stretching open the gashes until the metal assumes the form shown by Fig. 233. The material and the method of manufac-

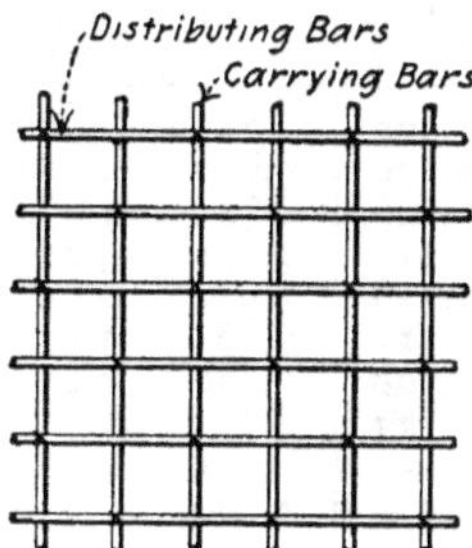

FIG. 232.—Monier Netting.

FIG. 233.—Expanded Metal.

turing it were invented by the J. F. Golding. The metal used is a high-grade low-carbon Bessemer steel. From tests made on commercial stock by Mr. George Hill, M. Am. Soc. C. E., the following figures are available: Carbon content 0.008 per cent. Unannealed specimen tested in tension parallel with the grain showed a tensile strength of about 65,000 lbs. and an elastic limit of about 30,000 lbs. per square inch, with an elongation of 15 per cent. in 8 ins. Tests made of strands of metal, which had been embedded in a floor tested to rupture, showed an average ultimate strength of 54,240 lbs. per square inch. Expanded metal is commercially designated by giving the gage of the steel and the amount of displacement between the junction of the meshes; thus No. 10, 3-in. mesh, designates an expanded metal made from No. 10 steel, in which the displacement of the bridge amounts to 3 ins., the axis of the diamond being 2¾ ins. and 6 ins. When expanded this size and mesh weighs 0.56 lb. per square foot and has a sectional area per foot width of 0.168 sq. in. and per inch width of 0.014 sq. in. The largest size of expanded metal is made from No. 4 plate with 5×12 in. meshes, and the smallest size is made from No. 27 plate with ⅝-in. meshes. The expanded plates are uniformly 8 ft. long and vary in width from 1 ft to 6 ft.

Electrically Welded Wire Fabric.—This form of reinforcement consists of longitudinal and transverse galvanized steel wires welded to each other at intersections. It is manufactured in continuous sheets of any desired length and varying in width up to 8½ ft. by the Clinton Wire-cloth Company of Clinton, Mass. The following information regarding the material is furnished by the manufacturers:

Size per W. & M. Gage.	Diameter of One Wire, Inches.	Weight per Lineal Foot of One Wire, Pounds.	Tensile Strength of One Wire, Pounds.	Tensile Strength of Longitudinal Wires Only, in 1 Foot of Width of Fabric, when spaced as follows:				
				2-in., Pounds.	3-in., Pounds.	4-in., Pounds.	5-in., Pounds.	6-in., Pounds.
No. 3	.243	.158	2,799	16,794	11,196	8,397	6,717	5,598
" 4	.225	.135	2,392	14,352	9,568	7,176	5,740	4,784
" 5	.207	.114	2,019	12,114	8,076	6,057	4,845	4,038
" 6	.192	.098	1,738	10,428	6,952	5,214	4,171	3,476
" 7	.177	.083	1,476	8,856	5,904	4,428	3,542	2,952
" 8	.162	.070	1,237	7,422	4,948	3,711	2,968	2,474
" 9	.148	.058	1,036	6,216	4,144	3,108	2,486	2,072
" 10	.135	.048	.859	5,154	3,436	2,577	2,061	1,718
" 11	.120	.038	.684	4,104	2,736	2 052	1,641	1,368
" 12	.105	.029	.525	3,150	2,100	1,575	1,260	1,050

Longitudinal strands may be spaced on centers of 2 ins. or wider, the increase running in steps of ½ in. Transverse strands may be spaced on centers of 1 to 20 ins. inclusive, in steps of 1 in. When longitudinal and transverse strands are of the same size, No. 6 wire is as heavy as it is advisable to specify; but if heavy wire is required in one member only (either longitudinal or transverse strands), can furnish No. 3 to No. 12 inclusive in combination with No. 8 to 12 inclusive. Can furnish various widths, the maximum in a fabric of 5 ins. or wider spaced longitudinals being 8½ ft.; but if longitudinals are spaced on centers of 4½ ins. or narrower, it is 21 meshes of the desired size plus whatever overhang appears in the transverse strands. Transverse strands must project at least ½ in. beyond the two outside longitudinals, and they may be as much longer as required.

Lock Woven-wire Fabric.—This form of reinforcement consists of a netting of transverse and longitudinal galvanized steel wires with a lock-wire tie at each intersection. The character of this lock tie is shown by Fig. 234. The fabric is made in rolls 56 ins. wide and 300 ft. long, of any gage, wire and any size of mesh required. The component wires have an ultimate strength of 116,000 lbs. per square inch. The material is furnished by W. N. Wight & Co. of New York City.

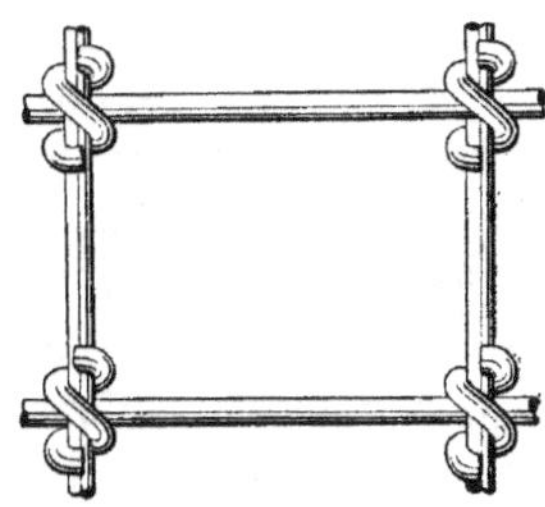

FIG. 234.—Lock Woven-wire Fabric.

Twisted Steel Rods.—Steel bars of square section cold-twisted to the form shown by Fig. 235 is the form of reinforcement patented by Mr. E. L. Ransome and used in the various systems of Ransome concrete-steel construction. The purpose in twisting the rods is twofold: first, to give the metal a mechanical hold on the surrounding concrete like that of a screw in wood, and, second, to increase the strength and decrease the ductility. The usual practice is to purchase commercial squares of the proper size and twist them on the site of the work, but the bars already twisted may be purchased if desired. The following figures show the approximate number of twists per foot usually given the bars:

FIG. 235.—Ransome Bar.

Size, Inches.	Number of Twists.	Size, Inches.	Number of Twists.	Size, Inches.	Number of Twists.
$\frac{1}{8}\times\frac{1}{8}$	7	1×1	1	$1\frac{3}{4}\times 1\frac{3}{4}$	$\frac{1}{8}$
$\frac{1}{4}\times\frac{1}{4}$	5	$1\frac{1}{4}\times 1\frac{1}{4}$	$\frac{1}{2}$	2×2	$\frac{1}{16}$
$\frac{1}{2}\times\frac{1}{2}$	3	$1\frac{1}{2}\times 1\frac{1}{2}$	$\frac{3}{8}$		

The effect of cold twisting on the strength is indicated by recent comparative tests of mild-steel bars and cold twisted bars from the same lot of steel. These tests were made at the testing laboratory of Columbia University, and each of the results given is the average of four or more tests. The results in pounds per square inch are as follows:

Size of Square Bar, Inches.	Elastic Limit,		Ultimate Strength.	
	Plain.	Twisted.	Plain.	Twisted.
¼	46,170	62,350	57,800	86,700
¾	34,840	56,720	59,200	85,240
⅞	39,700	56,150	62,100	84,730

Corrugated Bars.—Steel bars of square section with corrugations on all four sides, as shown by Fig. 236, are the invention of Mr. A. L. Johnson, M. Am. Soc. C. E. These bars are rolled from high-carbon steel breaking at about 105,000 lbs. per sq. in. and having an elastic limit of between 65,000 and 67,000 lbs. per sq. in. They are made in the following sizes and weights:

FIG. 236.—Corrugated Bar.

Size in Inches.	Net Section, Square Inches.	Weight, Pounds per Foot.
¼	0.19	0.78
¾	0.38	1.56
⅞	0.55	2.25
1	0.70	2.90
1¼	1.10	4.56

The following explanation of the design and construction of the bar is given by the inventor:

By numerous practical examples, as well as by laboratory tests, it has been determined that it is very easy to break the adhesion between the concrete and the surface of the embedded metal, and it is therefore not safe to use plain bars, or any form of bar depending upon this adhesion for its efficiency. The bar should be provided with some form of ribbing device, the sides of which ribs should be nearly at right angles to the axis of the bar, varying therefrom by an angle not exceeding the angle of friction between the materials used. Our corrugated bar is designed on this principle, and the practical results and tests that have been made by others, as well as by ourselves, have proved the correctness of the theory. Theory and practice do not always agree, but they do in this case. This bar is rolled hot, like any other shape, from special material, and has double the amount of work put on it that is given ordinary structural steel. This results in a very high elastic limit, which is desirable in steel-concrete construction, the load producing this stress in the steel being the ultimate load of the combination in all cases.

Thacher Steel Bars.—The special form of reinforcing-bar patented by Mr. Edwin Thacher, M. Am. Soc. C. E., is shown by Fig. 237. These bars are rolled from medium steel, and the nominal sizes in

FIG. 237.—Thacher Rolled Bar.

which they are made, together with the sectional area, weight, and mean ultimate strength of each size, are given in the following statement:

Nominal Diameter, Inches.	Area, Square Inches.	Weight per Lineal Foot.	Mean Ultimate Strength Medium Steel.
1/4	0.047	0.16	3,000
3/8	0.10	0.34	6,400
1/2	0.18	0.61	11,500
5/8	0.28	0.95	17,900
3/4	0.41	1.39	26,200
7/8	0.55	1.87	35,200
1	0.71	2.42	45,400
1 1/8	0.90	3.06	57,600
1 1/4	1.10	3.74	70,400
1 3/8	1.32	4.49	84,500
1 1/2	1.56	5.30	99,800
1 5/8	1.81	6.16	115,800
1 3/4	2.08	7.07	133,100
1 7/8	2.35	8.00	150,400
2	2.65	9.02	169,600

The inventor presents the following claims for the bar and gives the following advice concerning its use:

This bar is specially designed for the work it has to do, and numerous tests have shown that it will not slip in or disrupt the concrete. The sectional area is practically uniform throughout, and all changes in shape of section are made by gradual curves, as all sharp corners, fins, knife-edges, or sudden changes of section in concrete or any other non-fibrous material are liable to cause cracks, and are therefore not desirable. The bars are laid flat, the widest dimension transversely, and the distance from the center of the bars to the outside of the concrete should not be less than one and one-half times the diameter of the bar. The bars should be held in their proper position until they are firmly embedded in the concrete. Bars can be furnished from ¼ in. diameter up to 2 ins. diameter, but it will not often be necessary to use bars larger than 1¼ ins. or 1½ ins. diameter, and as a rule, when the bars are subjected to tension only, as in slabs or girders, it is better to use small bars spaced from 3 in. to 8 in. apart, depending upon size of bar and depth of member, than to use large bars spaced a greater distance apart.

In slabs, transverse rods of small diameter, spaced two or three feet apart, should be used to prevent cracks due to shrinkage. For deep girders, requiring large areas of steel, it is better to use small bars in two or more rows if necessary, than to concentrate the metal. For arches subject both to compression and bending stresses, it is not desirable to use small rods or close spacing, and heavy concentrations of metal give the best results. Tests made by the Austrian Society of Engineers and Architects on the Melan arch, using heavy concentrations of metal spaced 39 ins. apart, gave much higher results than tests on the Monier arch with rods spaced only a few inches apart.

In much of his heavy bridge work Mr. Thacher has used a wholly different form of reinforcing-bar. This is a flat bar of structural steel with rivets through it at close intervals and generally staggered, whose

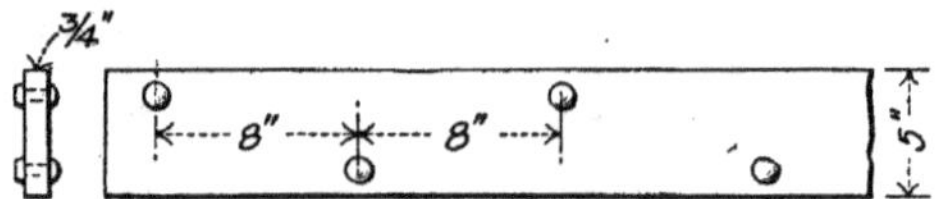

Fig. 238.—Thacher Riveted Bar.

projecting hands furnish a mechanical hold on the concrete. The sketch Fig. 238 shows one of these bars designed for an arch span of 122 ft. For smaller arches the bars are made narrower, but seldom less than 3 ins. wide.

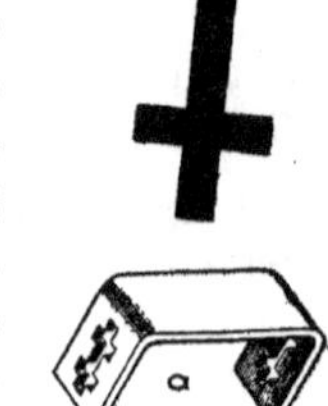

Fig. 239.—Columbian Bar.

Columbian Bars.—The form of reinforcing-bar employed for concrete-steel floors of the Columbian type up to and including spans of 15 ft. is shown in cross-section by Fig. 239. These bars are rolled from structural steel and are made in sizes running from 1 in. to 5 ins. in depth. For longer spans a deeper bar with additional ribs is employed. The smaller sizes of bars are connected to the floor-beams by means of stirrups of the form shown by Fig. 239-*a*. The manner of using these bars in floor construction is described on page 145.

Kahn Steel Bars.—The form of reinforcing-bar invented by Mr. Julius Kahn, Assoc. M. Am. Soc. C. E., and illustrated in Fig. 240, is designed with the assumption that: (1) Concrete should be reinforced in a vertical plane as well as in the horizontal plane; (2) The reinforcement should be inclined to the vertical, preferably with varying upward curvature approximating the line of principal tensile stress; (3) The metal should be distributed in proportion to the strains existing at

FIG. 240.—Kahn Bar.

any place; (4) The shear members should be rigidly connected to the horizontal members. The bar is constructed by shearing a rolled bar of the cross-section shown by Fig. 240 and bending up the sheared strips to the positions shown. These sheared bars are manufactured in several sizes by the Trussed Concrete-steel Company of Detroit. Mich.

Cummings Bars.—The special form of reinforcement designed by Robert A. Cummings of Pittsburg, Pa., is shown by Fig. 241. It is

FIG. 241.—Cummings Bars.

constructed of round steel bars bent and arranged as shown by the drawing.

Commercial Shapes.—The use of reinforcing-bars of special form has had its greatest development in the United States. In Europe commercial shapes are employed almost exclusively. Bars of circular section are most used; other forms are T bars, + bars, channels, and straps. A number of examples of each of these forms of reinforcing-bars are given in the preceding section, and they will serve to explain in detail the various methods of combination. The material in these bars is usually wrought ron or soft steel in Europe and soft or medium steel in the United States.

Efficiency of Mechanical Bond.—A number of the forms of bar reinforcement described have been especially designed to increase the bond of the bar beyond the normal adhesion of concrete to metal.

This object is sought to be accomplished by providing the bar with mechanical anchors in the shape of ribs, notches, or projections of one sort or another. Tests to determine the efficiency of this mechanical bonding are rare, and the results of those which have been made permit only the most general conclusions. In a series of tests of Johnson, Thacher, and Ransome rods by Prof. C. W. Spofford of the Massachusetts Institute of Technology the concrete split or the bar broke in all cases where Johnson and Thacher rods were used. The result usually obtained with the Ransome rods were as follows: The rod slipped at a certain pull, then pulled through sometimes several inches with a smaller force, and finally took another fast grip which required a harder pull than the first to break and which generally split the concrete block. Substantially the same results were secured with Ransome rods in tests made at the Case School of Applied Science by S. W. Emerson and George A. Peabody. In the latter tests the average resistance to initial slip of four ½-in. bars was 1,259 lbs. per linear inch embedded. This same series of tests furnished interesting information on the adhesion of 1-in. plain square bars to mortars and concretes of different compositions. In all twenty-four bars were tested, four for each composition; and the average results were as follows:

Composition of Mixture.	Pounds per Square Inch of Surface in Contact.
Neat cement	278
1 cement, 3 sand	411
1 cement, 2 sand, 4 broken stone	587
1 cement, 3 sand, 6 broken stone	478
1 cement, 2 sand, 4 gravel	547
1 cement, 3 sand, 6 gravel	516

Preservation of Iron in Concrete.—It has generally been assumed that iron or steel embedded in a concrete does not corrode, and many instances are cited of embedded steel being removed from concrete quite as clear and bright after a long period of exposure to the elements as it was when first embedded. It should be noted, however, that an occasional instance is cited to show that under certain circumstances metal embedded in concrete will corrode. As the durability of concrete-steel required that the steel shall be permanently protected from corrosion, this question is an important one and it has received consideration from a number of experts. The commonly accepted theory accounting for the protection from rust of iron embedded in concrete has been recently stated by Prof. Spencer B. Newbury as follows:

The rusting of iron consists in oxidation of the metal to the condition of hydrated oxide. It does not take place at ordinary temperatures in dry air or in moist air free from carbonic oxide. The combined action of moisture and carbonic acid is necessary. Ferrous carbonate is first formed; this is at once oxidized to ferric oxide and the liberated carbon dioxide acts on a fresh portion of metal. Once started the corrosion proceeds rapidly, perhaps on account of galvanic action between the oxide and the metal. Water holding carbonic acid in solution soon, if free from oxygen, acts as an acid and rapidly attacks iron. In lime-water or soda solution the metal remains bright. The action of cement in preventing rust is now apparent. Portland cement contains about 63 per cent. lime. By the action of water it is converted into a crystalline mass of hydrated calcium silicate and calcium hydrate. In hardening it rapidly absorbs carbonic acid and becomes coated on the surface with a film of carbonate, cement mortar thus acting as an efficient protector of iron and captures and imprisons every carbonic-acid molecule that threatens to attack the metal. The action is, therefore, not due to the exclusion of the air, and even though the concrete be porous, and not in contact with the metal at all points, it will still filter out and neutralize the acid and prevent its corrosive effect.

The use of cement washes and plasters for the specific purpose of protecting iron and steel from rust is quite common and has extended over a long period of time. Cement paint is largely used by the railway companies of France to protect their metal bridges from corrosion. Two coats of liquid cement and sand are applied with leather brushes. After investigation and careful tests the engineers of the Boston Subway adopted Portland-cement paint for the protection of the steel beams of that structure. Iron spirit-tanks for European distilleries are universally painted on the inside with Portland-cement paint to prevent corrosion. In the United States it is a frequent practice to coat the inside of steel salt-pans, sulphate digesters, etc., with cement plaster to prevent corrosion. Regarding the damage from corrosion by the sulphur in the cinders of cinder concrete Prof. Newberry expresses himself as follows:

The fear has sometimes been expressed that cinder concrete would prove injurious to iron on account of the sulphur contained in the cinders. The amount of this sulphur is, however, extremely small. Not finding any definite figures in this point, I determined the sulphur contained in an average sample of cinders from Pittsburg coal. The coal in its run state contains a rather high percentage of sulphur, about 1.5 per cent. The cinders proved to contain only 0.61 per cent. sulphur. This amount is quite insignificant, and even if all oxidized to sulphuric acid it would at once be taken up and neutralized in concrete by the cement present, and would by no possibility attack the iron.

In connection with this statement it may be noted that in the demolition in 1903 of a tall steel-frame building in New York City, which was built in 1898 and had practically all of its framework except the columns embedded in cinder concrete, the steel removed showed practically no rust which could be considered as having developed after the metal was embedded.

Tests of a reliable character, made to determine the efficiency of concrete in protecting embedded metal from corrosion, are comparatively few. The most important ones which have been published are those of Mr. Breuillie of France and those of Prof. Charles L. Norton of Boston, Mass. Mr. Breuillie's tests were extended in character and the conclusions drawn from them by the experimenter were: (1) That the cement attacked the iron; (2) that water dissolved the composition which formed at the contact of the two materials; (3) that the adhesion of the steel to the cement disappeared when water passed through the concrete for a certain time; (4) that the weight of the iron salts which adhere to the steel and the normal adhesion between the steel and the concrete increased with time; (5) in all cases the action of the cement on the iron prevented rust and removed the rust from metal which had been allowed to corrode before being embedded.

The tests conducted by Prof. Charles L. Norton of the Massachusetts Institute of Technology, Boston, Mass., were of a somewhat different character from those of Mr. Breuillie. Briquettes or blocks were made of neat cement; of 1 part cement and 3 parts sand; of 1 part cement and 5 parts broken stone, and of 1 part cement and 7 parts cinders. Portland cement was used, and was tested chemically and physically and found good. The cinders when washed down with a hose-stream and dried tested alkaline, and analysis revealed very small amounts of sulphur. In each block there was embedded a $\frac{1}{4}$-in. rod, a piece of soft sheet steel $6 \times 1 \times \frac{2}{32}$ in., and a 6×1-in. strip of expanded metal. These blocks were exposed as follows: one-quarter of them in sealed chests containing an atmosphere of steam, air, and carbon dioxide; one-quarter in a similar chest with an atmosphere of air and carbon dioxide; one-quarter in a chest with an atmosphere and steam, and one-quarter on a table in the open air of the testing-room. At the end of three weeks the blocks were carefully cut open and the steel examined and compared with specimens which had lain unprotected in the corresponding chests and in the open air.

The results of the examinations were as follows: The unprotected specimens consisted of rather more rust than steel. The specimens embedded in neat cement were perfectly protected. Of the remaining specimens hardly one had escaped serious corrosion. The location of the rust-spot was invariably coincident with either a rod in the concrete or a badly rusted cinder. In the more porous mixtures the steel was spotted with alternate bright and badly rusted areas, each clearly defined. In both the solid and the porous cinder concrete many rust-spots were found, except where the concrete had been mixed very wet, in which case the watery cement had coated nearly the whole of the steel, like a paint, and protected it. The following are Prof. Norton's conclusions from his tests:

(1) Neat Portland cement, even in thin layers, is an effective preventative of rusting.

(2) Concrete, to be effective in preventing rusting, must be dense and without voids and cracks. It should be mixed quite wet when applied to the metal.

(3) The corrosion found in cinder concrete is mainly due to the iron oxide, or rust, in the cinders and not to the sulphur.

(4) Cinder concrete, if free from voids and well rammed when wet, is about as effective as stone concrete in protecting steel.

(5) It is of the utmost importance that the steel be clean when bedded in concrete. Scraping, pickling, a sand-blast, and lime should be used, if necessary, to have the metal clean when built into a wall.

At first sight the conflicting testimony which has been quoted appears to have but little solid ground upon which the practicing engineer can base a decision as to the probable damage from rust of iron or steel embedded in concrete. A brief analysis will show, however, that this is not actually the case. In the first place there are many instances where steel embedded in concrete has shown no signs of rust upon removal. None of the evidence presented disputes this fact. Secondly, steel removed from concrete which contained cracks or voids has in many instances shown rust, always at the points where the cracks and voids were located. None of the evidence presented disputes this fact. Thirdly, the theory that the concrete covering filters out and renders innocuous the corrosive elements so completely as to protect the steel even where it is not in contact with the concrete is disputed by the results of Prof. Norton's tests. Fourthly, Prof. Norton's tests show that where the concrete is so closely in contact with the steel as to completely cover it with cement there is no corrosion. This fact is not disputed by any of the other evidence. Fifthly, Prof. Norton's tests show that wet concrete mixtures more certainly insure the close contact of the steel and concrete at all points than do dry mixtures. This fact is not disputed by any of the other evidence.

Sixthly, Prof. Norton's contention that the steel should be perfectly cleaned before it is bedded in concrete is controverted by the tests of Mr. Breuille, which show that bedding in concrete will remove the rust from previously corroded steel. Seventhly, all the evidence presented indicates that the sulphur content of the cinders is not a serious element of danger in cinder concrete and that, other conditions being the same, cinder concrete and stone concrete are about equally efficient in preventing the rusting of embedded steel. The useful conclusion which the practicing engineer can draw from all this is that, so far as danger from subsequent rusting is concerned, he can confidently embed steel or iron reinforcement in either cinder or stone concrete if he secures a close contact between the concrete and steel at all points, and if no cracks develop in the concrete to expose the metal to attack.

CHAPTER XIII.—METHODS OF CONSTRUCTION IN FOUNDATION WORK.

THE mode of procedure followed in constructing foundation work in reinforced concrete varies with the character of the foundation structure and the character and arrangement of the reinforcement. It may be very simple, as is the case in building beam grillages, or it may be quite as complex as floor and column construction, as is the case in casting and driving reinforced-concrete piles or in building footings with elaborate rod and stirrup reinforcement. The methods followed can best be explained by considering the different forms of foundations separately.

Spread Foundations.—Spread foundation, in concrete-steel are either isolated column footings or extended footings covering the whole building area. The mode of procedure followed is the same in constructing both, the work being merely conducted on a larger scale in one case than in the other. Attention can therefore be confined to the differences in procedure introduced by the differences in the form of reinforcement employed. The most common form of spread-foundation reinforcement that is employed in the United States is the steel-beam grillage; and the form next in favor is the mesh or bar reinforcement. Bar and stirrup footings are mostly confined to foreign practice.

The manner of using beams in grillages is illustrated on p. 132. The footing course of concrete is first laid by leveling off the ground and constructing a boxing of the exact size of the footing. This boxing is accurately centered and its top edges brought to exact level by transit and Y level. The concrete is then filled in and tamped in 6- or 8-in. layers, and the top layer is accurately struck off level with the top of the boxing. A coating of mortar of 1 cement and 2 sand is then spread over the top as a bed for the beams forming the first course of the grillage. These beams are set one after another, inserting the spacers as the work proceeds. When all set, the beams are filled between with concrete. Enough space between beams is usually left for this purpose in designing the grillage. A space of at least 2 ins. is desirable to admit of easy work. When the beams are very close together grouting is resorted to, since the spacers prevent filling from the ends. The filling should be carefully done, with particular care to crowd the concrete up under the top flanges and into the angle between web and flanges. A wet concrete is necessary to get good results without excessive labor. When filled between, the ends and sides of the beams are encased with concrete to the dimensions designed, a boxing like that used for the footing being employed to give the necessary form and face. The tops of the beams are then plastered to form a bed for the second tier of beams, which is laid and embedded exactly as was the first. The final task is the setting of the stone or metal column footing in a thin bed of mortar, and the plastering of the exposed surface, to give it a smooth finish.

When mesh or bar reinforcement is employed instead of beam grillages, the general method of building up the footing is the same. The concrete is deposited in boxing, or forms, in layers, and as the proper levels are reached the reinforcement is laid and bedded on the soft concrete and the filling continued to the next level for reinforcement. The boxing or forms used are somewhat more elaborate than those required for beam grillages, but are in most cases comparatively simple structures. The footing is given a final surfacing of plaster.

In constructing extended footings covering the whole building area the work of procedure is substantially the same as that followed in constructing isolated footings, but the work is on a larger scale and can seldom be prosecuted continuously. This makes necessary the joining of old and new work, in which task the construction must conform to the usual requirements. If the footing is of the ribbed-plate type, such as the Hennebique illustrated on p. 137, the construction calls for special methods. The foundation-bed has to be smoothed off and trenched so as to form a mold for the ribs and plate. In this mold the concrete is deposited and the reinforcement placed exactly as is done in building floors as described on p. 359.

Piles with Reinforced-concrete Caps.—In constructing reinforced concrete caps for timber piles the earth is first excavated the proper distance below the pile-heads and brought to a level surface. The concrete is then deposited in layers in suitable boxing or forms, and the reinforcement is placed and covered exactly as described for spread-foundation work.

Casting Concrete-steel Piles.—The casting of reinforced-concrete piles is analogous in all respects to the molding of columns or beams in place; when the mold is set upright the molding of piles resembles that of columns, and when the mold is set horizontally the molding resembles that of beams. Each method possesses advantages and disadvantages. In vertical molding there is the advantage of having the layers in which the concrete is tamped at right angles to the line of pressure on the pile, and the disadvantage of having to compact the concrete in a narrow space with long runners. In horizontal molding the planes of the layers are parallel to the direction of the pressure, but the molding is performed under much easier conditions. The forms for molding piles are similar to the forms employed for constructing columns and beams. As they have to be used over and over they should be so constructed that they can be rapidly assembled and taken apart. The piles can be removed from the forms in one or two days, but they are not sufficiently hardened to be driven for about thirty days.

The following is the method in detail which was adopted in constructing the triangular pile, illustrated on p. 142. The molds or forms were set up vertically and the concrete placed in them in 8-in. layers. The concrete was mixed very carefully in machine mixers and was made with a minimum amount of water. Of the three sides of the mold, two sides were set up to the full height, but the third side was built up as the concrete was placed. The reinforcing-rods were placed and secured in position before any concrete was placed. Each 8-in. layer of concrete was rammed down as hard as could be done with hand rammers, then the top surface was roughened with a trowel and a set of spacing-ties placed on the rods. Another 8-in. layer was then placed. Piles were built in this way to lengths varying from 16 to 26 ft. After being molded the pile was left to set for about a day, without water, and was then kept in the form for about a week longer with continual sprinkling. By that time the piles had hardened sufficiently to allow them to be lifted vertically out of the forms and set away in a vertical position for another period of a week or so, during which time they were kept constantly wet. After this they could be removed for transportation or storage.

Driving Concrete-steel Piles.—Reinforced-concrete piles may be driven by hammer, pile-drivers, or by water-jet, or in any other manner in which timber piles are driven. In driving by hammer, however, the fall of the hammer is usually decreased and some form of driving-cap designed to distribute the impact stress evenly over the top of the pile is usually employed. The weight of the hammer, on the other hand, is usally considerably increased. In driving the triangular piles described on p. 142 a hammer weighing 5,000 lbs., with a drop of 5 ft., was employed. The 16-in. square piles of Hennibique construction employed for the Brumath Bridge and illustrated on page 139 were driven with an 8,800-lb. hammer with a maximum drop of 8 ft., which was decreased as the penetration decreased. Mr. Charles F. Marsh is authority * for the statement that Hennebique piles 12 ins. square are frequently driven by a 30-cwt. hammer with a drop of 4 ft. In driving sheet piling 0.3×0.4 in. (11.8×14.8 ins.) in cross-section at Chantenay sur Loire in 1898, a hammer weighing 3,520 lbs. and having a drop of 6.25 ft. was employed. In constructing the wharf at Novorossisk, Russia, the pile illustrated on p. 142 was driven 2 ft. into hard bottom with from 10 to 14 blows of a 3,375-lb. hammer falling from 12 ft. to 14 ft. Except that stronger tackle is required to handle the heavier concrete-steel pile, the ordinary pile-driving rig for timber piles is all that is essential.

To protect the heads of concrete-steel piles during driving use is made of a cap or false pile to distribute the shock. The most carefully worked out device of this kind is the cast-steel cap devised and used by Mr. Hennebique. This cap is shown by Fig. 45. It is a cap-shaped casting which fits over the neck of the pile loosely. The annular opening at its lower edge is calked with clay and tow and the hollow space above is filled through a hole in the top of the cap with sand or, preferably, with sawdust. A separate cap is required for each size of pile. In driving a large lot of Hennebique sheet piling at Southampton, England, a false pile was used in addition to the metal cap. A rather more complicated arrangement which was employed in driving sheet piling at Chantenay sur Loire, France, is shown by Fig. 242. Here the piles were cast with reinforced tops, which were designed to be broken off enough after driving to expose the reinforcement for bonding to the work above. This allowed the top of the pile to be clamped loosely to the guides of the pile-driver as shown. A metal cap with an upper and lower socket was set on the neck of the pile with a filling of sawdust over the pile top, and its upper socket received the end of a hard-wood false pile clamped to

* Trans. Inst. C. E., Vol. CXLIX.

the pile-driver guides as shown. The relative positions of the pile, its cap, and the false pile and hammer are clearly shown by the illustration. In driving the triangular piles shown on p. 142 a cap was used which was made of alternate layers of iron plate, wood, and lead. This cap was clamped to the pile head and also served to guide the pile in the leads. Another device used consists of a metal cap like

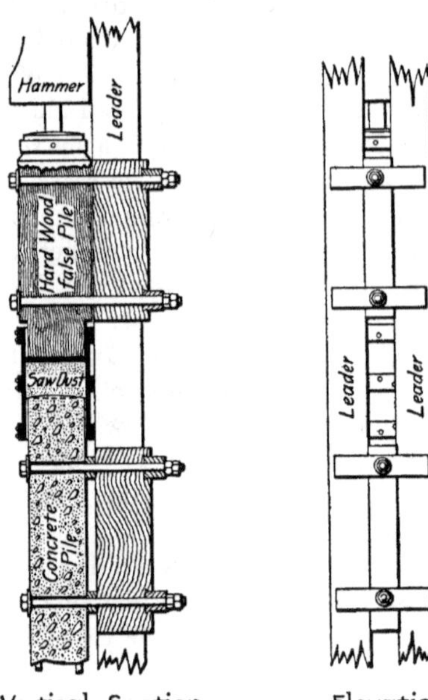

FIG. 242.—False Pile and Driving Cap Employed at Chantenay sur Loire.

that shown by Fig. 45 but having the socket on top, in which is set a disk of hard wood to take the direct blow of the hammer.

When the nature of the so.l permits, a water-jet is always a safe and sure method of driving concrete-steel piles. Piles designed for this method of driving are cast with a hollow center or have jet-pipes cast into the concrete, as shown by several of the preceding illustrations of Chapter VI. The method of driving and the apparatus employed are the same as are used in timber-pile work.

The most notable example of the use of water-jet in driving concrete-steel piles of which the author has definite knowledge is furnished by the construction of the River Aisne Bridge at Soissons, France, described on p. 210. In constructing the abutments and two piers for this bridge, 168 sheet piles and 134 bearing piles were employed. Both the sheet piles and the bearing piles were of Hennebique construction; 40 of the bearing piles were 25 c.m. (9.84 ins.) square, and 92 were 30 c.m. (11.81 ins.) square, and all of the sheet piles were 15×35 cm. (5.9×13.78 ins.) in cross-section. The material penetrated by the piles was sand mixed with a few pebbles. An attempt was made at first to drive some of the 25-cm. (9.84-in.) square piles with hammer drivers, using the arrangement of driving pile and cap shown by Fig. 243, with a 600-kg. (1,320-lb.) hammer. The first pile driven reached a depth of 2.5 m. (8.2 ft.), but its head was completely shattered for about 0.6 m. (1.97 ft.); the second pile had its head shattered at the first blow, and by the time it had reached a

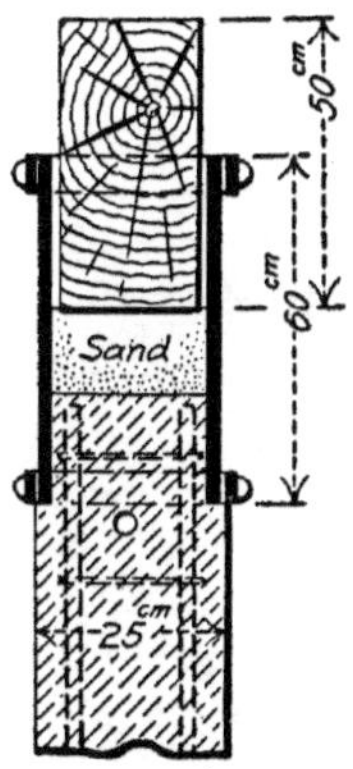

FIG. 243.—False Pile and Cap for Driving Piles in Bridge Foundations at Soissons, France.

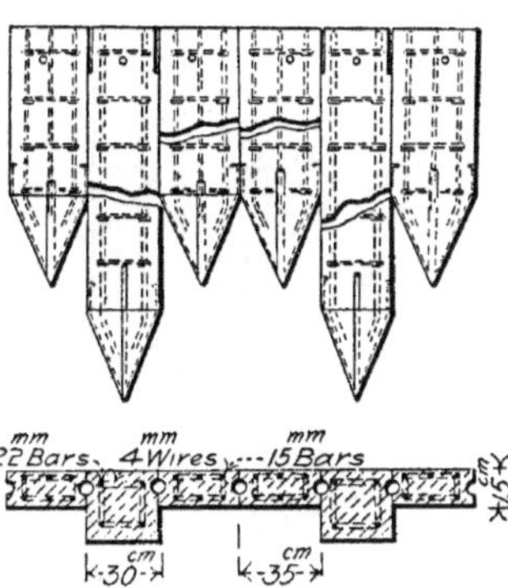

FIG. 244.—Elevation and Section of Piles in Soissons Bridge Foundations.

depth of 1 m. (3.28 ft.) the metal was bare for a length of 0.7 m. (2.29 ft.). Owing to this poor success with hammer driving, the engineers substituted a water-jet apparatus for driving the remainder of the piles. It is important to note, before giving the results of the new mode of procedure, that the engineer in charge of the work, Mr. Ribaud, Ingénieur des Ponts et Chaussées, attributes the failure of hammer driving to the insufficient age (1½ months) of the piles and to the fact that they were molded on their sides instead of vertically.

For sinking the remaining 38 piles under the abutments use was made of a hand-pump, which gave a pressure too low to loosen the material efficiently. The mode of procedure was to operate the pump until the material around the point of the pile was loosened, and then to give the pile a few light blows of the hammer to settle it into the loosened soil. This process was repeated until the piles had a penetration of from 2.5 to 3 m. (8.2 to 9.8 ft.). Six days were required to drive 18 piles for the right abutment.

In placing the piles for the two piers, power pumps were employed. Eeach pier comprised 46 bearing piles and 84 sheet piles, arranged to form a coffer-dam pointed at both ends and 20 m. (65.6 ft.) long from point to point. The arrangement of the piles in this coffer-dam is shown by Fig. 244. The bearing piles were 5.45 m. (17.8 ft.) long and the sheet piles were 5.1 m. (16.7 ft.) long; a bearing pile weighed 1,150 kgs. (2,530 lbs.) in air and 880 kgs. (1,936 lbs.) in 3 m. (9.84 ft.) of water. To guide the sinking of the piles use was made of an interior framework constructed as shown by Fig. 245. A panel of

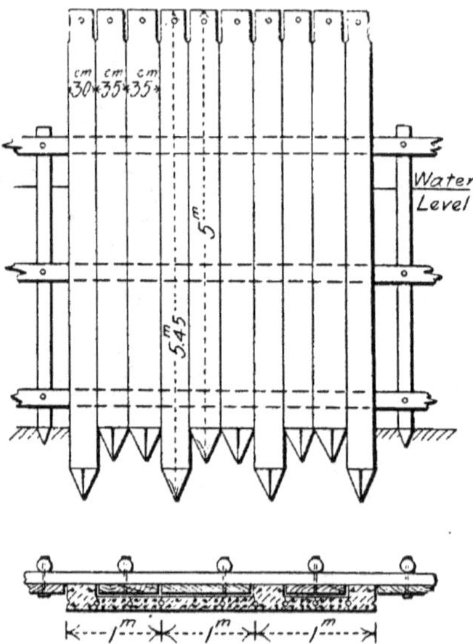

FIG. 245.—Timber Frame to Guide Piles During Driving, Soissóns Bridge.

eight or ten piles was erected in the guides and driven, then another panel, and so on. A steam-pump supplied the jet which was delivered against the soil by an armored hose terminating in an iron tube 35 mm. (1.38 ins.) in diameter and 7 m. (22.96 ft.) long. This tube or nozzle was manipulated by a workman so as to deliver the jet underneath the point of the pile being driven, and at the same time the pile was forced down by a lever over its head. With a force of six men a panel of five or six piles could be sunk 1 m. (3.28 ft.) in an hour; in one instance 45 piles were driven in a day's work. Eight days were required to drive the piles for the right pier, and 4½ days to drive the piles for the left pier. A section of the pier foundation is shown by the drawings of Fig. 130.

Constructing Piles in Place.—In the preceding section two forms of reinforced-concrete piles designed to be constructed in place are described. To perform this task use is made in both cases of a pointed metal cylinder or casing for forming the hole to be occupied by the pile and for retaining the walls of this hole in place until the concrete is placed. Fig. 246 shows the arrangement adopted for the so-called Simplex pile construction. A wrought-iron cylinder of the diameter and length which the pile is required to be is slipped over the neck of a previously cast concrete point. This cylinder, with its point, is driven by means of an ordinary pile-driver to the depth required, there being an oak driving cap or head applied to the cylinder to receive the blow of the hammer. When completely driven the cylinder of expanded metal reinforcement is slipped inside the driving cylinder until its lower end bears on the concrete point. Concrete is then shoveled into the driving-cylinder, which is gradually withdrawn as the filling progresses, until it is entirely free. Another concrete point is then attached and a second pile completed. Only the single driving-cylinder is employed when the pile penetrates earth throughout, but when there is a depth of water overlying the earth, as is the case in driving piles in a stream or lake bottom, a second cylinder or shell is clamped outside the driving-cylinder as shown in the illustration. This shell is shorter than the driving-cylinder, and in fact is just long enough to penetrate the bottom a sufficient distance to prevent the free access of water to the hole. It is driven with the driving-cylinder, which is withdrawn, leaving it in place, by simply unclamping the collar. This shell forms the mold for that portion of the pile above the bed of the stream, and is left permanently in place.

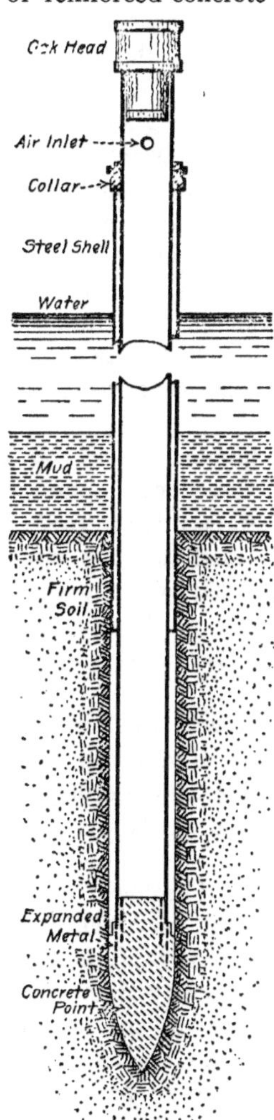

FIG. 246.—Driving Shell and Cap for Constructing "Simplex" Pile in Place.

The Simplex pile was used in constructing the extensive foundations for the new Engineer School at Washington Barracks, District of Columbia, and a rather careful study of its action was made by the army engineers in charge, Capt. Sewell and Lieut. Clarke S. Smith. From a description furnished the writer by Lieut. Smith the following data of interest are taken:

The ground where the concrete piles were driven is about 100 ft. back from the river wall along the Potomac. The character of the material penetrated was determined by borings and pits and proved to be quite variable. The surface of the ground when work began was yellow clay mixed with loam. From four to five feet below this is a stratum of blue silt or clay moderately well dried out and compacted. The thickness of this stratum is variable and in some places is 5 or 6 ft. It is underlaid by wet sand and gravel. The blue silt or clay was no doubt formed by deposit from the river, since a half-century ago the shore line extended about 150 ft. further inland.

The reference of the tops of the piles at different places varies between 11 and 14 ft. above mean low water, and the change in the surface of the river due to tides is between 4 and 5 ft., and due to freshets reaches 11 or 12 ft. From tests it was determined that the upper yellow clay would not stand 500 lbs. pressure per sq. ft. without serious settlement. This small bearing power was caused by the soft layer of blue silt or clay. As the ground is alternately wet and dry, wooden piles could not be used.

For concrete pile the apparatus consisted of a wrought-iron shell, an interior driving form, a concrete point, a pile-driver, and frame and tackle for drawing. The wrought-iron shells were made of $\frac{3}{8}$-in.-thick boiler-steel and were 11 ft. long. The same shells were used throughout the work with occasional repairs to rivets, etc. The interior driving-form was a 10-in. cast-iron pipe. The lower end of this pipe fitted over the tenon of the concrete point, and the upper end held a driving-head of black ash, so arranged as to drive both the form and the shell. The shell, form, and point were driven with a 2,200-lb. hammer falling 10 ft., and the penetration at the last few blows averaged $\frac{1}{3}$ in., the point of the pile being then embedded in the sand. When the driving-form was raised it generally brought up the tenon of the point, which had been broken off by the shock of driving. This left a rough top on the point favorable for a good bond between the concrete point and the concrete which was then dropped into the shell.

Enough concrete was placed in the shells at one time to make a layer about 12 ins. thick. Each layer was rammed and the shell was gradually withdrawn, so that there was always about 6 ins. of concrete in the bottom of the shell. Since the piles were constructed during the wet season, this process prevented the water from entering the shell and kept the earth from caving in. Work progressed during rainy weather almost as well as in good weather.

It was found that the piles could be constructed so that their axes centered well upon the positions staked out. The variation to one side was sometimes 3 or 4 ins. On account of an occasional small boulder, the pile would sometimes glance off to one side and take up a position oblique to the vertical, but this defect could usually be remedied when the shell was being withdrawn. Difficulty was experienced at first in drawing the shell from the ground, but after the apparatus was perfected there was little trouble.

To determine the bearing value of these piles, one of them was constructed at some distance from the buildings; the concrete was allowed to set two weeks and the pile was then tested and dug up. The result of the test was as follows:

Date.	Pounds.	Change in Level Reading.
April 9, 1903....	578	0.000 ft.
" "	5,859	0.000 "
" "	7,674	0.000 "
" "	9,357	0.000 "
" "	11,749	0.000 "
April 10, 1903...	15,135	0.000 "
" " ...	16,000	0.000 "
April 17, 1903...	18,604	0.000 "
" " ...	21,760	0.000 "
" " ...	25,951	0.002 "
" " ...	29,949	0.004 "
" " ...	32,525	0.008 "
" " ...	34,123	0.008 "
April 18, 1903...	34,123	0.009 "
" " ...	37,000	0.009 "
" " ...	41,095	0.009 "

The pile when dug up showed a uniform cross-section and the point was intact, but not very well bonded to the upper part of the pile. However, the pile set squarely on the point and was thus furnished with a good bearing. The length of the pile over all was 13 ft.; the length of the point being 3 ft.

At the site of another building the ground consisted of an upper layer 5 ft. thick of mud pumped in from the river during the last year. Below this is a stratum of soft clay and loam about 3 ft. thick which gradually becomes harder yellow clay as the distance below the surface increases.

There were also some piles constructed 17 ins. in diameter and about 35 ft. long. They penetrated very dry soft blue silt and landed

FIG. 247.—Driving Core Expanded for Constructing "Raymond" Pile in Place.

up in sand. The principle was entirely similar to that used for the short pile, but a shell of ⅜-in.-thick boiler-steel was employed. On account of the great friction on the outside of these shells, they were very difficult to draw until the plan of painting the outside with grease or heavy oil just prior to driving was adopted.

FIG. 248.—Raymond Driving Core Collapsed and Being Withdrawn from Shell.

The concrete points were made of 1 part vulcanite Portland cement, 2 parts concrete sand, and 5 parts broken stone (run of crusher) all by volume. The concrete for the pile was one part Old

Dominion Portland cement, 3 parts concrete sand, and 5 parts gravel by volume.

The apparatus employed in constructing Raymond concrete-steel piles is of rather elaborate character. In brief, a core jacketed with a shell of thin steel plate is driven into the ground and is then withdrawn, leaving the shell in place. Figs. 247 and 248 show the con-

FIG. 249.—View Showing Heads of Raymond Concrete Piles Ready for Wall Footing.

struction of the core. It is made in two longitudinal halves of cast steel between which is fitted a cast-steel key, with wedge-shaped lugs. When the core is assembled for driving these lugs spread the two halves apart as shown by Fig. 247, and the three parts are secured by a horizontal key through the head. The steel casing is slipped over the assembled core, and the whole is driven by any ordinary pile-driver by means of a driving-cap which sets over the head of the core and has horizontal wings fitting into the guides of the pile-driver. When the driving is completed the horizontal key is taken out and the central key-piece of the core is lifted until the lugs slip into the recesses in the side pieces. This position is shown by Fig. 248, and it allows the core to close together or collapse so that it can be easily withdrawn from the shell, the end of which is shown projecting above the ground in Fig. 248.

The reinforcing-bars are inserted inside the shell in their proper position and the remaining space is then filled with concrete. Piles constructed in this manner, but without reinforcement, have been employed successfully in a number of structures. In constructing a library building at Aurora, Ill., 142 piles 20 ins. in diameter at the top and 13 ins. in diameter at the bottom were driven by an ordinary pile-driver with a 2,400-lb. hammer having a drop of from 20 to 25 ft. The penetration at each blow ranged from ½ in. to 2 ins. The shell used was made of No. 20 sheet iron and the concrete was 1 part Portland cement, 2 parts sand, and 4 parts broken stone. The material penetrated was filled ground overlying rock at a depth of 14 ft. The heads of the piles were enclosed in concrete foundation walls as shown by Fig. 249.

CHAPTER XIV.—METHODS OF CONSTRUCTION IN BUILDING WORK.

In directing building work in concrete-steel the engineer has a long list of traditions to overcome. Foremen and masons whose training has been wholly in the use of concrete as a filler over brick arches or trough sections of steel do not readily perceive that a reinforced concrete floor-slab is a radically different structure and must be fabricated according to entirely different standards of workmanship. To meet this difficulty the engineer has to arm himself with a rigidly-drawn specification and careful supervision and inspection. The requirements for materials and workmanship in concrete-steel construction and the imperative necessity of enforcing them have been discussed in a preceding chapter. Negligence in these matters is frequently punished by disaster in concrete-steel building work.

Cost of Forms.—The successful construction of buildings in reinforced concrete is quite as much a matter of intelligent planning and expert carpentry in constructing and erecting the forms as it is a matter of careful manipulation and workmanship in setting the reinforcement and molding the concrete around it. A building of the factory type of reinforced concrete throughout, including footing, outside and inside columns, walls, girders, beams and floor-plates, roofs and stairs, will cost the contractor seldom less than $20 per cubic yard of concrete placed, and of this cost from 25 per cent. to 35 per cent. will be for forms, including materials, erection, and removal. These are, of course, round figures, since it is evident that the cost of forms must vary between wide limits depending upon the design, the times the forms can be reused, and other obvious variables. For example, the cost of forms for the 16-story Ingalls building at Cincinnati, O., including foundations, columns, walls, floors, stairs, and all other parts, was $5.85 per cubic yard of concrete in place, whereas in constructing a 4-story shoe-factory in the same city, including only footings, beams, and floor-plates, the forms cost the same contractors about $6.25 per cubic yard of concrete in place.

The following statements as to the cost of forms in building construction have been secured from a number of experienced contractors and designers of this class of work. The costs given include materials, erection, and removal:

(1) For plane floor-slab construction, such as is common with expanded metal, International metallic sheeting, and Clinton fabric, the cost of forms may go as low as $2.20 per cubic yard on a 16-ft. span with a 400-lb. load, and as high as $3.25 per cubic yard on a 16-ft. span with a 100-lb. load. For beam and plate construction, such as is used by Ransome and others, the cost of forms will run from $5.50 per cubic yard on a 16-ft. span with a 500-lb. load, to $10.50 on the same span with 100-lb. load.

(2) The forms for concrete-steel floors will cost from 4½ cts. to 6 cts. per square foot, including everything. The forms for a concrete wall 4 ins. thick will cost from 8 cts. to 11 cts. per square foot of wall. The forms for a column cost 22 cts. per lineal foot, the quantity of concrete being 3.7 cu. yds. per 100 lin. ft. of column.

(3) For floor work the cost of the forms will range from 10 cts. to 20 cts. per square foot, for wall work the cost should not exceed 5 cts. per square foot based upon the area of one side of the wall only. Where forms can be used several times over in a single building the cost often runs below the above minina.

(4) Experience on about 30 buildings shows that it is rarely possible to furnish the centering and remove it for much less than $4 per cubic yard, and that only by very bad management or under unfavorable circumstances can the cost exceed $6 per cubic yard.

Floors Supported between Beams.—The construction of floors of the types designed to be carried by steel I beams is a comparatively simple process. With rare exceptions such floors are plain flat plates or plain arches, and they are built on centers consisting of plain lagging boards carried on joists or arched ribs. Fig. 250 shows a form of center designed for flat plate floors carried on the top flanges of the beams. For similar floors resting on the bottom flanges the center shown by Fig. 251 is employed; this center is usually hung from the beams by hook-bolts or other suitable devices. A common form of center for plain arch floors is shown by Fig. 252. All these centers are built of wood, and are within the ability of any ordinary carpenter to construct. The lagging used should be smooth dressed, chiefly to prevent the adhesion of the concrete to it, but perfect smoothness is not essential otherwise, since the underside of the plate is usually

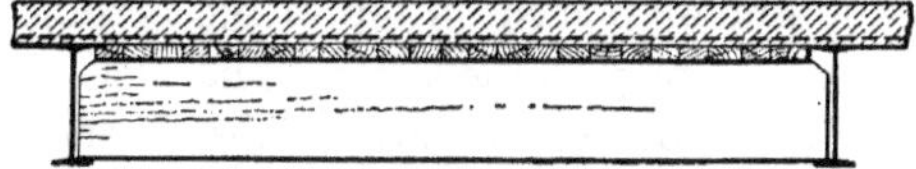

FIG. 250.—Center for Slab Floor Carried on Tops of Beams.

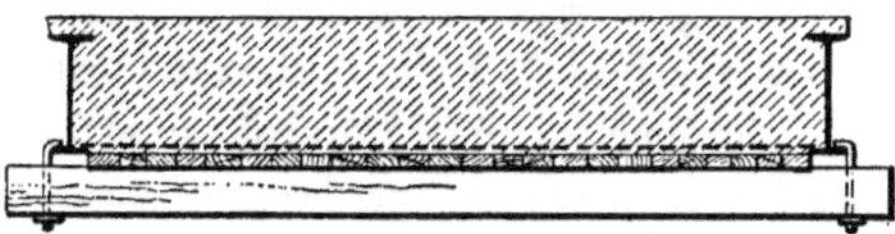

FIG. 251.—Center for Slab Floor Carried on Lower Flanges of Beams.

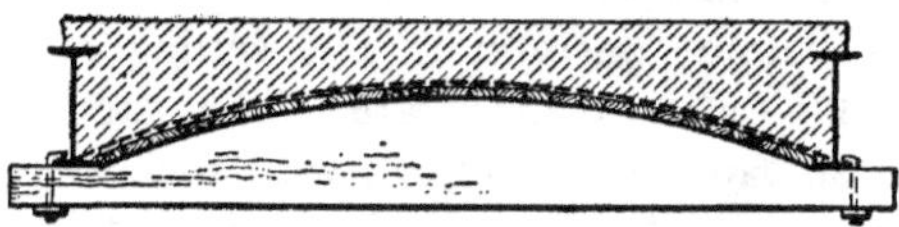

FIG. 252.—Center for Arch Floor Between Beams.

concealed by a suspended ceiling or else is covered with a plaster finish. To reduce still further the chance of adhesion the lagging can be greased or soaped or it may be covered with a sheet of paper or cloth. Ample means to prevent the adhesion of the lagging to the concrete should be provided, otherwise the striking of the centers is certain to subject the floor to injurious shocks and to tear away portions of the underside of the slab. The practice of jarring the lagging loose by blows of a sledge or crowbar is wholly reprehensible.

The method of placing the reinforcement and the concrete which embeds it depends largely upon the character of the reinforcing element. It should always be such that the reinforcement occupies the position in the slab which the designer intended it to occupy and that the metal is thoroughly embedded in the concrete. The concrete slab should be a continuous piece of construction. These are essential requirements to the strength and stability of the construction, and if they are not met perfectly the structure is to that degree imperfect and unable to perform the duty its designer intended it to perform

With network reinforcement, when a single bottom net is used, it is usually laid directly on the centering and the concrete is deposited in a single layer on top. This is the normal mode of procedure with expanded metal reinforcement, but a better practice is to deposit a thin bed of concrete first, or, at least, to make certain that the concrete gets under all portions of the netting by lifting it with hooks as the concrete is placed. With Monier netting a bed of mortar is always laid first on the centering; the netting is laid on this and then the remainder of the mortar is placed on top of the netting. When upper and lower nettings are employed, the latter is placed as already described and the concrete filled above to the level of the second netting; this is then laid on the concrete and covered above to the depth required. The concreting of floors with bar reinforcement is a somewhat more particular task. Usually a bed of concrete is placed on the center and on this the bars are laid in their designed locations; concrete is then filled above to the level of the second layer of bars, which is laid exactly like the first, or in case only one layer of bars is used the filling is carried up to the top of the floor-plate at once. With certain forms of bar reinforcement like the Columbian, the bars are hung in place from the beams before concreting. The setting of bar and stirrup reinforcement calls for particular care. The bars are placed on a bed of concrete spread on the centers with the stirrups under them at the proper points. To keep these stirrups in position a small mound of concrete is packed around the foot of each one. The filling above them proceeds with care not to disarrange the stirrups and to get the concrete in thorough contact with them. With all dry mixtures this is practically a task of the fingers. With rib reinforcing members, like the Melan and Wünsch, the mode of procedure is exactly the same as in constructing bridge arches of these types, except, of course, that the concrete is deposited in one layer and for the full arch at once. Certain forms of network reinforcement, the most noted example of which is the Roebling woven-wire netting, are sufficiently stiff when arched between I beams and of such fine mesh that they themselves serve as a centering upon which the concrete is deposited. In such cases the metal is never thoroughly embedded in the concrete and the arch has to be plastered on the underside.

Monolithic Floors.—The ordinary form of monolithic floor is a flat plate stiffened underneath by ribs which serve the purposes of girders and floor-beams. Generally the ribs are beams of uniform depth, but sometimes they are arch ribs. In either case the center has to be so shaped that it will mold both ribs and plate. A somewhat elaborate construction is required to accomplish this, and it must furthermore be supported from below by means of a staging of some sort. Generally also the task of placing the reinforcement and depositing the concrete requires more care than in the case of floors between steel-beams. A few examples from practice will illustrate both features of work.

Forms.—The drawings of Fig. 253 show the centers commonly

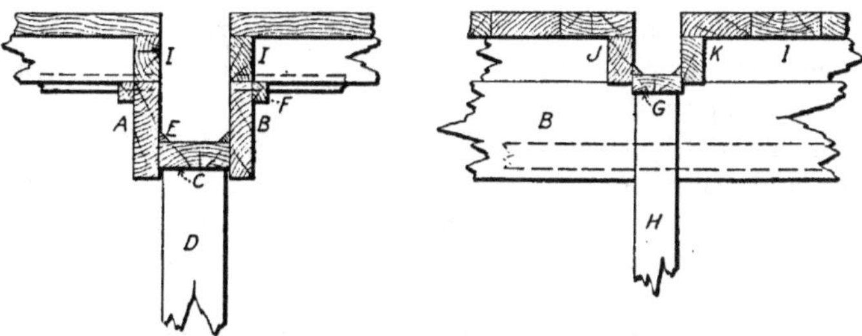

FIG. 253.—Forms for Hennebique Slab and Girder Floor.

employed in constructing ribbed floors of the Hennebique type, when the construction of the centers and the concreting are in a measure simultaneous processes. The molds for the main girders or ribs are first built up to the level of the bottoms of the secondary beams by erecting the pieces *A*, *B*, and *C*, and the bottom board is carried by the post *D*, and the side boards *A* and *B* are clamped to it by means of metal clamps of the form illustrated by Fig. 254. To give the beveled

FIG. 254.—Clamp for Girder Forms, Hennebique Construction.

FIG. 255.—Floor Girder Form Employed at Ingalls Building, Cincinnati, Ohio.

edges to the girders, triangular strips *E* are nailed to the side boards. In the mold thus formed the bottom portion of the main girder is concreted nearly to the tops of the side pieces *A* and *B*. The construction of the molds for the secondary ribs or the floor-beams is then commenced. Battens or cleats *F* are nailed to the side pieces *A* and *B*, to carry the ends of the bottom boards *G*, which are also supported at intermediate points by posts *H*. The upper boards *I* of the main girder forms are then placed and the side boards J and K of the beam forms are clamped to the edges of the bottom boards G. The placing of the lagging L completes the center. The concreting of the remainder of the main girder, the floor-beams, and the plate then proceeds. It is, of course, possible to erect the forms complete instead of in sections, and in that case the concreting is a continuous process.

A modification of the form just described is illustrated in Fig. 255. This form was employed for molding the floors for the Ingalls office building at Cincinnati, O. The clamp consists of a bent iron rod on the tail of which slides a washer, with an eccentric hole a little larger than the diameter of the rod. By driving this washer toward the hook it becomes "bound" on the rod and tightly clamps the side boards to the bottom board of the mold. A blow with a hammer loosens the washer and consequently the clamp. These forms were carried on timbers between the column molds and by posts at the center of span.

The drawings of Fig. 256 illustrate a form of centering employed

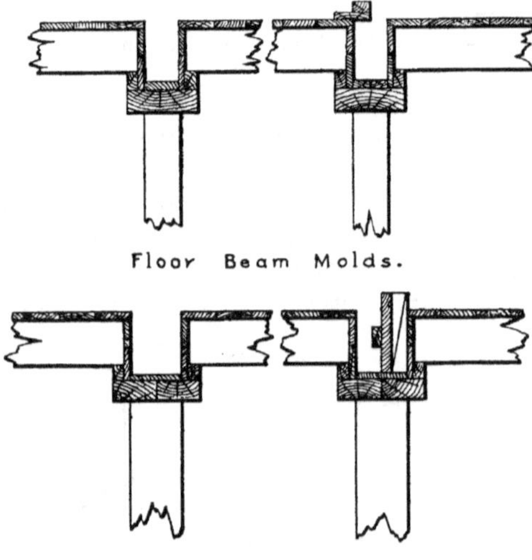

FIG. 256.—Forms for Slab and Girder Foundry Floor, Eastwood Company, Paterson, N. J.

in building the floors of the Eastwood Company's foundry and described on p. 164. The construction resembles that of the Hennebique centering previously described, and only the devices employed for producing the expansion joints in the floor-slabs and the planes of weakness in the main girders call for description. The expansion rebates in the floor-slabs are formed against the temporary strips A nailed over the beam mold. These strips are removed after the concrete has hardened and before the joining slab is built. The lines of weakness in the main girders are produced by building them in two parts in the same mold. One half of the mold is first shut off by the partition board *B* held in place by wedges and the other half is concreted. When this concrete has hardened the board and wedges are removed and the second half is concreted.

Fig. 257 shows the mold employed in building the floors for the works of the Pacific Borax Company, which are described on p. 165.

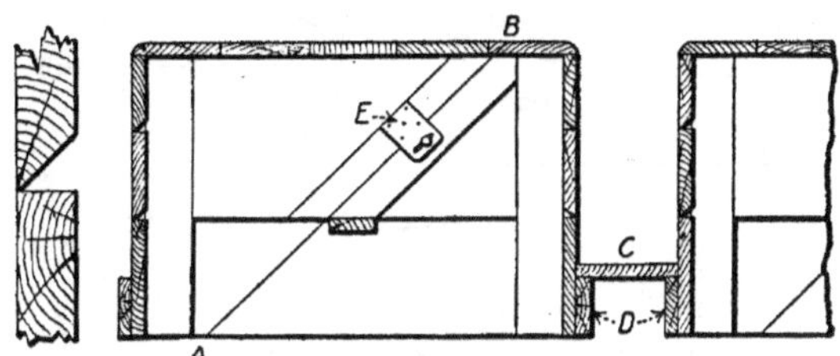

FIG. 257.—Form for Slab and Girder Factory Floor for Pacific Coast Borax Company, Constable Hook, N. J.

These molds consist of boxes having solid sides and top but no bottom, and exactly corresponding in shape and dimensions to the cavity of the panel between each pair of floor-beams and the girders with which they connect. A number of these molds are set side by side with their ends resting on stringers carried by the column molds and with an interval between each two equal in width to the thickness of the floor-beam. The bottoms of these intervening spaces are closed by boards *C* resting on cleats *D*, and they thus form molds for the floor-beams. To facilitate the removal of the mold it is cut into halves along the plane *AB*. These two halves are clamped together by the slotted plates *E*, which when loosened allow the two parts to slide and so partly collapse the mold. The drawings of Fig. 258 show a similar mold employed in constructing the floors of the Kelley & Jones brass factory which are described on p. 167. In these molds collapsibility is secured by a longitudinal hinged joint in the top lagging which allows the bottom edges to close together when the cross braces are removed. In constructing this building enough molds were made to construct one entire floor, and they were usually left in place seven days after the concrete was placed.

Fig. 259 shows the girder and slab mold for building the floors of the Central Felt & Paper Company's factory which are described on pp. 182 and 183. In this mold the bottom piece is separate and easily detached from the vertical side pieces. Both bottom board and side pieces are independently supported on transverse cap planks carried by posts spaced 6 ft. to 8 ft. apart along the center line of the girder. The bottom pieces were single 3-in. planks and the side pieces were made of $4\times\frac{7}{8}$-in. boards nailed to 2×4 in. top and bottom timbers. A third 2×4-in. horizontal piece near the top received the loose ends of the transverse joists which carried the lagging for the floor-slab. Ver-

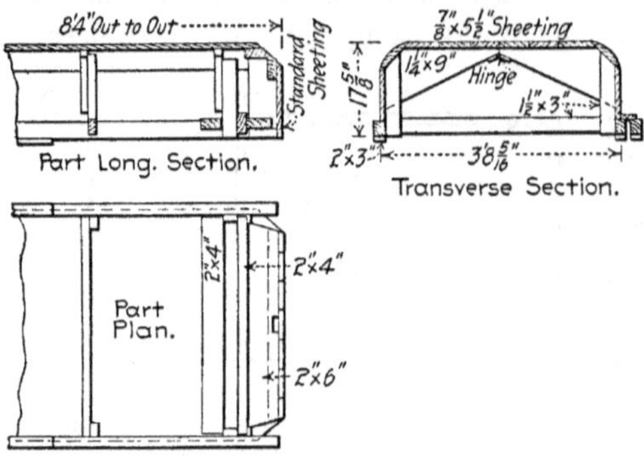

FIG. 258.—Forms for Floors for Factory Building at Greensburg, Pa.

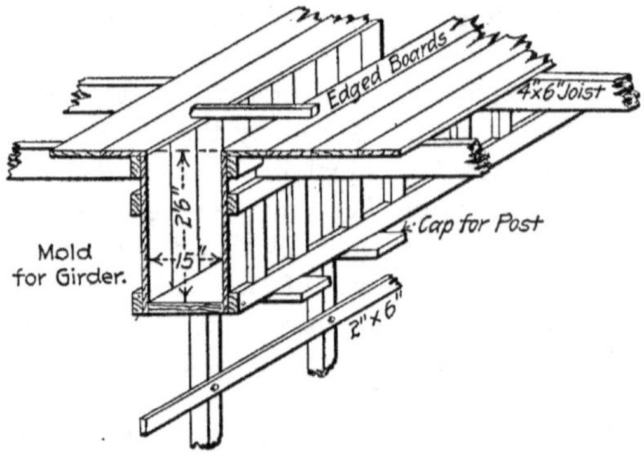

FIG. 259.—Form for Slab and Girder Factory Floor for Central Felt and Paper Company, Long Island City, N. Y.

tical pieces under the ends of the joists transmitted their loads to the bottom piece and to the post-caps. The lower edges of the side pieces were temporarily screwed to the bottom plank and their upper edges were spaced by occasional cross pieces from which the reinforcing-rods were suspended prior to placing the concrete. The first lagging boards for the floor-slabs were nailed to the side pieces and to the ends of the joists but the others were laid loose. After the floor-slab had set about seven days, the joists were revolved 90° to lay on their flats on the side pieces and the lagging followed them down to a level about 4 ins. below the ceiling. A few days later the side pieces were unscrewed and removed leaving only the bottom board supported in place by the false works posts. The bottom boards were left in place about three weeks and were then dropped about 1 in. by removing fillers from the post-caps. The centering remained in this position for another week. The advantage claimed for this form of mold arrangement is that it allows the concrete to be quickly freed to the action of the atmosphere while remaining supported until it is well hardened.

Staging for Forms.—The floor forms are supported in position for the concreting by staging or false work carried on the finished floor beneath the one being constructed. This false work is of simple construction; it consists usually of uprights resting on double wedges on the finished floor and bearing against the bottoms of the beam and girder molds at the tops. Diagonal boards connect the successive uprights and brace them together into a stable false works. The principal precautions to be taken are to have the false work strong enough to carry its load and to use uprights with sufficient frequency to make sure that the beam and girder molds do not deflect materially between supports. Generally speaking the uprights should be not over 5 ft. apart under girder and beam molds. Unless the span of the floor-slab is considerably over 5 ft., the lagging will not need to be supported between beams, but at least a 2-in. lagging is necessary and one considerably thicker will be preferable. Where intermediate support is required for the slab-lagging it may be secured by uprights carrying plank caps. The entire falsework should be so designed that its removal can be accomplished rapidly and easily and without shock and stress to the concrete floor which it supports. The amount of timber required for falsework is considerable, since it is seldom possible to count on reusing any part oftener than once a fortnight in cold weather and once a week in warm weather, and the contractor can profitably spend considerable thought in designing his stagings so as to economize material and to preserve as much of it as possible for future use.

Concreting.—Careful workmanship is essential in placing the reinforcement and the concrete in floor construction. The process of concreting is in other respects simple. Usually the placing of the reinforcement and the concrete proceed simultaneously, but in some cases they are made separate operations. In placing the concrete and reinforcement for Hennebique floors the following mode of procedure is the generally accepted practice: A bed of concrete is first placed in the bottoms of the beam and girder molds, and on it are laid the tension-bars with the stirrups slipped under them and held in position and upright by small mounds of concrete packed around their bottoms. The remaining concrete is then placed and rammed in 2-in. layers until the level of the floor-slab is reached. The reinforcing-rods and stirrups of this are then blocked up on the floor-lagging or laid on a bed of concrete exactly as in constructing girders. Care has to be exercised to make certain that the stirrups are kept in contact with the tension-bars and that neither is distorted or shifted in position by the ramming. The practice of Mr. Hennebique, and of European engineers generally, is to use a moderately dry concrete, and this, with the stirrup reinforcement, calls for very thorough and careful ramming to secure a perfectly embedded skeleton and a homogeneous and compact concrete. With bar reinforcement alone the process of construction is much more simple; this is particularly the case if a wet concrete is used, as is the usual American practice.

Instead of placing the reinforcement piece by piece as described above, the practice is sometimes adopted of erecting the skeleton complete before any concrete is placed. An excellent example of this procedure is presented by the construction of the girders in the factory of the Central Felt and Paper Company, at Long Island City, N. Y. In this work the girder and beam skeletons were connected up completely, as is indicated by the drawings of Fig. 260, and suspended in the molds from cross-pieces resting on their top edges, as shown by Fig. 259. The concrete was made very wet, so that little ramming was required to settle it well around the metal.

In concreting the floors for the Kelley & Jones factory at Greensburg, Pa., the concrete was compacted in the beam and girder molds by cutting it with a spade having a $6\times18\times\frac{1}{8}$-in. blade attached to a 5-ft. gas-pipe handle. The concrete of the floor-slab was compacted by rolling, using first a 3×3-ft. 250-lb. wooden roller, then a $2\frac{1}{2}\times2\frac{1}{2}$-ft. 500-lb. iron roller, and finally a $2\frac{1}{2}\times2\frac{1}{2}$-ft. 700-lb. iron roller. This mode of procedure is reported to have given excellent results.

When the concreting of the floor cannot be made a continuous process from start to finish, which is usually the case, care must be exercised in stopping off one day's work and in joining the next day's work to the break. If the floors are divided into panels by expansion and shrinkage joints, like, for example, the floor for the Eastwood foundry illustrated by Figs. 75 and 256, the usual practice is to make the concreting of each section continuous and in each day's work to begin and finish a new set of sections. When these natural stopping-places are not available it is the best practice to stop each day's work with an inclined face over the beam and girder molds. In no case should the concreting be stopped before the full depth of the floor has been completed over the entire area of the section chosen for a day's work.

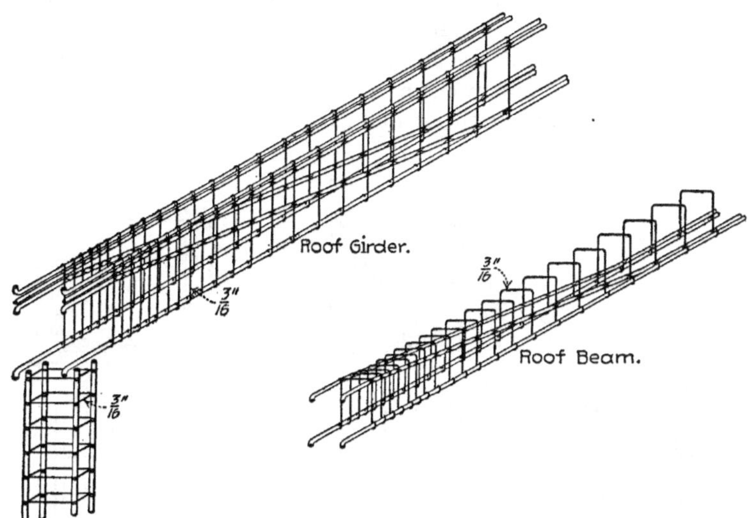

FIG. 260.—Girder and Beam Reinforcement Assembled for Erection, Central Felt and Paper Company, Long Island City, N. Y.

In warm weather, and particularly when the floor receives the direct sunlight, it is advisable to keep the surface well sprinkled, or covered with a cloth kept constantly wet, to prevent the formation of surface cracks. The finish of concrete floors calls only for brief mention. When they are to be covered with wood flooring the usual practice is to place nailing-strips in the slab and fill between them with a cement of cinder and mixture. Mosaic slate or marble slab flooring is of course laid directly on the concrete with the usual plaster bed. When granolithic or other special finish is employed it should always be laid if possible while the concrete is still soft. In case this procedure is impossible the concrete surface should be roughened and well washed with water and then a grout wash used to join the finishing coat to the slab.

COLUMNS.

Columns are molded in place in vertical forms which are usually built upward from the bottom in short sections as the concreting progresses. Sometimes, however, the form is erected complete the full height of the column and the concrete is filled in from the top. With the first construction the ramming of the concrete in layers is quite practicable, and it is preferred by many engineers for that reason. When the forms are built to their full height at the start the ramming of the concrete

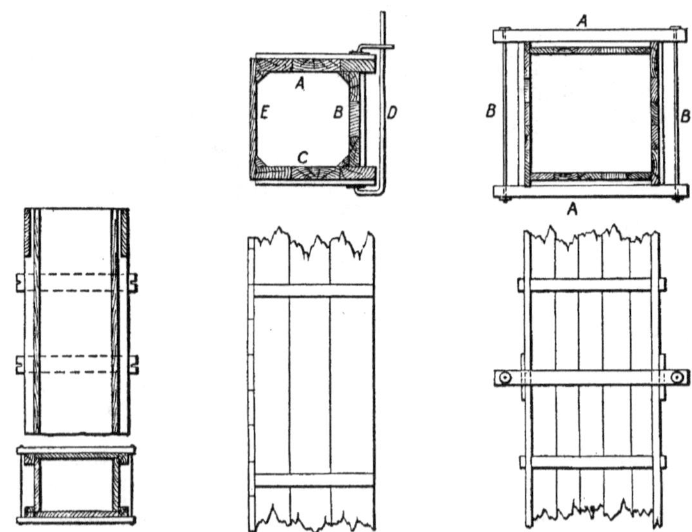

FIG. 261.—Form for Column for Factory of Pacific Coast Borax Company, Constable Hook, N. J.

FIG. 262.—Standard Column Form, Hennebique Construction.

FIG. 263.—Column Form Employed at Ingalls Building, Cincinnati, Ohio.

can be accomplished only by means of long-handled rammers much like cannon-rammers, and then only with danger of disturbing the reinforcement. In such cases it is the more common practice to use a very liquid concrete and trust to gravity and its fluidity for the complete filling of the mold and the thorough embedding of the metal.

Forms.—Fig. 261 shows the form used in constructing the 16-ft. columns for the factory of the Pacific Coast Borax Company. In this construction four corner uprights are erected and are covered with short lengths of vertical boarding with the joints under the horizontal cleats.

These cleats are notched at the ends to provide for tie-rods with nuts which screw up and clamp the form together. At the top of the form the column uprights are notched to carry stringers which reach from column to column and serve to carry the floor-molds. This form is open for concreting on all four sides at the top of each section of vertical boarding.

The part elevation and transverse section of Fig. 262 show the form usually employed in molding Hennebique columns. The vertical side boards *A*, *B*, and *C* are erected to the full height of the column and are secured by cleats and metal clamps *D*. The fourth side *F* is built up from the bottom with horizontal boards as fast as the concreting progresses. This form is quickly and easily removed, but it permits concreting from only one side.

The column forms employed in building the 16-story Ingalls Building at Cincinnati, Ohio, were constructed substantially as shown by Fig. 263. Vertical boards secured by horizontal cleats form the four sides of the form. Clamps with two yokes *A* and two spacing-pieces *B* hold the sides of the form in place after erection. To hold these forms in place longitudinal and transverse X-bracing was fastened between them. The other details are indicated by the drawings.

Fig. 264 shows the forms for interior columns for the factory of the Central Felt and Paper Company, Long Island City. It has three continuous sides and one sectional side, the latter being built up of horizontal boards as the concreting progressed. It will be noted from this sketch that the horizontal ties were all temporarily assembled on the vertical bars near the top before concreting was begun. This enabled the ties to be slipped down and located without delay to the concreting.

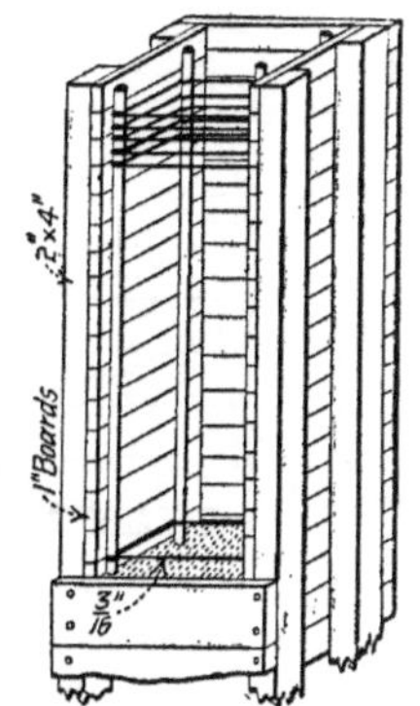

FIG. 264.—Column Form Employed at Factory of Central Felt and Paper Company, Long Island City, N. Y.

In constructing the Kelley & Jones factory at Greensburg, Pa., use was made of the column molds shown by Figs. 265 and 266. Fig. 265 shows the form construction for the interior columns. This is built with four separate side pieces of $5\frac{1}{2}\times\frac{7}{8}$-in. planed boards having one beveled edge and one square edge. On two opposite sides the boards are vertical and are nailed to the edges of horizontal cleats spaced 20 ins. apart; the other two sides are made of horizontal boards nailed to a pair of outside vertical cleats. The horizontal cleats project beyond the sides of the mold and are slotted to receive the tie-rods which clamp the four sides of the mold together. The topmost cleats project

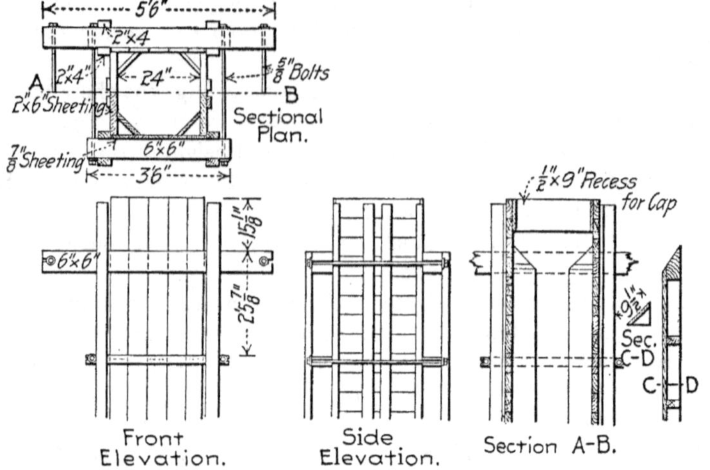

FIG. 265.—Form for Interior Column, Kelley & Jones Factory, Greensburg, Pa.

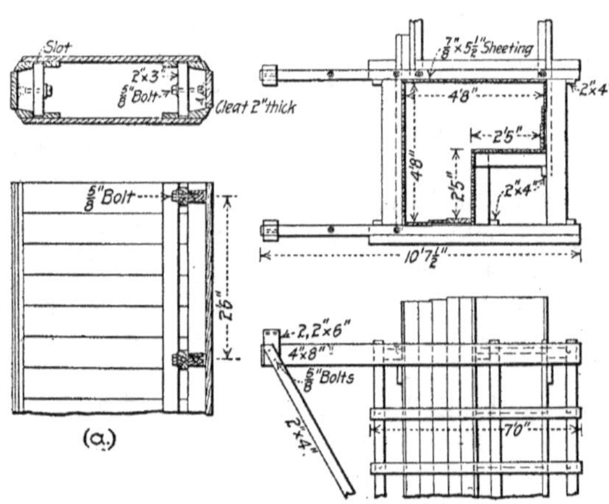

FIG. 266.—Form and Core for Wall-column, Kelley & Jones Factory, Greensburg, Pa.

about 2 ft. beyond the sides of the mold and are made heavy to carry the main girder forms, as is shown by Fig. 265. Vertical struts reaching to the floor below support the ends of these long cleats. The other details of the construction are shown by the drawings. The outside column molds were similar in a general way to those for the interior columns, except that all the boarding was set vertically, and as the columns were hollow a core or mandrel was used inside the form to shape the hollow interior. Fig. 266 is a drawing of one of the corner piers or columns, and Fig. 266 (*a*) shows the construction of the core. To collapse these cores the cross-piece resting in the slots was revolved on the bolt as an axis until it slipped free of the slots and allowed the sides to close together.

Figs. 267 and 268 show the construction of the molds employed in building the architectural columns for the Nassau County Court-house at Minneola, N. Y. Fig. 267 is the mold for one of the twelve-sided columns in the court-room. It is made of wooden staves doweled together and hooped with adjustable iron straps. Fig. 268 is the mold

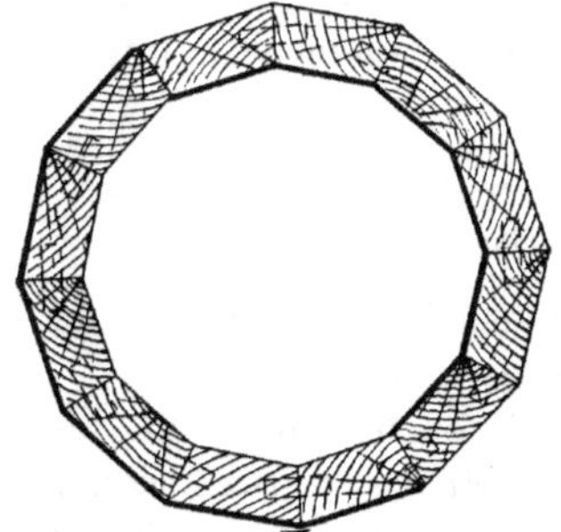

FIG. 267.—Form for Twelve-sided Column, Nassau County Court-house, Minneola, N. Y.

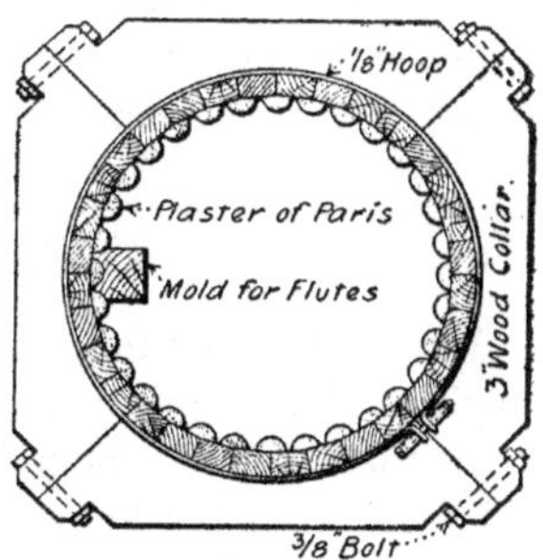

FIG. 268.—Form for Fluted Column, Nassau County Court-house, Minneola, N. Y.

for the fluted portico columns. These columns have nominally circular perimeters, but they are really 24-sided polygons. The mold consists of 24 staves doweled together and hooped around by adjustable bands; the flutes are of plaster of Paris secured to the inside of the mold at the angles of the staves by screws. In removing the molds the screws were loosened and the staves stripped off, leaving the plaster strips in place as a protection to the column.

Concreting.—The first step in molding a column after the form is erected is to set up the reinforcement. This, as already stated, generally consists of four or more vertical rods lengthwise of the column and of horizontal ties which bind these rods in position at intervals vertically. In case the column is concreted entirely from the top of the full-height form this reinforcement has to be erected complete before the last side of the form is finally closed in. When the form is built up in short vertical sections as the concreting progresses, the horizontal ties can, if desired, be placed just ahead of the filling, but the main rods have of course to be erected complete before concreting begins. When the filling is accomplished from the top of a full-height form ramming is a rather difficult process and the concrete must be made very wet to insure the thorough embedding of the metal. In constructing the columns of the Ingalls Building at Cincinnati, Ohio, enough concrete was deposited at one time to fill the mold about 12 ins., and this was then compacted by long-handled rammers before another layer was deposited. When the forms are built up just ahead of the filling the deposition and ramming of the concrete in layers is much simplified; thin layers can be employed and a fairly dry concrete used if desired. For ramming the concrete in the columns for the Kelley & Jones factory at Greensburg, Pa., a rammer of the following construction was used with notable success: The handle was 5 ft. long and was attached to a head shaped like an elongated plumb-bob with a point 1 in. in diameter, a top 3 ins. in diameter, and 18 ins. long.

Walls and Partitions.

Walls and partitions are molded in place between vertical forms which are built upward in sections as the concreting progresses.

Forms.—The principal problem in wall construction is the design and erection of the forms so as to reduce their cost to the lowest possible amount. The simplest and most obvious construction for a wall-form is to fix posts or standards at intervals on both sides of the intended wall and to sheath them inside with boards. The posts are fixed in place by outside struts and by connecting ties through the wall-space. Forms constructed in this manner consume a large amount of timber, which is usually fit only for the roughest of uses after it is removed, but they are very frequently employed, and where lumber is cheap they compare favorably in cost with any of the various sectional forms designed to be built upward by taking away the rear sections and erecting them ahead as the concrete hardens.

A sectional wall-form in familiar use in England is shown by the drawings of Fig. 269. The posts are erected and held the proper distance apart by tie-bolts passing through conical spacers. On both sides of the wall-space between the posts are set shutters made up of boards nailed to battens. These shutters are set flush with the inner side of the post and are held in that position by strips nailed to the sides of the posts. The battens project at the top and are there tied together through the wall-space by bolts and spacers. In building a wall the lowermost shutters are removed from the hardened concrete and set on top for concreting another "lift" of wall. To remove the shutters the holding-strips are removed, the tie-bolts unscrewed, and the taper spacers driven out. The holes left in the wall by the spacers are ultimately filled with mortar. The boarding for the shutters should be narrow, about 6 or 7 ins. wide, and the joint should be as tight as is practicable. Many patented forms of sectional wall-molds exist, but on the whole they offer few advantages over the simpler forms. Those who are interested will find a number of these molds described in Thomas Potter's "Concrete: Its Use in Building" (London, 1894).

A form of wall-mold successfully employed in America is shown by the sketch Fig. 270. It consists of slotted standards placed in pairs

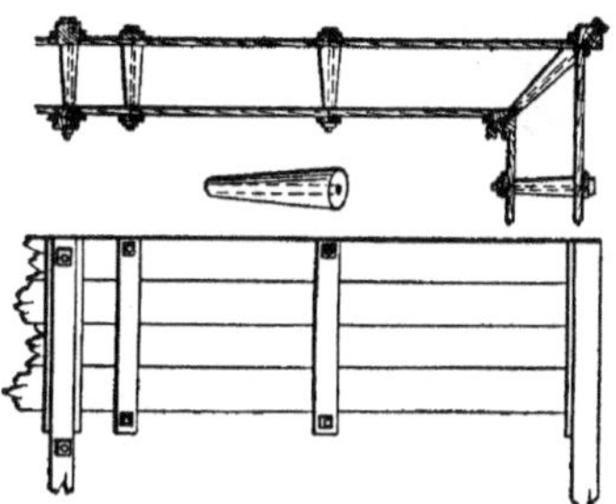

FIG. 269.—Sectional Wall-form Supported by Independent Staging.

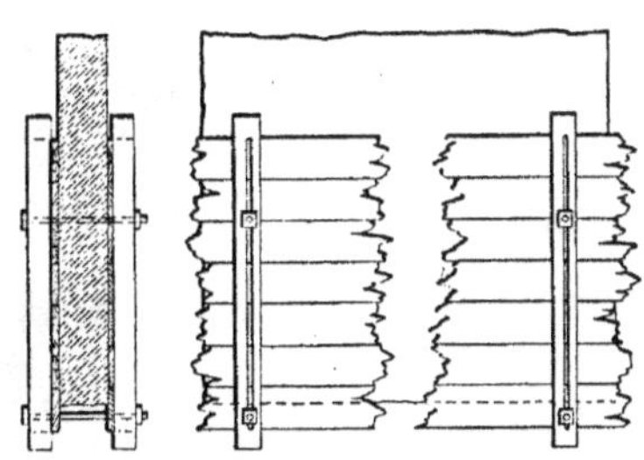

FIG. 270.—Sectional Wall-form Supported by Completed Wall.

at intervals along the wall and one on each side of the wall-space. The two standards of each pair are connected through the wall-space by bolts fitted with hand-nuts. These bolts pass through the slots in the standards, and the molding-boards are set up against their inside faces. In operation the molds are filled to about the top of the uppermost board, the bolts are then slackened and the standards are pushed up a space and again clamped by tightening the bolts. The advancement of the standards liberates some of the bottom boards, and these are placed on top to extend the mold. When the bottom bolts have reached the lower end of the slot they are taken out and inserted at the top. By a repetition of their operations the concreting of the wall is continuous.

When corners are turned a special mold construction is necessary. One form of corner mold is shown by Fig. 269. This has the serious objection that the concrete is perforated by the bolt- and spacer-holes at a point where these are particularly objectionable. A corner form which has been employed with entire success in America in a number instances is shown by Fig. 271. The tie-bar *B* is left permanently in the wall.

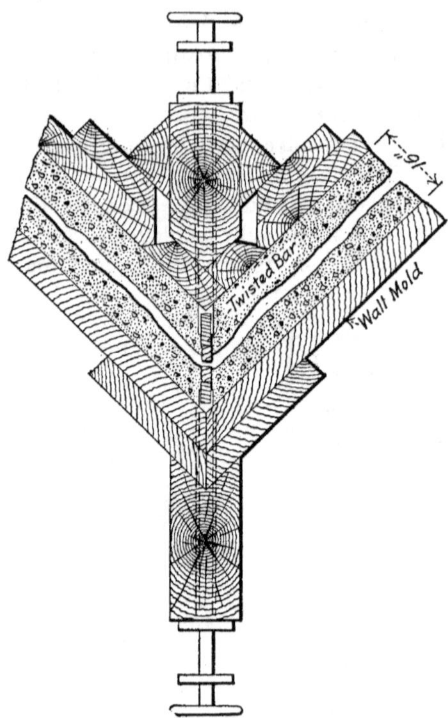

Fig. 271.—Details of Corner Form for Walls, Pacific Coast Borax Company, Constable Hook, N. J.

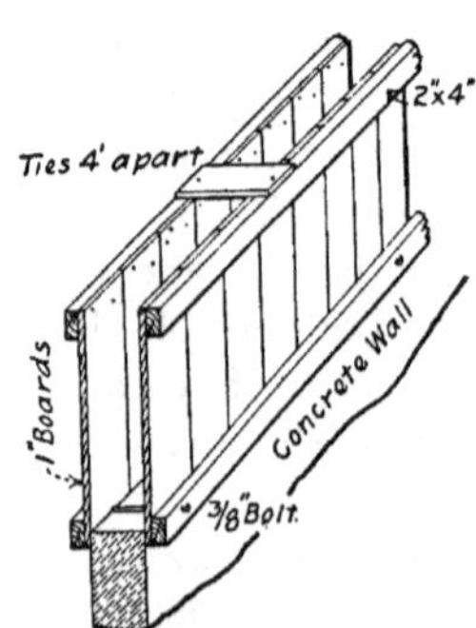

Fig. 272.—Wall-form for Factory of Central Felt and Paper Company, Long Island City, N. Y.

Fig. 272 shows the wall form employed in constructing the wall of the Central Felt and Paper Company, which is illustrated by Fig. 97. Each panel of the mold consists of two vertical side pieces 3 ft. high and 16 ft. long. For the first course they were seated on the ground and braced by props on both sides. After the concrete had set for three days, the molds were loosened and lifted until the lower edges were 2 inches below the top of the concrete. In this position they were supported by bolts passing through sleeves resting on the top of the concrete, and their upper edges were held by transverse boards nailed 4 ft. apart. The sleeves used consisted of pasteboard tubes, it being found that these were just as efficient and much easier to point and trim than iron pipe. In using this mold care must be taken to start the wall to exact line and dimensions and to finish each section flush with the top of the mold. Bond between the old and new concrete was secured by washing the surface with broom and hose and flushing it with neat cement.

Wall-mold Ties.—It is usual in thin walls to use a tie between standards which is removed completely, leaving a hole entirely through the concrete. These holes, while they are not seriously objectionable, are better avoided if possible, and an endeavor is usually made to limit their number and to have them located in the least objectionable places. Where the walls are thick it is always difficult and sometimes practically impossible to withdraw the ties entirely, and if they are cut off flush with the face the exposed ends rust and discolor the masonry. To overcome these difficulties use has been made in a number of instances of detachable wall-ties. Two examples of such devices are shown by Figs. 273 and 274.

The device shown by Fig. 273 is a combination of rods, castings,

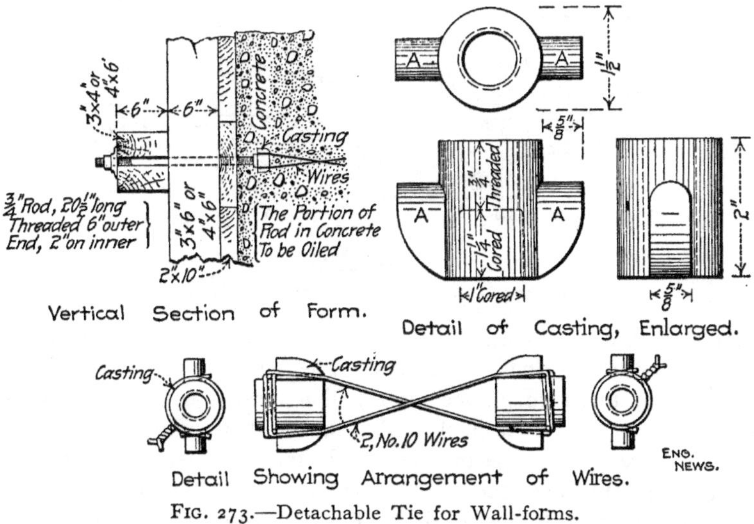

FIG. 273.—Detachable Tie for Wall-forms.

and wire. The ¾-in. rods, 1 ft. 9 ins. long and threaded on both ends, are placed through the form and bracing-timbers so as to allow sufficient clearance for tightening. On the inner end is fitted a casting, and it will be seen that the threads inside of the casting extend but half the depth, and the remainder is cored to 1 inch diameter. It may allow mortar to surround the ends of the rods, but that is provided for by oiling or greasing the rods before placing, which makes their removal easy. The castings are simply rough castings, but the rounded surface of the wings (*A*) is cast or filed smooth so as to leave no sharp projections to cut into the wires when they are tightened. The castings are screwed onto the inside ends of the rods to their final positions and remain permanently in the concrete. The wire used is No. 10 annealed fence-wire; four strands securely tied, as shown, to two opposite castings and tightened by turning the nut on the outside ends of the rods until about 2 inches of each rod will project into the concrete. After the concrete has hardened sufficiently the forms are removed, the rods withdrawn, and the depressions left by them grouted, thus making a clean face, with no exposed metal to cause discoloration. The use of wire also eliminates the necessity for any previous calculations for lengths of ties, and enables a stock of castings, short rods, and wire to be kept to anticipate the demand, thus avoiding delays.

The advantages claimed for the device shown by Fig. 274 are

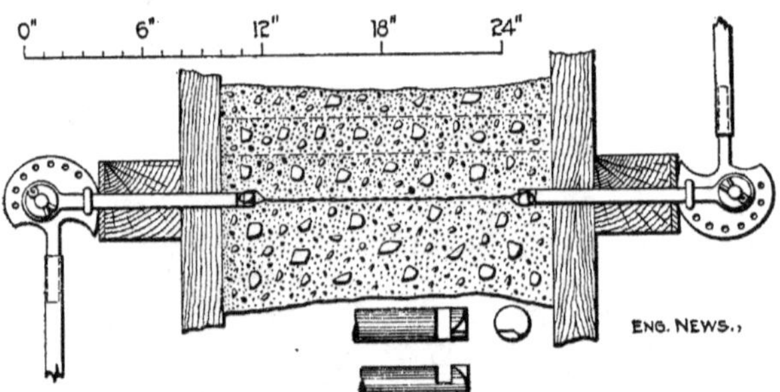

FIG. 274.—Detachable Tie for Wall-forms.

that there are no threaded bolts or nuts or wrench to use in putting together or taking down the forms. The clamp is made of malleable iron and is put through the form, and a twisted wire of the desired strength is placed on the hook. The cam is then tightened, by placing a 2-ft. piece of gas-pipe on the handle for leverage. To release the forms it is only necessary to reverse the cam, when the I bolt is at once freed from the wire tie and the clamp can be immediately removed, leaving the tie in the concrete. The form is then taken down and the hole left by the bolt is filled with cement.

Concreting.—In building walls the concrete is deposited and rammed in layers. The practice of French engineers is to make these layers quite thin, seldom over 4 inches. In America 6-inch layers are commonly employed, and often the thickness is increased to 8 inches. The usual practice in concreting is to carry up the walls uniformly all around the building, making each day's work a continuous operation. The concrete is mixed wet and is not heavily tamped, but rather joggled into place.

Roofs.

Roof construction does not differ materially from floor construction in reinforced concrete. When the construction is of previously cast plates the task consists simply in attaching these to the roof framing and closing the joints with cement. The exact mode of procedure varies with the form and design of the cast plate. The corrugated plate roofing shown by Fig. 101 is constructed by attaching the plates to the rafters and purlins by means of clips and then plastering it on both sides with cement mortar. Roofs built in place over an iron framework and monolithic roofs require forms for their construction. These may be quite simple in the case of plane-roofs, or very complicated in the case of domed roofs with ribs and panels. For plane roofs the forms are substantially the same as those used for floor construction of the corresponding character. In illustrating the dome roof for the Nassau County Court-house, New York, described on page 190, the form used for molding the concrete is shown in place. As will be seen, a circular working floor was built at the height of the column tops, supported on light vertical studs, and on it were built the molds shown in the half-sections. Three horizontal scarf-board rings were set to hold the vertical paneling for the inner surface of the drum, and from this was blocked and braced the paneling for its outer surface and for both vertical surfaces of the outside concentric cylindrical wall. Radial scarf-boards in vertical planes were supported on 3×4-in. studs to carry a continuous surface of horizontal courses of paneling, on which 3×4-in. circular horizontal pieces cut to horizontal radius were set and received the continuous planed curve of the intrados. The extrados of the dome shell was formed against sectional plaster molds spaced and supported from the outside paneling by separator-lugs. The plaster molds were reinforced with wire netting, and were made in horizontal strips 18 inches wide, with a lug cast on each corner.

After the ceiling dome had set, the upper drum and roof dome were built in similar molds supported at the circumference on the concrete structure and at the center on vertical studs set in the central light-well. As the intrados of the roof dome is not exposed to view, it is made on straight paneling, which gives the horizontal section a polygonal instead of a circular outline, but makes the vertical section curved. This enables a single set of horizontal paneling-boards, nailed directly to the radial scarf-boards, to suffice for the mold for the intrados. The plaster molds for the extrados of the roof-dome shell are similar to those for the ceiling dome, but have no integral lugs and are separated from the inside paneling by wooden spacers, to which they are secured by horizontal circular wires on the outside. These blocks are removed as the concrete is built up to them. The horizontal joints of the mold-blocks are covered by 2-in. iron straps. Spacing-blocks are nailed to the inner paneling, as shown in the detail section, to locate the radial rods.

CHIMNEYS.

In constructing chimneys of concrete steel, the task to be performed is to build two concentric shells of concrete with an annular space between them which is intersected radially by ribs or buttresses that spring from the opposite shells and nearly meet along the median line of the annular space. The universal mode of procedure is to build the shells and buttresses in vertical sections formed by a mold which is moved upward as the concreting progresses. A staging or tower of timber is erected inside the chimney, its top being kept well above the highest level of the concreting, and from this tower the molds are hung by adjustable suspender-rods. In the accompanying drawings Fig. 275 shows the assembled apparatus in elevation and in plans; Fig. 276 shows the shell-molds; Fig. 277 gives the details of the core-molds for forming the annular space, and Fig. 278 shows the construction of the hoops for the shell-molds. This apparatus was employed in constructing the 180-ft. chimney for the power-house of the Pacific Electric Railway Company at Los Angeles, Cal., and its manipulation is best explained by a description of that work.

A timber tower about 6½ ft. square was built up on the center line of the chimney a little in advance of the work. This tower was built with four legs braced together on all four sides by horizontals and intersecting diagonals. Across the topmost horizontals there were placed two pairs of transverse beams about 16 ft. long which cantilevered about 5 ft. beyond the sides of the tower and supported the molds. The inner and outer molds were each suspended from the cross-beams by four equally spaced vertical rods having threaded tops which engaged screw-wheels bearing on the beams. A light working-platform was bracketed out from the outside mold near the top, and a concrete platform was hung from this mold and a little below

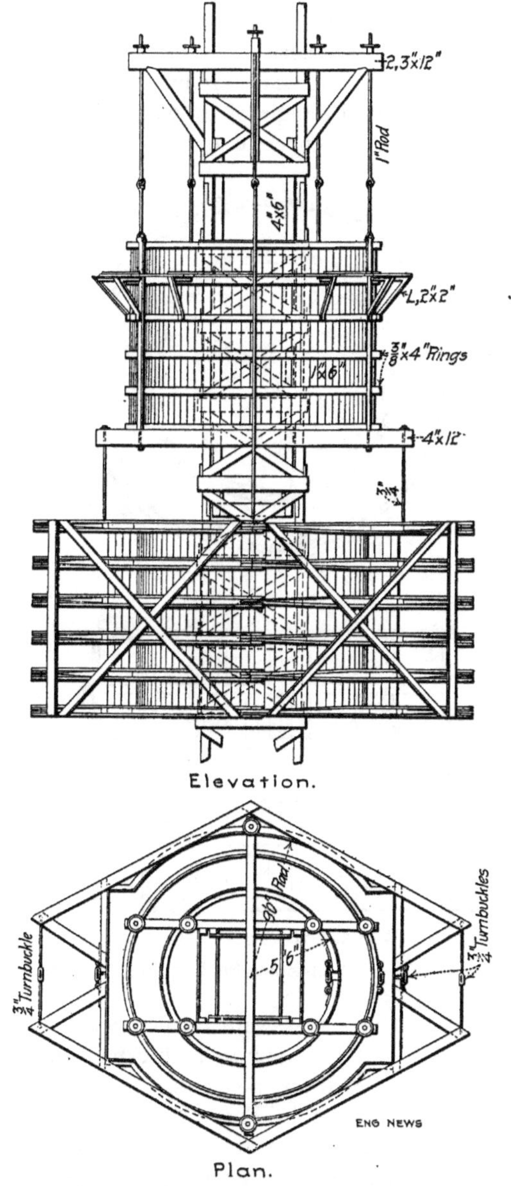

FIG. 275.—Details of Wall-molds for Stack of Electric Power-house, Los Angeles, Cal.

it to catch dropping material or a fallen workman. Inside the chimney a staging was supported from the tower on which the workman stood in placing and tamping the concrete.

The shell-molds shown by Fig. 276 were made with vertical wooden staves 12 ft. long beveled to an angle of 10° on both edges so as to be in contact on the face next the concrete and have a V-shaped opening

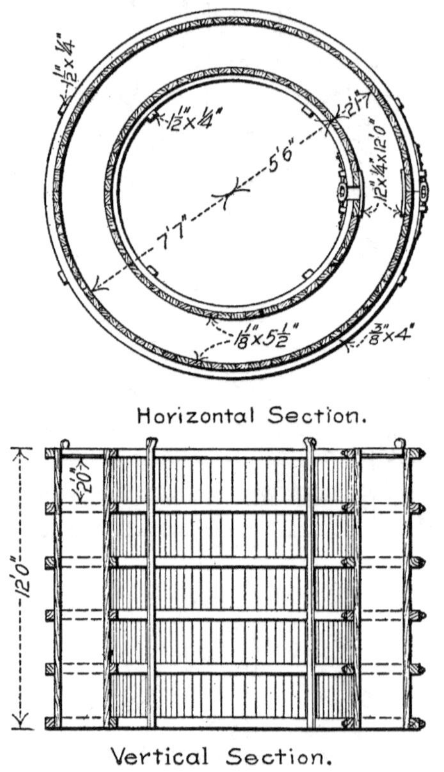

FIG. 276.—Details of Wall-molds for Los Angeles Chimney.

on the opposite face. The staves were hooped together with bands built up of $\frac{3}{8}$-in. strips of Oregon fir to a thickness of 5 ins. and a width of 4 ins. The ends of these bands were connected by the jaw-forgings and sleeve-nuts shown in Fig. 278. Six of these bands or hoops were used for each mold, they being located outside the staves in the outside mold and inside the staves for the inside mold. To attach the molds to the suspenders a $1\frac{1}{2}\times\frac{1}{4}$-in. bar with an eye at the top and a length of 12 ft. was bolted to each hoop in a vertical position at the four points on the circumference directly beneath the suspender-bars.

The space between the inner and outer shells was formed by

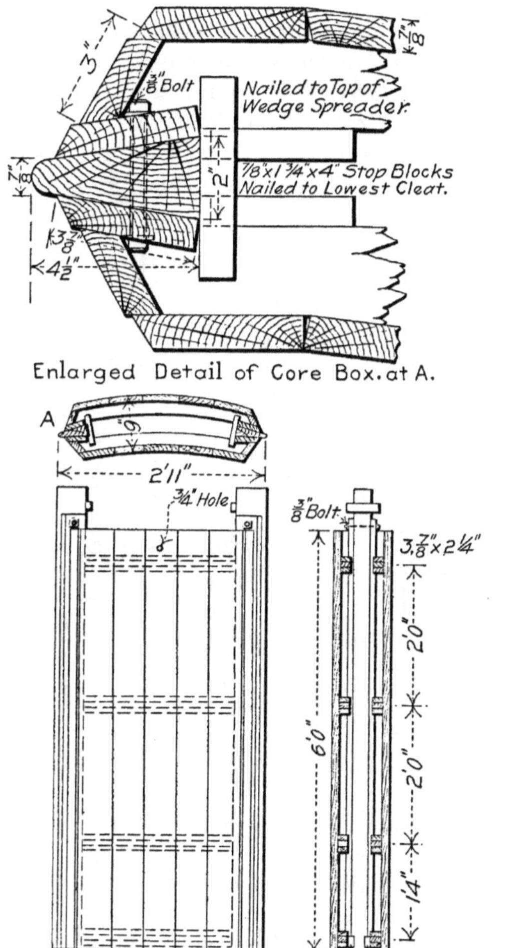

FIG. 277.—Details of Core-molds for Los Angeles Chimney.

means of core-boxes inserted between the faces of the molds. The construction of these core-boxes is shown by Fig. 277. Each consisted of two parts separated by a wedge-shaped timber. To collapse the cores the wedges were drawn inward, and to spread them they were forced outward. The edges of the wedges of adjacent cores in place nearly touched, but to insure perfect separation between the opposite shells at the ribs a ½-in.-square strip was tacked to the edge of each wedge.

The concreting was done in 5-ft. sections, and one section was completed every day. All material was raised through the falsework tower by means of an electric hoist, and the tamping and placing of the concrete was done from the inside staging. To make the extension of the falsework tower possible, a telescopic section 24 ft. long

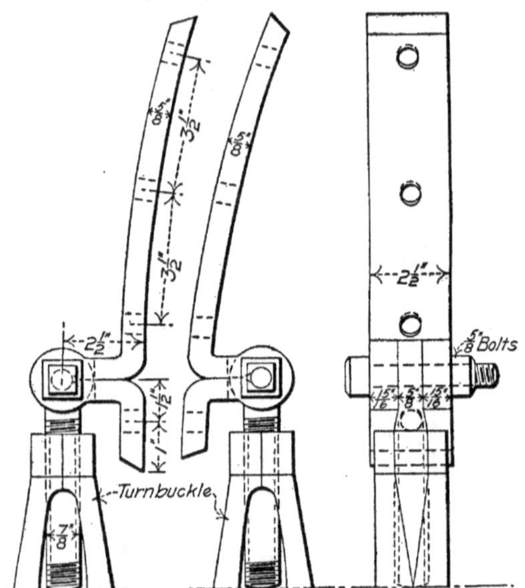

FIG. 278.—Details of Hoops for Wall-molds.

was built at the upper end. When it was desired to build up the main tower this section was advanced and clamped and the bearings of the cantilever beams were transferred to it. Two panels, 5 ft. high, of the main tower were then erected.

The sequence of daily operations was about as follows: In the morning the hoops of the outside mold were loosened and the form was raised ten feet by screwing up the suspender-rods with the hand-wheels. The hoops were then tightened, clamping the bottom of the mold to the upper 2 ft. of chimney finished the night (about 14 hours) before. The cores were then collapsed and pulled out in pieces, first the wedges and then the shells. They were then cleaned and assembled and set in position for the next length of work. The inside molds were then raised and the whole form was leveled up and secured firmly. Concreting was then begun and the 5-ft. section was completed at about five o'clock in the afternoon. To keep the concrete properly wetted during construction a circular pipe pierced with numerous holes was attached to the bottom of the outside mold and connected with the city water-mains. This device kept the concrete below the molds thoroughly sprayed with water.

CHAPTER XV.—METHODS OF CONSTRUCTION IN BRIDGE-WORK.

THE sequence of operation in constructing an arch bridge of reinforced concrete is substantially the same as that adopted in masonry arch construction. Girder bridges of concrete-steel are built like floors of that material, and they will not be considered further. Arch-bridge construction involves the following items of work: (1) Arch centers; (2) arch-ring construction; (3) spandrel construction; and (4) parapet-wall construction.

Arch Centers.—In general design and construction the centers employed for building concrete-steel arches are substantially the same as centers for masonry arches, and for equal loads the same rules and practice serve for both. There is this essential difference, however, between centers for concrete-steel arches and for masonry arches: the former serve as molds for shaping the soffit of the arch ring and as a support for the forms for molding the arch-ring face and the spandrel walls. This difference calls for certain obvious modifications of construction, but necessitates no radical change in general design and arrangement. In certain forms of concrete-steel arches in which exceptionally strong reinforcing-ribs are used the practice is sometimes adopted in Europe of carrying a portion of the weight of the center and its load of concrete by suspension from the reinforcing-ribs. This practice has not gained recognition in the United States, and would seem to be of doubtful expediency under all ordinary conditions. For most forms of arch-reinforcement, like the Monier, Thacher, Hennebique, it is of course impracticable.

Mr. Edwin Thacher, M. Am. Soc. C. E., specifies the following requirements for the general design of arch centers:

The lagging shall be dressed to a uniform thickness, so that when laid it shall present a smooth surface, or it shall be made smooth by plastering or other efficient means. In framing the centers allowance shall be made for settlement of centerings, deflection of arch after the removal of centerings, and for permanent camber. The centers shall be framed for a rise of arch greater than the rise marked on drawings by an amount equal to one eight-hundredth of the span.

In constructing plain arches the necessary soffit-mold is secured by using smoothly dressed lagging or by giving the lagging a smooth surface by a suitable covering. The smooth dressing and perfect fit of the lagging accomplishes all the purposes of a special covering, and usually at a cost which is no greater, and this is the only means usually sought for securing a smooth soffit-mold. When conditions render necessary the use of undressed lagging-timber a plastering of wet clay, or of mortar covered with thick paper or with cloth, have been found to give a satisfactory covering. It is common practice even with dressed lagging to employ a paper or cloth covering for the purpose of preventing adhesion between the concrete and the lagging which might disturb the ready striking of the centers. Smearing the lagging with soap or with oil is practiced for the same purpose. It is recommended by European authorities that a mineral oil be used for this purpose, as the disintegration of the vegetable and animal oils produces acids which attack the concrete.

The construction of the soffit-mold for ribbed arches is a more particular task; it involves the construction in wood of a negative of the paneled soffit, and its difficulty depends upon the intricacy of the paneling. The illustrations previously given of forms for ribbed floors indicate the general character of the work required.

A requisite of an arch center second only to its strength and stability is ease of removal, and the provisions for striking centers for concrete-steel arches should be quite equal to those for stone-arch centers. The devices usually provided for this purpose are wedges or sand-boxes or both. Sand-boxes, when they work as they are supposed to work, allow of a very gradual and uniform lowering of the arch, but they sometimes fail to operate perfectly because of the packing of the sand, due to wetting with silt-bearing water, or to dirt or cement finding its way into the box. The careful sealing of the boxes against the entrance of foreign matter, and their frequent inspection, are obvious precautions against trouble from this source. Sand is not entirely incompressible, and in a succeeding paragraph an incident is cited where serious trouble arose from the compressibility of the sand-cushions used to carry the centers.

Examples from Practice.—The following examples of arch centers taken from recent work will serve to illustrate current practice in connection with the discussion on pp. 124 to 127:

The illustrations of the Jacaquas River bridge given by Figs 279

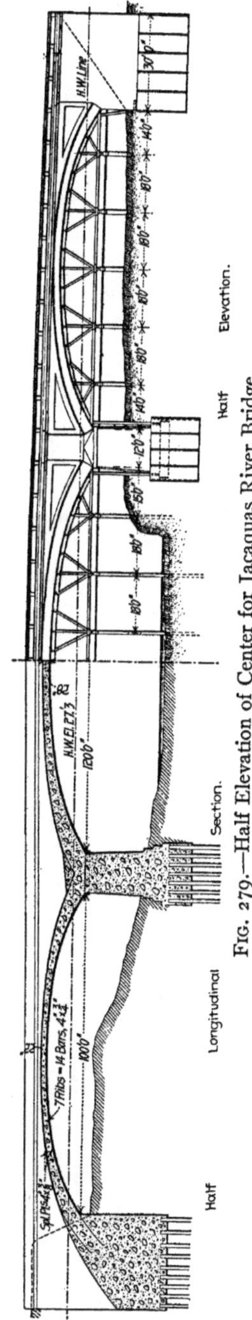

FIG. 279.—Half Elevation of Center for Jacaquas River Bridge.

and 280 show the form of centering used for the arches of that structure. The centering proper was carried by transverse pile-bents in the channel span and on timber bents in the side spans. The piles for the channel-span bents were driven to refusal with a 2,240-lb. hammer, and the sills of the timber bents were carried on concrete footings built in pits excavated from 2 to 6 ft. deep according to the nature of the ground. The ribs of the centers were spaced 6½ ft. center to center and were covered with 4×6-in. lagging set on edge. Sand-boxes, 10 ins. in diameter and 12 ins. high, were used at all points of support. It was observed that the centering settled under the weight of the concrete much more than was anticipated, and after considerable search the greater part of this settlement was traced to the sand-boxes, some of them going down 1½ ins. or more. The plungers were as close-fitting as would work. An experiment was made by carefully packing a sand-box

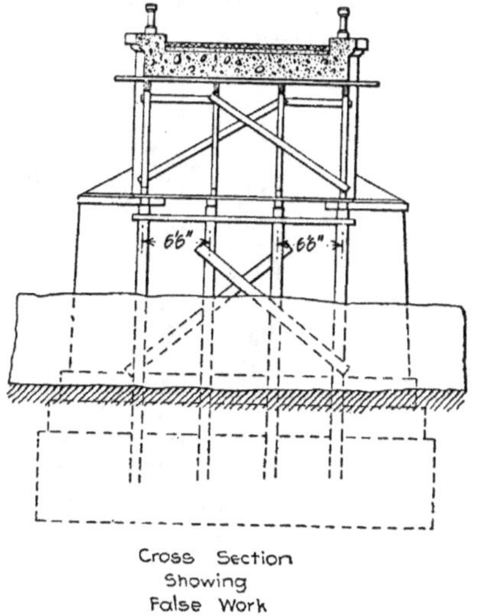

FIG. 280.—Cross-section of Center for Jacaquas River Bridge.

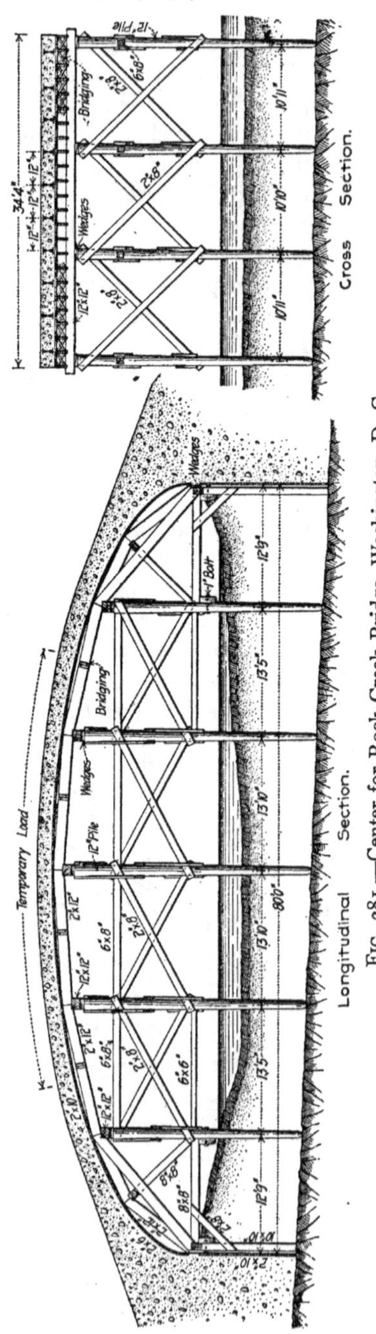

FIG. 281.—Center for Rock Creek Bridge, Washington, D. C.

and then sledging down a plug the size of the plunger employed. This plug was sledged down about 3 ins.

In planning the construction of the arch centers and the spandrel-wall forms for the Maryborough Bridge, Queensland, which is described on page 208, the centers were made of pine timber and consisted of six ribs 4 ins. thick supported on caps carried on piles driven into the river-bed at the center of each span, the outer ends being carried on the staging of the piers. Each rib was 2 ft. deep at the center and 1 ft. deep at the ends, and was formed of two flitches bolted together as shown. The lagging was of 1½-in. pine planking. The spandrel molds were formed of 1½-in. planking on studs, and were braced against piling. Wedges between the ribs and pile-caps provided means for striking the centers.

The centering for the Rock Creek bridge built in the National Park at Washington, D. C., in 1901 is shown by Fig. 281. Transverse

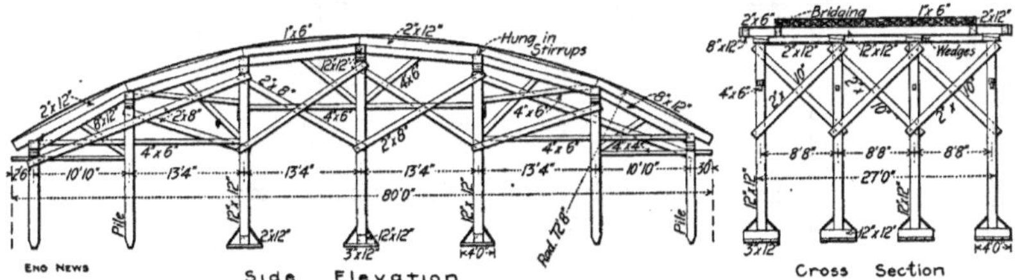

FIG. 282.—Center for 80-ft. Melan Arch, Washington, D. C.

four-pile bents were driven to hardpan and sawed off about 2 ft. below the arch soffit. On these bents transverse caps were laid with double wedges on each pile-head. These caps carried the ribs of the center, which were placed quite close together and braced apart by transverse X-bracing over each bent. The ribs were covered with transverse lagging. The other details of the center construction are shown by the drawings of Fig. 281. A center for an 80-ft. Melan arch built over the same stream in 1902 is shown by Fig. 282. In this center the posts carried timber shoes which rested on concrete footings. Wedges over each post provided means for striking the center. The other details of construction are clearly shown by the drawings. Fig. 283 is a transverse sectional elevation through one of the arch centers for the Zanesville, O., Y bridge. A timber trestling carries each rib of 6×14-in. oak and 8×14-in. pine spaced 36 ins. apart and covered with dressed and matched lagging of Norway pine 2 ins. thick and 4 ins., 6 ins., and 8 ins. wide.

An unusual form of center, which furnished particular advantages in the erection of spandrel-wall forms, was employed in constructing two 100-ft.-span arches near Mechanicsville, N. Y., in 1903. Four-pile bents were driven through the mud of the river-bottom to rock; the piles being spaced 6 ft. between centers, and the bents 10 ft. apart. The piles were capped with 10×12-in. pieces drift-bolted to place, and four lines of 10×12-in. stringers were drift-bolted to the caps, bringing the structure about 4 ft. above the elevation of the spring-

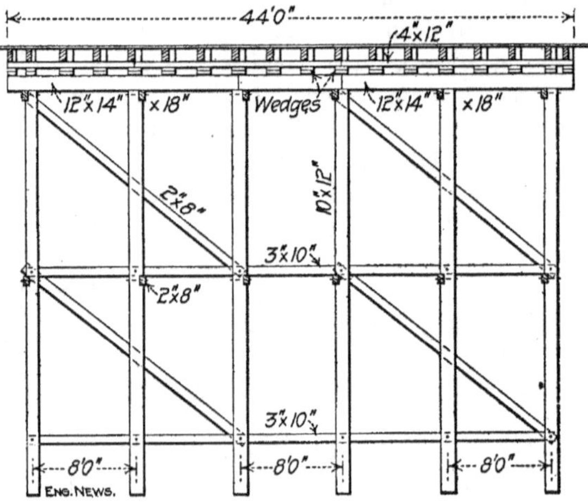

FIG. 283.—Cross-section of Center for Bridge at Zanesville, Ohio.

line of the arches. Posts 8×10 ins. spaced 3 ft. apart were erected on the stringers and cut at such lengths that the caps on top of the posts conformed to the curve of the arch. These caps were 8×10 ins. in section and 23 ft. long, projecting 3 ft. beyond the faces of the arch ring for convenience in bracing the forms for the spandrel walls. The top faces of the caps were beveled to conform to the curve of the arch. The lagging was then put on parallel to the center line of the structure, and sprung down and nailed to place over the caps. The lagging was put on in two thicknesses of inch boards. Rough boards were used for the first course, and good stock, planed on the upper surface and carefully jointed and laid, for the top course, making a smooth surface for the concrete. Hard-wood wedges were used under the posts for striking the centers.

Fig. 284 shows the form of center employed in constructing the arch of the Laibach Bridge in Austria. The center had eight ribs spaced about 7 ft. apart and braced together by diagonal battens. Each frame rested on seven piles, each carrying a sand-box. The upper members of the ribs were cut in the middle of each panel, and

TABLE XXXV.—SHOWING DEFLECTION OF ARCH CENTERS FOR BIG MUDDY RIVER BRIDGE.

Conditions of Work when Elevations Were Taken.	Elevation taken on Under Side of Keys of Main Arches.					
	North Arch.		Center Arch.		South Arch.	
	West Side.	East Side.	West Side.	East Side.	West Side.	East Side.
Centers ready for concrete	385.06	385.06	385.95	385.93	386.32	386.33
Concrete finished in main arches	384.92	384.92	385.77	385.70	386.21	386.22
Compression, etc., of centers	0.14	0.14	0.18	0.14	0.11	0.11
Alternate centers removed	384.92	384.92	385.77	385.70	386.12	386.15
Further yield of centers	0.00	0.00	0.00	0.00	0.09	0.07
Rest of centers removed	384.69	384.70	385.70	385.71	386.03	386.06
Deflection of arches *	0.23	0.22	0.07	0.08	0.09	0.09
Total change from original height of centers †	0.37	0.36	0.25	0.23	0.29	0.27
Figured elevations on plan	384.95	384.95	385.60	385.60	386.24	386.24
Allowed for compression, shrinkage, etc.	0.11	0.11	0.35	0.33	0.08	0.09
Final elevations as compared with original plan	−0.26	−0.25	+0.10	+0.10	−0.21	−0.18

* The greater apparent deflection of the north arch includes all changes due to the variation of 50° to 70° in temperature, from July 25, 1902, to Jan. 19, 1903.

† The centers were loosened on dates as follows: North arch, Aug. 8–15, 1902, wedges were loosened; when removed, Jan. 19, 1903, there was no further deflection of the arch; rib of arch was finished July 25, 1902. South arch was finished Oct. 5, 1902, wedges were out Jan. 15, 1903. Center arch finished Dec. 5, 1902, wedges were out Jan. 15, 1903. Note.—The bridge is on a grade of 0.004 per cent., rising to the south.

‡ The centers of the middle arch were set high purposely: 1st. To make this arch higher at the crown than its position on the grade; 2d. To allow for a slight movement of the north pier, which had apparently yielded under the thrust of the north arch, which was finished first, centers being loosened before any centering was in place in the center arch, or anything in place to assist this pier in resisting the thrust.

here rested on cross-timbers A, A hung from the steel reinforcing-ribs by means of hangers as shown by Fig. 285. By this means about half

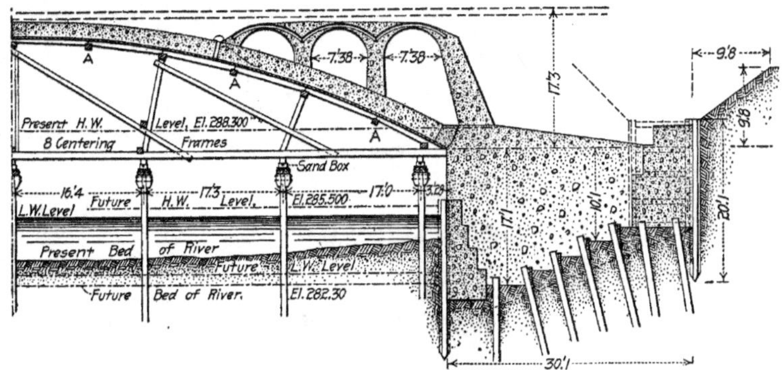

FIG. 284.—Arch Center for Laibach Bridge, Austria.

of the weight during erection was carried by the steel ribs and half by the centering.

The center illustrated by Fig. 286 was employed in constructing the Illinois Central Railroad bridge crossing the Big Muddy River in southern Illinois. This bridge has three arches of 140 ft. span each. The two end arches were built first, and two complete sets of centering were provided for their construction. To provide a center for the middle span alternate ribs were removed from the side-span centers and re-erected with new lagging. The centers for the side-span were set 1 in. higher than the calculated elevation for finished concrete work. A careful record was kept of the deflection of the centers and of the arch, and the results obtained are shown by Table XXXV. The load coming on each center from the completed arch ring was about 3,000 tons, or from 18 to 20 tons per pile in the falseworks carrying the centers.

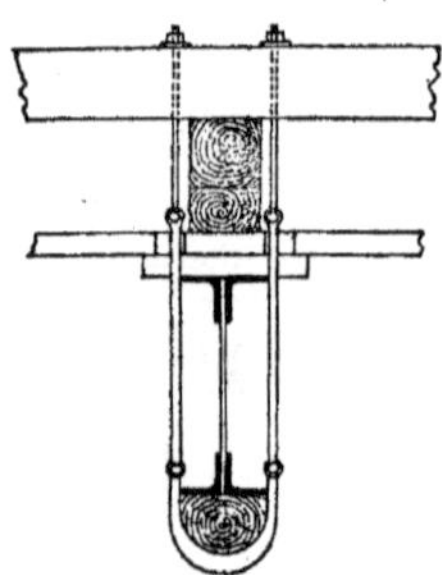

FIG. 285.—Hanger to Carry Centering from Melan Ribs, Laibach Bridge.

Arch-ring Construction.—The construction of the arch ring involves two processes which may or may not proceed simultaneously.

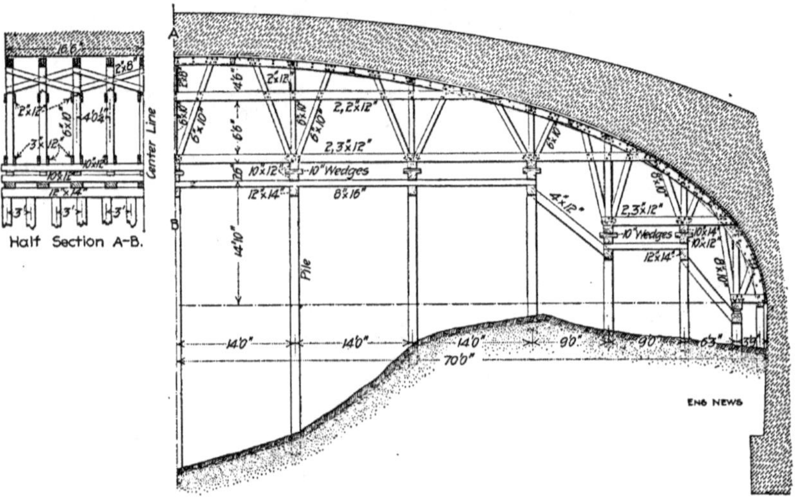

FIG. 286.—Center for Big Muddy River Bridge, Illinois Central R.R.

These processes are the erection of the metal skeleton or reinforcement and the placing of the concrete which embeds the skeleton.

Erection of Reinforcement.—The method of erecting the reinforcement varies with its form and arrangement. Certain forms, of which the Melan may be taken as an example, are self-supporting or require but little support from the arch center; other forms, of which the

Monier will serve as an example, have practically no rigidity by themselves and require complete support by the center. Again, in certain forms the erection units are few in number and of large size, and may be erected complete before any concreting is done, while in others the units are small and numerous and have to be placed and embedded one by one as the concreting progresses. Thus while certain general rules of practice are common to the erection of all forms of reinforcement, the mode of procedure in detail is different for each form.

In Melan construction the reinforcement consists of parallel arch-ribs, which are either rolled I beams or lattice girders bent to the curve of the arch-ring. For short spans with I-beam ribs each rib is in a single piece, and the only ironwork done in the field is that required to attach the ends of the ribs to the cross-angles in the abutments. For longer spans, with lattice-girder ribs, the ribs are usually received in two or more segments and these have, of course, to be joined together in the field. When hinged ribs are employed, as in some of the European examples of Melan bridges which have been described in Chapter VIII, the work of assembling the separate parts is still more extensive. Generally all the ribs of each arch are accurately set in position before any of the concrete is placed. In wide spans in which the concreting is done in longitudinal sections the placing of all of the ribs before any of the concrete is placed is not necessary. The ribs are first fixed to elevation and distance at the ends. They are then lined in longitudinally and blocked up from the lagging of the center to the proper height to get the desired thickness of concrete underneath them. As thus set up, the ribs are not extremely rigid against lateral displacement, and they are usually braced at several points in their length by diagonal struts from the top flanges to the lagging. These braces are removed one after another as the concreting approaches them.

Forms of reinforcement analogous to the Melan, like the Wunsch and Bonna forms, are erected in essentially the same manner as the Melan. For erection purposes the Thacher flat-bar reinforcement may be considered as consisting of Melan ribs with the webs eliminated, leaving only the top and bottom chords. Generally the two bars of each rib are erected at the same time and before any concreting is done, the bottom bar being blocked up from the center the requisite distances from the arch soffit and the top bar being similarly supported in position near the extrados of the arch ring. When the concreting is done in transverse sections, planks set on edge across the arch center at suitable intervals serve both as end molds for the sections of concrete and as supports for the reinforcing-bars.

Monier arches are usually reinforced by an intradosal and an extradosal network. The center having been erected, the first operation is to lay down the intradosal reinforcing-net. This is made up of rods laid parallel to each other and to the longitudinal axis of the bridge, and of wires crossing these rods at right angles and traversing the arch ring transversely. The distance apart of these wires and rods varies, of course, with the span and load, or with the

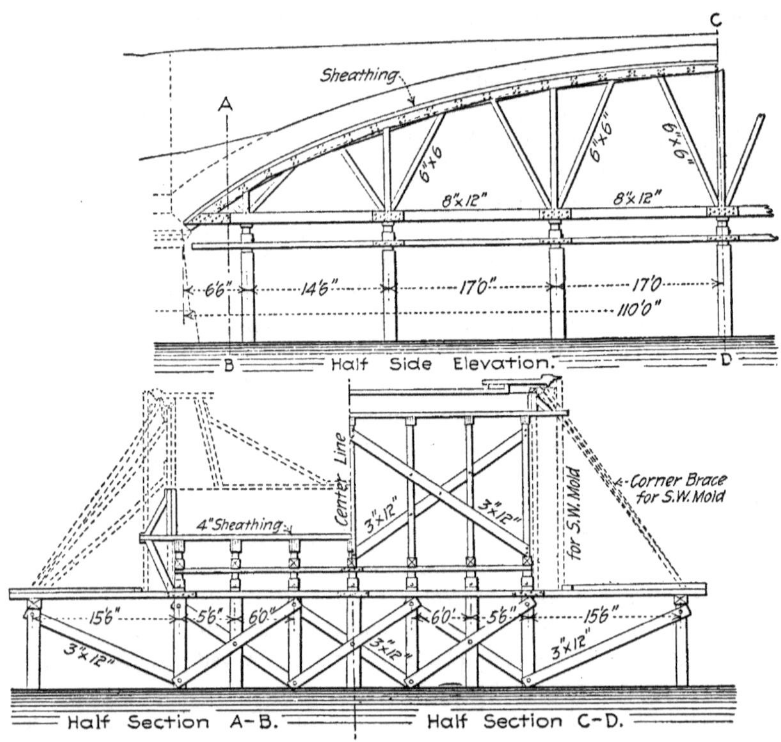

FIG. 287.—Centers for Topeka Bridge, Topeka, Kan.

strength of reinforcement required. When the cross-wires have been placed they are tied to the rods at alternate intersections by means of wire ties. So far as is practicable splices are avoided, but when a splice is necessary it is made by lapping the rods or wires a distance of at least 20 times their diameter and wrapping the lap with fine wire. An effort is always made also to stagger the splices of adjacent bars and to locate all splices at those points where the tension in the arch is least. In erecting this network the longitudinal rods are laid first and then the cross-rods are placed and tied. To insure the accurate spacing of the rods and bars, their alinement is frequently marked on the paper covering the lagging of the center. When the netting has been completed it is blocked up from the center and concreting is begun and continued in concentric layers until the plane of the extradosal netting is reached. This netting is then laid onto the concrete in one piece, it having been built beforehand ready for this operation.

In erecting expanded-metal and other forms of reinforcing network which are furnished in sheets, the sheets are simply spread over the center with overlapping edges so as to cover completely the lagging. In some cases the overlapping edges are fastened together by means of fine wire ties. The network being completed as described, it is blocked up from the centering and the concreting begins. The extradosal net, if one is employed, is erected in practically the same manner except that the sheets are laid out upon the concrete after it has been brought up to the proper level.

The erection of bar and stirrup forms of reinforcement, like the Hennibique, is done in smaller units and as the concreting progresses. Taking, for example, a Hennebique ribbed arch, the sequence of operations is substantially as follows: The proper depth of concrete is first placed in the bottoms of the longitudinal rib molds and in this the first bar or group of bars are laid with the stirrups hooked under them at the proper intervals. Mounds of concrete packed around the bottoms of the stirrups serve to hold them vertical unless they are of considerable length, when an overhead framework is built to support them. The concreting then proceeds until the level of the next layer of bars is reached, when these are placed exactly like the first. This alternate placing of the concrete and the metal of the reinforcement continues until the main ribs, the transverse ribs, and the arch ring are completed in order.

Bar reinforcement, like the Ransome, Thacher, and corrugated bars, is erected in different ways. Usually the arrangement of the bars is in pairs, each consisting of an intradosal and an extradosal bar in the same vertical plane. The task is to place and support these bars in position for the concreting. Usually temporary blocking is employed. This blocking is constructed in different ways. For the bottom rods simple cleats of the proper thickness are laid on the lagging and the bar laid on them. For the upper bar boards set on edge, short strips set endwise, or small A-frame horses are employed. Another form of support consists of boards set on edge transversely across the lagging and notched to receive the bars. When the concreting is done in transverse segments this is an especially convenient arrangement, since the transverse boards will serve the secondary purpose of forming the ends of the segments. Considerable rigidity is required of the blocking, since the bars have to be held bent and from shifting laterally, and a long bar of steel will develop surprising contortions oftentimes while being fastened into position. For spans of much length the bars have to be spliced. This is usually done by simply lapping the ends and wiring the laps together. In place of temporary blocking sheet-metal spacers can be employed and left permanently in the work.

Placing the Concrete.—The important thing in placing the arch-ring concrete is to secure its perfect contact with every part of the metal reinforcement. This object must also be accomplished without displacing or distorting the reinforcing members from their designed positions and forms. Finally, the concrete must be so deposited as to insure its intended monolithic homogeneity and unity.

Theoretically the proper procedure is so make the construction of the arch ring a continuous process. Except in the case of foot-bridges of narrow width, however, it is seldom attempted to construct the arch ring in one piece. The usual practice is to divide it into two or more sections which may be either longitudinal rings or transverse segments. In either case each section is built up in thin concentric layers. Practice seems to have settled upon no well-defined rule in choosing between longitudinal rings and transverse segments, but it is unanimous in requiring each ring or segment to be of such dimensions that it can be finished in one continuous operation. In his specifications for concrete-steel bridges Mr. Edwin Thacher specifies the following requirements:

For square arches the concrete shall be laid in transverse sections of the full width of the arch between timber forms normal to the center line of the arch, the length of sections being such that the center section or a pair of intermediate or end sections shall constitute a day's work. Work shall be started at the center section and carried toward the end sections, which shall be laid last. For skew arches the concrete shall be started simultaneously from both ends of the arch and be built in longitudinal sections at least 5½ ft. wide and wide enough to constitute a day's work. The concrete shall be deposited in thin layers, each layer being well rammed in place before the previously deposited layer has had time to partially set. The work shall proceed continuously day and night if necessary to complete each longitudinal section. These sections while being built shall be held in place by substantial vertical timber forms, parallel to the face of the arch and to each other, and these forms shall be removed when the section has set sufficiently to admit it. The sections shall be connected by cleaning and roughening the surfaces and mopping them with a 1 cement

1 sand mortar, and also by steel clamps spaced about 5 ft. apart connecting the adjacent reinforcing-ribs.

Too great emphasis cannot be given to the necessity of securing the most perfect bond possible between adjoining sections. The Melan concrete-steel arches partly wrecked by flood at Paterson, N. J., in the fall of 1903 failed by shearing their full length along the plane of junction between adjoining sections of the concreting. In view of this experience some method of interlocking the sections by clamps, dovetailing, or rabbeted joints would seem to be always advisable. In case longitudinal sections are employed, the dividing plane between them should be always located between consecutive reinforcing-ribs or bars. This is particularly essential when rib-like forms of reinforcement like the Melan and Wunsch are employed.

Substantially the following mode of procedure is followed in Europe in concreting arches with Monier reinforcement. The intradosal netting having been woven and blocked up from the center as previously described, the concreting is begun by filling around and beneath the wires and carefully tamping the filling. When the metal has been thoroughly buried, the remaining concrete is placed in concentric layers from 4 to 6 ins. thick, and each layer is thoroughly rammed. This filling is continued until the plane of the upper or extradosal reinforcement is reached. This upper network is then placed and the covering layer is deposited and tamped so as completely to embed the metal. The top surface of the ring is finally troweled to a smooth surface, and generally plastered with a rich mortar. In arches of considerable span the depositing of each layer is usually begun simultaneously at four points, the two springing lines and midway of each haunch, and from each point the progress is toward the crown. Each layer is fully completed before the next layer is begun. It is deemed desirable also that the entire concreting of the arch-ring should proceed continuously.

In Melan arches the placing of the concrete is begun at the ends of the span, and it is a frequent practice to embed the ends of the ribs to their full depth for the disance they project over the piers or abutments before carrying the concrete up the haunches. The purpose of this preliminary filling is to fix rigidly the ends of the ribs and thus to increase their stability against lateral displacement. The arch-ring concreting is done in concentric layers, each layer being constructed from the ends toward the crown and from both ends simultaneously. When the concreting is done in longitudinal sections of the ring each layer is carried entirely over the span, but when the ring is built in transverse sections each layer is carried only from the lower to the higher ends of the sections in which work is progressing. In either case the first layer is usually made deep enough to cover the tops of the bottom-chord flanges of the reinforcing-ribs, and is laid and tamped with particular care to insure a complete and compact filling underneath and around the chord. Special small tamping-bars are advisable for this work, and the ordinary iron railway tamping-bar has been found a suitable tool for this particular purpose. Succeeding layers are deposited and tamped until the full depth of the arch ring is attained. In placing all of these layers the most particular care is essential in compacting the concrete around the ribs so as to fill tightly against the web and up into the angles between the web and the top-chord flanges. There is great liability of voids being left at these angles unless the filling is carefully done. In case the ribs have latticed webs special effort is necessary to fill closely around the web members and to insure the transverse continuity of the concrete through the web openings. In tamping each layer of concrete it is considered an advantage to tamp the advancing edge by tangential blows in addition to the usual compacting from above and normal to the arch ring.

The concreting of Wunsch and Thacher arches follows substantially the mode of procedure pursued in Melan construction, while with expanded-metal and other network reinforcements the operations performed are much the same as with Monier reinforcement. The task of concreting in a bar and stirrup reinforcement is always complicated and tedious, owing to the number and small size of many of the reinforcing members. In Hennebique construction much of the concreting is a matter of depositing small amounts of concrete and of compacting it with small tools around a few members at a time. The work involves considerable care but no material difficulty.

Examples from Practice.—The following examples of arch-ring construction will serve to illustrate current practice in this class of work. The arches of the Zanesville, O., concrete-steel bridge described on p. 222 were built in 10-ft. segments extending the full width of the arch, 43 ft. A transverse form set radially was used to shape the end of each segment, and to this form a 3×8-in. plank was spiked to form a groove in the end of the segment into which the fresh concrete of the succeeding segment would enter and form a sort of dovetail joint. The lagging of the center was covered with about 1 in. of mortar on which the concrete was deposited. In building the concrete-steel arch at Oconomowoc, Wis., illustrated on p. 228, the arch ring was built in three transverse segments with radial joints so that the crown segment acted as a key. This key segment was built first, the ends being formed against plank forms set at an angle of about 30° with the vertical. Through these end planks holes were bored at intervals, and ¾-in. round rods 2 ft. long were inserted in them so as to project 1 ft. on each side. These rods served as dowels to connect the center and side segments at the joints. Each of the three segments was built in one day. The hinged Melan arch at Laibach, Austria, illustrated on p. 220, was concreted in the following manner: The concrete hinge-blocks were molded several months before the erection was begun, and they were first set in place with the lead cushion-strips in the joints. After the hinge-blocks were placed the concreting of the main arch proceeded from four points, being begun simultaneously on each side of the center and at each skewback. The steel reinforcing-arches were coated all over with a cement wash, and during concreting great care was taken to pack the concrete closely around the steelwork. Concreting was carried on uniformly along the full width of the arch, and at the end of each day's work a radial partition was placed against the fresh concrete. Before beginning work the next morning the radial face thus formed was roughed up with picks and coated with a mortar of 1 part cement to 3 parts sand, before fresh concrete was placed against it; this is thought to have secured a thorough joint between the sections placed on successive days.

Striking the Centers.—Practice varies in the time set for striking the centers of concrete-steel arches; and this operation is performed all the way from two weeks to two months after the concreting is finished. In American practice the centers are scarcely ever removed in less than 30 days, and they are frequently allowed to stand for double this time. To insure the center being of any practical use as a support at the end of these long periods, however, it must be kept well drenched with water, or else upon the drying of the ring concrete the lagging will shrink free from the arch soffit and the center will strike itself. This occurrence has taken place in a number of instances, and the fact that no serious consequences have resulted is urged by some engineers as warranting the striking of the center in much shorter periods of time. Except when several arches of the same span are to be built and the cost can be reduced by using the same centers for all of the arches, there is ordinarily not much to be gained by haste to remove them. On the contrary, the longer the concrete can be conveniently left to harden the stronger it is to resist deflection and distortion. The mode of procedure followed in striking the center of a concrete-steel arch is practically the same as in the case of stone arches. The lowering of the structure is performed gradually so as to allow the ring and its reinforcement to adjust themselves slowly to the strains which they have to withstand.

Spandrel-wall Construction.—Most concrete-steel bridges are built with solid spandrel walls, which, as stated in a preceding section, are either retaining-walls filled between with meagre concrete, earth, or cinders, or else are curtain-walls enclosing a cellular construction of arches or vaults on whose tops the roadway is laid. Walls of the retaining-wall type are the more common, and except that their faces are commonly paneled or otherwise ornamented and that they are surmounted by a moulded cornice, their construction is straightforward retaining-wall construction. When curtain-walls are employed they are made of uniform thickness and are reinforced.

Forms.—The first task in constructing spandrel walls is the construction and erection of the forms. The difficulties which arise in this task are due chiefly to the necessity of providing panels, molding, and other architectural embellishment in the front face and to the unsatisfactory conditions for bracing the face-lagging sturdily in place. Considering first the supporting framework for the lagging, the drawings of Figs. 288, 289, and 290 show typical arrangements. In Fig. 288 the form of centering is that used on the Mechanicsville, N. Y., bridge described on p. 386. The longitudinal lagging is carried on transverse timbers spaced 3 ft. apart and supported by posts resting on pile-bents. The advantage of the form of center from the standpoint of spandrel-wall construction is that the prolongation of the transverse caps give an excellent support for outside diagonal props for the face form. The tops of the facing-studs may be connected across the bridge by a horizontal tie, or they may be tied by diagonals to the sill timber of the rear-wall forms, as indicated by Fig. 289. Fig. 289 shows an arrangement for spandrel-wall forms with the ordinary arch-rib centering and without outside bracing; the drawing explains itself. The arrangement shown by Fig. 290 is one of the most satisfactory ones that can be employed, but it calls for a considerable extension of the falsework which cannot always be provided. The bracing of the rear-wall forms is simply accomplished by a cross-sill between the bottoms of opposite studs and by diagonals from this sill to the top and middle of the studs. The drawings shown give typical arrangements only, and can be modified in general and in detail according to the conditions. The chief desideratum is to secure an unyielding support for the lagging, and particularly for the lagging for the front face. To do this the studs should not be spaced more than 5 ft. apart and should be strongly braced and tied in position, and the lagging should be heavy enough not to bend between studs; lagging 2 to 3 ins. thick is none too heavy.

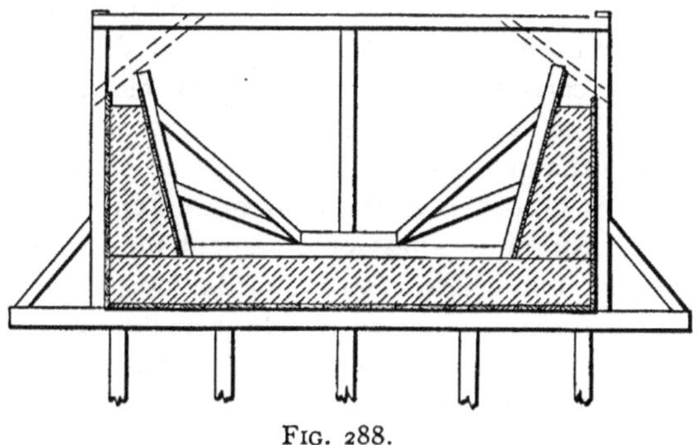

FIG. 288.

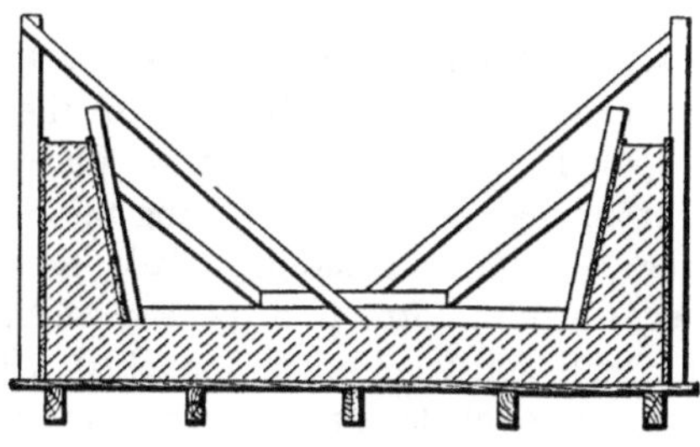

FIG. 289.

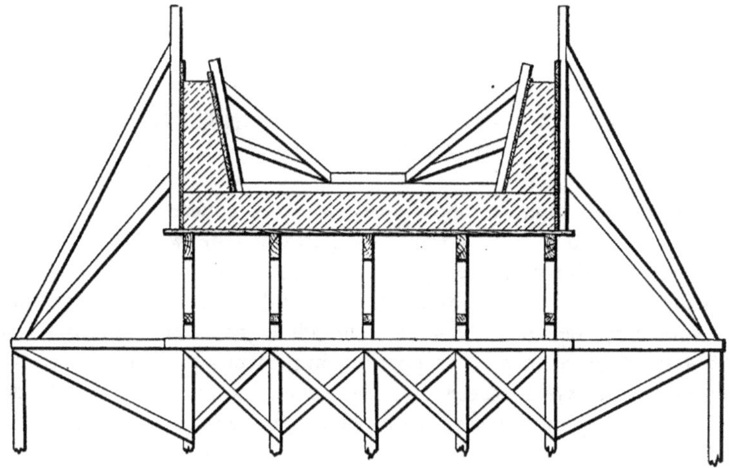

FIG. 290.

FIGS. 288–290.—Designs for Spandrel-wa'l Forms.

The lagging for the rear wall can be made of rough boards, but that for the face must be smoothly dressed and tightly jointed. The latter has also to carry the attached panels and molding or other ornamental forms. The task here is simply to construct on the lagging a negative of the ornamentation called for by the plans, and it may be simple or complex according to the amount and character of ornamentation desired. For most kinds of ornament the forms may be constructed of wood, but where rosettes, medallions, or similar complex forms are called for it is often preferable to make negatives in plaster of Paris from clay models and set these negatives in the lagging at the proper places. Another method is to cast these ornaments on the faces of concrete blocks which are molded separately and set into the concrete wall as the work progresses.

Placing the Concrete.—The first particular task in placing the concrete for the spandrel walls is to prepare the top surface of the arch ring so as to insure the perfect bonding of the wall to it. The surface being prepared, the next step is to place the concrete in horizontal layers from 6 to 8 ins. thick, using the means described in Chapter XVII for securing a smooth surface finish. Where stone facing is employed the face form is dispensed with, the stone being laid up as in any wall construction and backed with concrete built against a rear form of rough lagging-boards. As the concreting proceeds the cast ornaments, if any are used, are set in place and bonded into the wall.

CHAPTER XVI.—METHODS OF CONSTRUCTION IN CONDUIT WORK.

The construction of conduits of reinforced concrete is a task which requires particular care. This is especially true when the conduits carry water under pressure or are subjected when empty to an external pressure of considerable amount. A conduit carrying water under pressure must be not only strong enough to resist the bursting force, but it must be as nearly as possible impervious to water. The perfectness with which these two requirements are satisfied depends very largely upon the care with which the work of construction is performed.

Forms and Centers.—The principal requisites of forms and centers for conduits are: (1) They must be strong enough not to crush or become deformed under the load which they have to carry; (2) they must be so rigid that they will not easily warp and twist under the alternate soaking and drying to which they are subjected; (3) they must be so constructed as to give an even and smooth interior surface to the finished conduit; (4) they should be so constructed that they can be readily assembled and taken down and thus used over and over for succeeding lengths of conduit. The manner in which these requisites have been attained in practice is best demonstrated by actual examples.

The drawings of Fig. 291 show the forms and templates employed in constructing the arch culvert described on p. 247. The lower portion of the invert was brought to a true form by means of the invert template shown by Fig. 291. To form the curved-side portions of the invert two side forms constructed as shown by Fig. 292 were braced

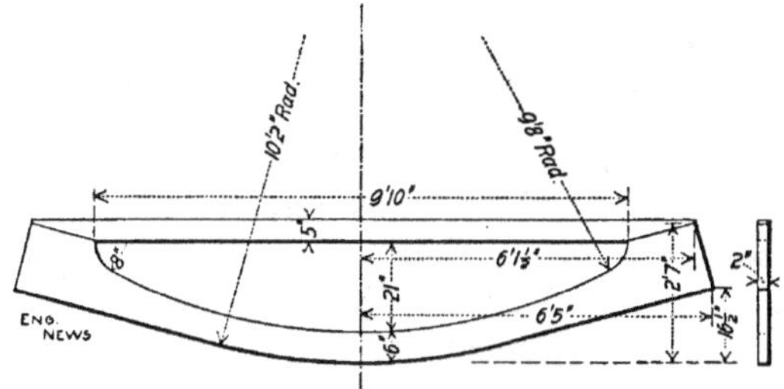

Fig. 291.—Invert Template for Culvert at Kalamazoo, Mich.

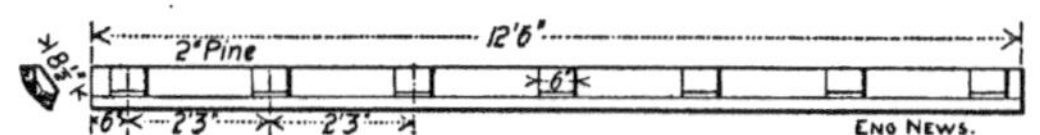

Fig. 292.—Side-wall Forms for Culvert, Kalamazoo, Mich.

in position by the hinged cross-struts shown in detail by Fig. 293. At this height the concrete was brought to a slightly inclined surface so as to properly receive the thrust of the arch, and was roughened so as to make a close bond with it. The arch center is shown by Fig. 294. This center rested on the invert forms at the sides and was supported

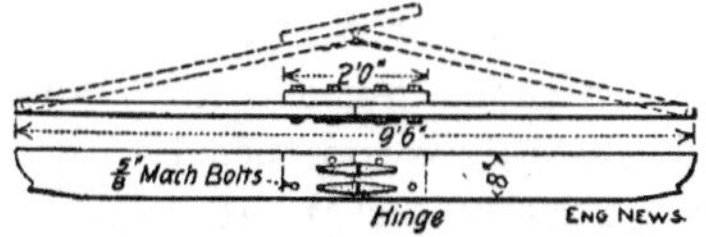

Fig. 293.—Cross-brace for Side-wall Forms for Culvert at Kalamazoo, Mich.

at the center by a vertical post. In placing the concrete forming the haunches planks were placed back of the centers to confine the concrete to the proper thickness, but for the upper portion of the arch where the concrete would stand alone these planks were omitted. To

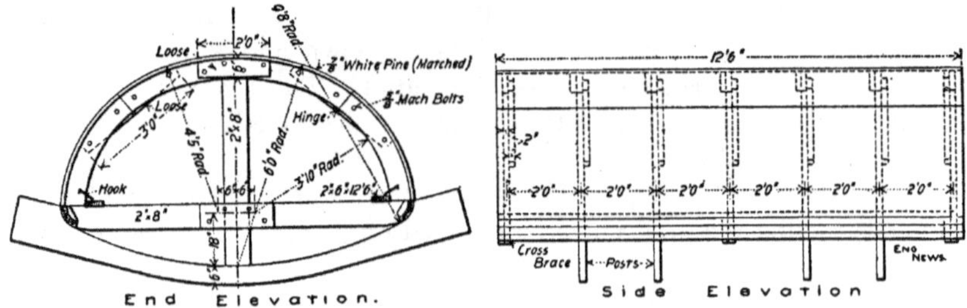

FIG. 294.—Arch Center for Culvert at Kalamazoo, Mich.

guide the concreting of the arch-ring, the template shown by Fig. 295 was used. The centers and the forms were all built in sections 12.5 ft. long, and a sufficient number were provided to lay twelve sections of invert and six sections of arch. The templates and centers are all

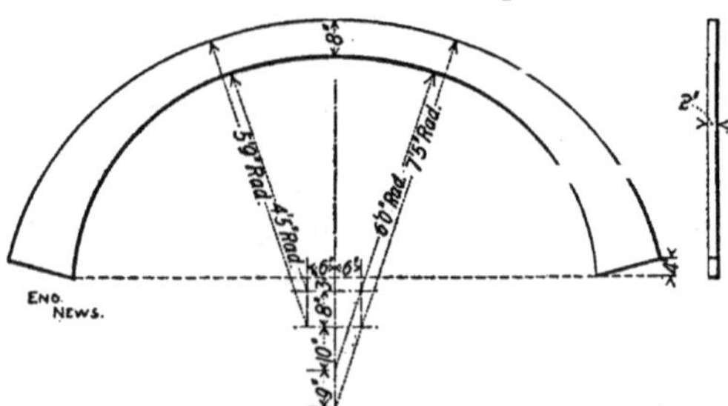

FIG. 295.—Template for Arch-ring of Culvert at Kalamazoo, Mich.

arranged so they can readily be taken down without any jarring of the concrete. The upper forms are hinged at the quarters and can be separated in two pieces and folded together so that those farthest back in the conduit, where the concrete has had time to set, can be brought forward through those that are still standing under the fresh concrete. All of the centering is loosened by unbolting a piece which spans a hinged joint in the center of the cross-brace, as is shown by Fig. 293. As the result of his experience in this work, the engineer, Mr. George S. Pierson, M. Am. Soc. C. E., makes the following statement:

In work of this kind it is very important to have the centering absolutely rigid so that it will not spring when concrete is being tamped against it and thus weaken the cohesion of the concrete. It is also important to have the arrangement such that all the centering can be removed without straining or jarring the fresh concrete.

The centers were generally removed in about three or four days after the concrete was in place.

Fig. 296 is the cross-section of the centers used in constructing the 60-in. conduits of the Forest Hill Reservoir at Newark, N. J. These centers were built in sections 16 ft. long and were supported on brick piers, as shown by the drawing. The footing course of concrete was laid first, and on top of this were built the brick piers, a pair

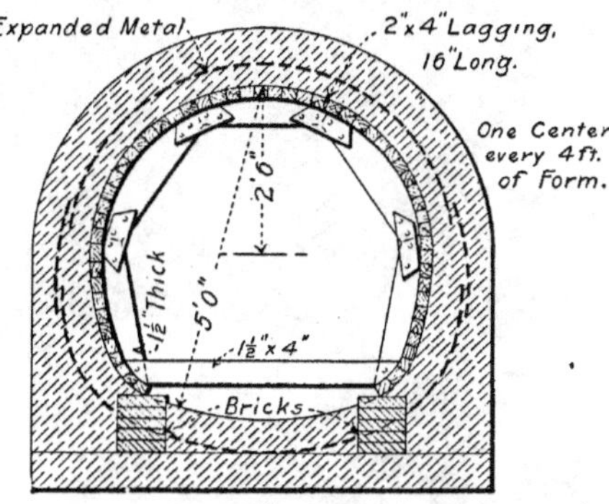

Fig. 296.—Center for 60-in. Conduit, Forest Hill Reservoir, Newark, N. J.

under each of the four ribs. The forms were then set up and lagged completely except for a manhole at the middle and crown of each. Through these holes the concrete for the inverts was passed to men working inside the form. The principal feature to be noted in the construction of these forms is the means by which they are collapsed. The bolts at one end of the bottom cross-bar are unbolted and the feet of the arch-rib are drawn together, thus allowing the center to be slipped free from the concrete. In order that the feet of the ribs may be easily drawn together the joints are made somewhat flexible. These centers were struck in about thirty-six hours after the concrete was laid.

A novel centering suitable for conduits of moderate size is shown by Figs. 297 and 298. This centering was used in constructing a 2½-ft. sewer at Medford, Mass. In this case the sewer had a crown of brick, but the centering should serve equally well for an all-concrete construction. The construction of the forms used for the invert is shown by Fig. 297. They were 10 ft. long and made in halves separating on a vertical line through their center, and were securely held together by means of malleable-iron clamps gripping the stringers along the inner edge of each. Proper width and stiffness were obtained by inserting an iron dog in the end ribs of each 10-ft. section. In using these forms they were first smeared to prevent the cement from sticking to the lagging; they were then set up in the trench in line and the concrete deposited around them. These forms were removed

FIG. 297.—Invert Form for Sewer at Milford, Mass.

from the concrete one after another by knocking out the iron dogs in each end and putting in place thereof turnbuckle hooks. After releasing the iron clamps the turnbuckles were turned so that the upper portions of the forms approached each other slightly, thereby separating them from the concrete in the most careful way. The arch centers, Fig. 298, were made in 10-ft. lengths, of $\frac{7}{8}$-in. lagging, 2-in. plank ribs and stout stringers on each side. All lagging was $1\frac{1}{2}$ ins. wide and was milled so as to have one beveled edge, in order that the exterior surface might be planed very smooth. The ribs were spaced 2 ft. apart between centers. On the forward end of each center there was securely fastened to each side a stout wooden wedge, which supported the rear end of the next center. It was then necessary to provide only one movable support for each 10 ft. of arch centering. In putting these arch centers in place a man stood upon the skewbacks, raising the center with an iron hook which passed

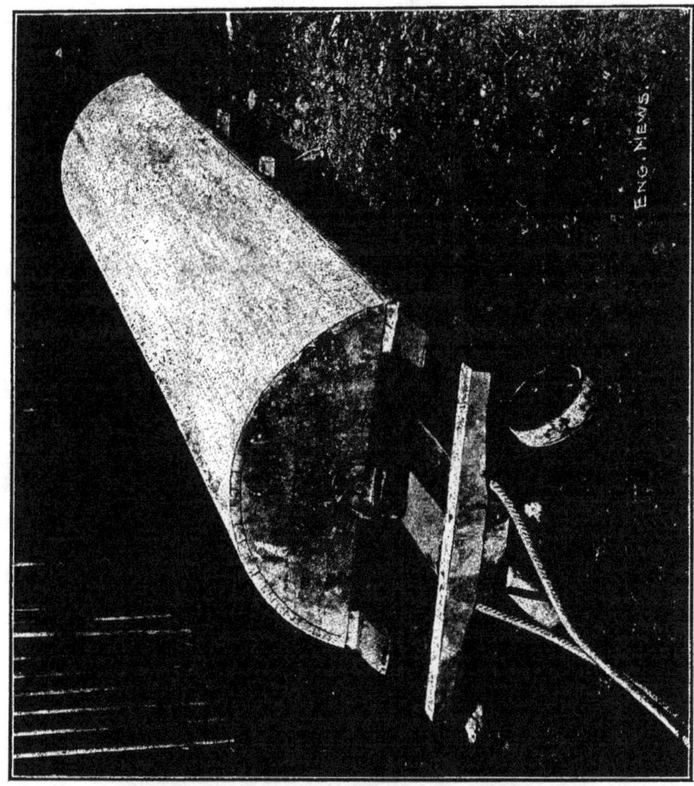

FIG. 298.—Arch Centers for Sewer at Milford, Mass

through a small hole in its top, while a man at the forward end pushed it into place upon the wedges of the preceding center. The forward end of each center was supported by a patent screw-brace or jack placed under a rib 2 ft. from the front end. After the arch had been turned, the arch centers were removed by the aid of a special truck, which had two bent axles and four cast-iron wheels so adjusted that they ran smoothly along the invert. At each end of this truck was a long rope. In the removal of the arch centering, an adjustable roller was first fastened by means of a thumb-screw to the outer rib, then the truck was pulled a foot or more under the center, the screw-jack was released, allowing the forward end of the center to drop about an inch upon the running-board of the truck, which was pulled into the conduit until the adjustable roller ran off the rear end of the truck, automatically locking itself there, whereupon the tail-rope was pulled by a man outside the finished sewer, drawing the arch center off its supporting wedges and allowing it to drop upon the truck to be wheeled to the outer end of the work. In this manner the arch centers were set up in place and removed with great rapidity and without injury to the concrete.

The centering shown by Fig. 299 was employed in constructing

FIG. 299.—Metal-wrapped Center for Sewer at Washington, D. C.

a 5-ft. egg-shaped sewer at Washington, D. C. A similar centering has been successfully used for constructing sewers down to 8 ins. in size. As shown the centering consists of two parts, an invert form and an arch center, each 16 ft. long and made up of lagging nailed to plank ribs spaced 2 ft. apart. To prepare the centering for use the arch center is set on top of the invert form with the wedge timbers *A* and *B* between, and the two are locked together by latches on the outside of the end ribs. A square gudgeon-timber is then passed lengthwise of the centering through the hole *C* in the ribs. This timber projects at each end and is rounded so that the centering can be mounted on bearings carried on two trestles. A crank-arm applied to one end of the gudgeon-timber enables the centering to be rotated so that a strip of steel 6 ins. wide and $\frac{1}{24}$ in. thick can be wound spirally around it. This wrapping-strip is shown in position by Fig. 299. The centering with the steel strip in place is set in the trench and the concrete is filled around it to the proper thickness. As soon as the filling is completed the wedge-timbers are driven in until the ribs drop into notches. This allows the upper and lower parts of the centering to close together when they are withdrawn, leaving the spiral strip of steel in place to carry the weight of the setting concrete while the centering is being rewound for shaping another section of conduit. After the concrete has set, the steel strip is removed by pulling on the loose end. To prevent the steel from adhering to the concrete it is smeared with oil.

Concreting.—The concreting of conduits is usally done in sections and is made a continuous operation for each section. This brings across the conduit all planes of weakness due to joining old and new concrete. The length of section chosen is usually that which can be completed in a day's work without any portion of the concrete becoming set before fresh concrete is joined to it, and this length varies with the size of the conduit and the conditions of the work. In the Cedar Grove Reservoir work about 32 ft. of the twin conduit was built in a day; the length of smaller conduits built in a day is often much greater. Upon the completion of a day's work the end to which the succeeding day's work is to be joined should be so finished as to insure the bonding of the new work with the old. Various ways of doing this have been practiced, but the most frequent practice is either simply to roughen the surface or to finish it with a groove or mortise.

In placing the concrete every effort should be made to secure a dense and compact product, particularly where the conduit has to carry water under pressure. This necessitates the use of a very wet mixture well joggled and churned into place, or if a dry mixture is employed, the tamping should be very thoroughly performed. Prevailing practice strongly favors a wet concrete for conduit work.

Examples from Practice.—Among the examples of reinforced concrete conduits given in Chapter IX two are of particular interest because of the magnitude of the work and the care with which it was performed. This is the conduit for the Jersey City water-supply and the conduits and sewers for the Torresdale Filters at Philadelphia, Pa. A brief description of the methods of constructing these conduits follows:

Jersey City Aqueduct.—Typical sections of the reinforced concrete aqueduct for the Jersey City water-works are given on p. 260. The centers for this aqueduct were made in sections 12.5 ft. long, consisting of eleven segments, each weighing about 200 lbs. The specifications required that the conduit should be built in monolithic sections not exceeding 20 ft. in length, but sections 50 ft. long were used wherever the top of the conduit was not over 18 ins. above the undisturbed ground. It was further specified that, if necessary to secure smooth surfaces, the concrete should first be deposited in the invert and the lower part of the side walls at least 1 in. back from the finished surface, and that immediately afterwards the forms for the invert should be placed and the space between them and the concrete poured full of a very wet mixture of equal volumes of cement, sand, and fine crushed stone. The urgency of the work did not permit of this procedure; the invert was formed by screeding, being finished by troweling while the mortar was plastic and the two bottom segments of the centering were not used. No outside forms were employed for a width of about 5 ft. on top of the arch. The centers were left in place about forty-eight hours, and occasionally they were greased with vaseline cut with kerosene.

The mode of procedure in setting the forms and building the conduit were in detail about as follows: The forms without the two bottom segments were set on 6-in. cubes of concrete previously built. When set, therefore, each form was open at the bottom for a width of 6 ft. and had in its top a manhole or scuttle about 2 ft. square. After being set, the lagging was greased and the reinforcing-rods were placed in position. Concrete was first deposited on the outside and forced under the forms, with tamping-bars, until it appeared at the inside opening and filled the annular space between the form and the ground. The invert concrete was then passed through the roof-scuttles and men inside the form deposited it and screeded the invert into shape. The scuttles were then closed and the concreting of the side walls and roof proceeded with from the bottom upward on each side. Back forms 12.5 ft. long and 2 ft. wide were employed to shape the exterior of the shell. These back forms were carried upward as the work progressed by adding segments until they reached a level about 2½ ft. each side of the crown. Here they were discontinued and the crown segment of the shell was built without forms to shape the exterior. Special care was taken to keep the forms clean, the sole duty of one laborer being to scrape and wash them for reuse as fast as they were At the end of each day's work a groove was formed in the face of the concrete to bond it with the succeeding day's work. Fine cracks have girdled the conduit at many of these joints, but nowhere else, and it is thought that these will silt up soon after the conduit has been put in service. Little or no finishing was required on the inside of the conduit, and it is as finished much smoother than the average brick lining. The speed of construction was rather remarkable. Between July 25 and November 14, 18,500 ft. of conduit were built. On one section 2,586 ft. of conduit were built in 65 days by a gang of 38 men, an average of 39.8 linear feet per day.

Torresdale Filter Conduits.—In the construction of the Torresdale filters for the Philadelphia water-supply the large sewers for waste water and the conduits for filtered water were built of concrete; all the conduits and the 8½-ft. sewers were built of concrete reinforced by expanded metal. The lengths of the several conduits were as follows: 300 ft. of 8½-ft. sewer reinforced by two layers of No. 10 3-in. expanded

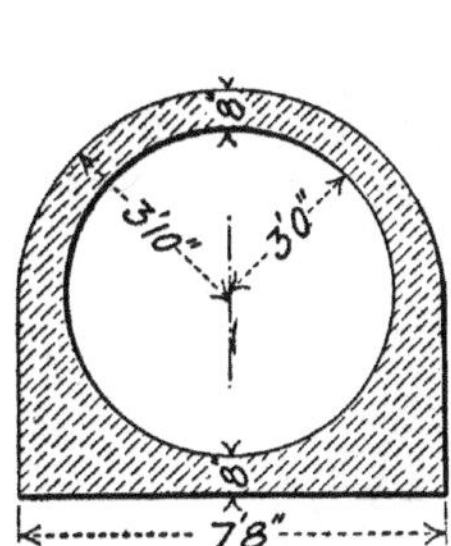

FIG. 300.—Section of 6-ft. Sewer, Torresdale Filters.

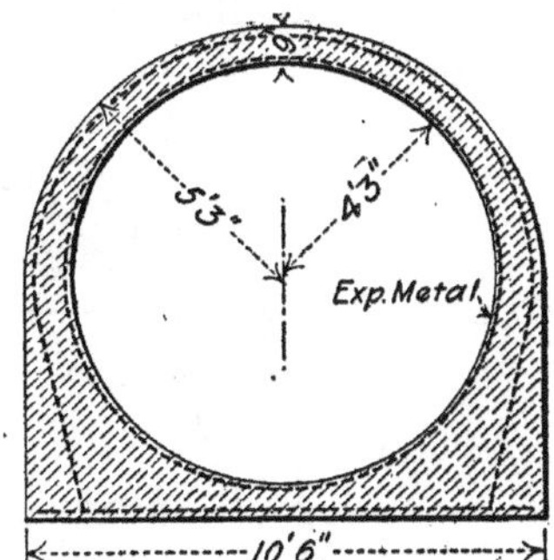

FIG. 301.—Section of 8½-ft. Sewer, Torresdale Filters.

metal; 576 ft. of 7½-ft. filtered-water conduit, 782 of 8-ft. by-pass conduit, 1050 ft. of 9-ft. conduit, and 1,430 ft. of 10-ft. conduit, all reinforced by expanded metal.

The 8 ft. 6 in. and 6 ft. sewers are built of concrete; the 6-ft. sewer is without reinforcement, and the 8 ft. 6 in. sewer is reinforced with two layers of No. 10 3-in.-mesh expanded metal.

The concrete was made of Portland cement, Bonneville, Lehigh, and Giant being the brands used, and the proportions were 1 part cement, 3 parts sand, and 5 parts ¾-in. stone. About one-third of the concrete for this work was mixed in a portable cubical mixer of ½ cu. yd. capacity and the remainder by hand. Owing to the relatively small amount used per day, about 20 cu. yds., it was found that there was practically no difference in the cost of mixing by machine or hand.

The specifications required that both the sewers and conduits should be built in monolithic sections; that is, the contractor could build as long a section as he could finish in one day,—so that this work con-

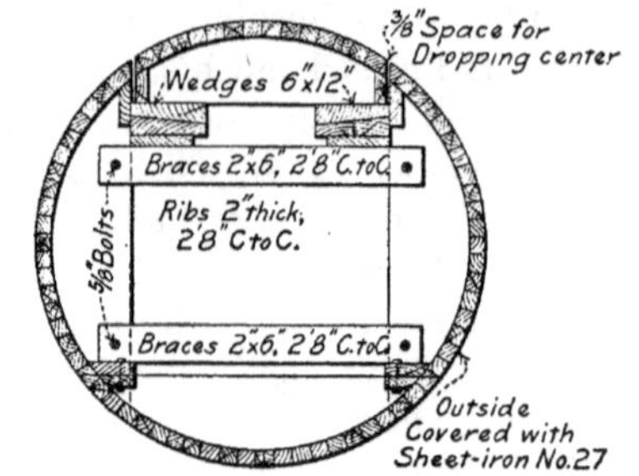

FIG. 302.—Center for 6-ft. Sewer, Torresdale Filters.

sists of monolithic blocks or rings from 12 to 16 ft. long keyed together. Figs. 300 and 301 show the section of the 6-ft. and 8-ft. 6-in. sewers, and Fig. 302 the center or form for building the 6-ft. sewer. The

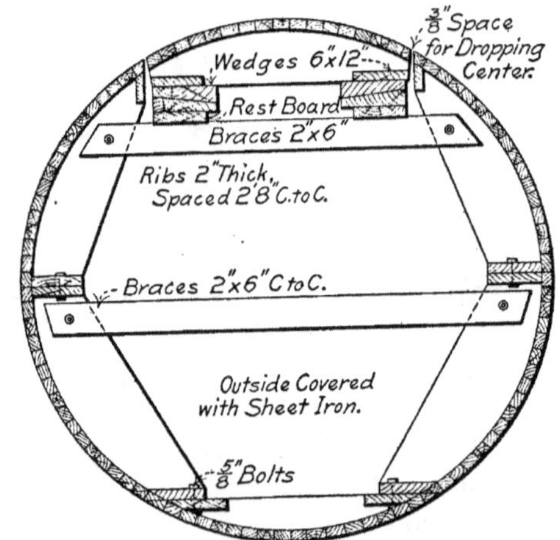

FIG. 303.—Center for 8½-ft. Sewer, Torresdale Filters.

forms were made 8 ft. long, and two sections or 16 lineal feet were built in from 8 to 9 hours, including setting of forms, by one foreman, one carpenter, and 15 laborers. This section contained 13 cu. yds. of concrete and required 80 bags of cement. The same gang built 14 lineal feet of the 8½-ft. sewers in 10 hours using the form shown by Fig. 303, and were somewhat delayed by lack of experience in handling the expanded metal, which was placed in two layers about 3 ins. apart. It was found that the small mesh and the two separate layers had a tendency to screen the mortar from the stone, and that better results could be obtained by using one layer of expanded metal with a larger mesh. Difficulty was also experienced in properly placing and ramming the concrete in the bottom and lower quarters of the circular section, and it was thought that better work could be obtained by using a "horseshoe" section or one with a comparatively flat bottom, as the metal could be kept closer to shape and the concrete more thoroughly rammed. The cross-section of the 10-ft. conduit, originally circular, was changed, for the above reasons, to that shown in Fig. 304. The size of the expanded metal was changed from

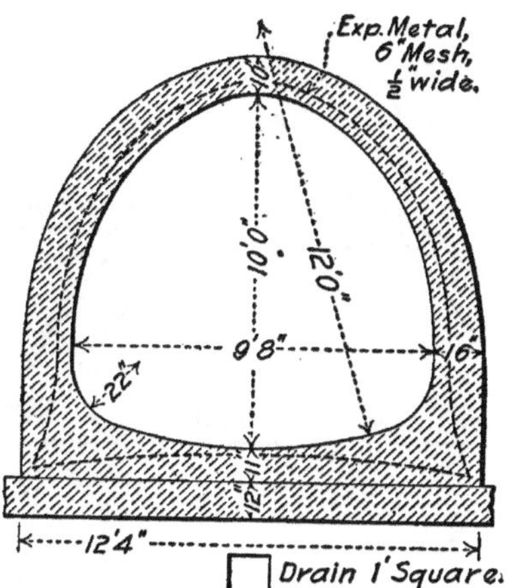

FIG. 304.—Section of 10-ft. Conduit, Torresdale Filters.

two layers of No. 10 cut double width 3-in. mesh to one layer of 6-in. mesh cut ½ in. wide from No. 4 steel, giving 6.77 sq. in. of metal per sq. ft. of area. It is believed that this is the heaviest expanded metal cut to date, and the Expanded Metal Company after cutting several carloads refused to furnish any more, claiming that it was breaking their machines, but that they were having heavier machines built that would do this work. In the mean time two layers of the No. 4 metal with 6-in. mesh were wired together and placed as shown, approximately 3 ins. from the inside of the conduit. This size mesh has done entirely away with any tendency to screen the stone from the mortar, and it was found that the cross-section was an improvement over the circular one and the 7½-ft. and 9-ft. conduits were changed accordingly. Figs. 305, 306, and 307 show the construction of the centers for building the 7½-ft., 9-ft., and 10-ft. conduits respectively. The general plan is the same, and the 9-ft. center, being

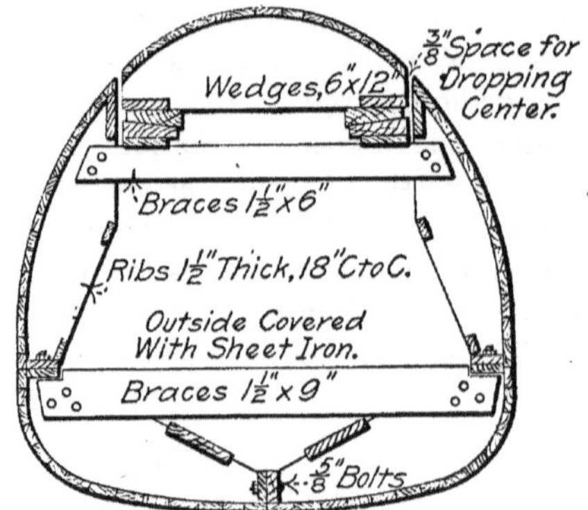

FIG. 305.—Form for 7½-ft. Conduit, Torresdale Filters.

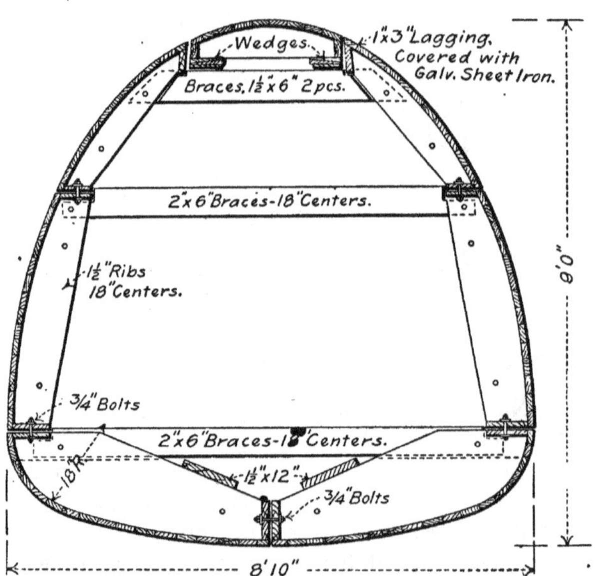

FIG. 306.—Form for 9-ft. Conduit, Torresdale Filters.

the last built, is the best. They were covered with No. 27 galvanized sheet iron, which not only gave a smooth surface to the concrete work, but added greatly to the life of the centers. Three sets of centers or forms were made for each side conduit, and while it was necessary to reiron them, they were in repairable condition at the completion of the work. The bracing inside the forms was arranged to allow a center to be taken apart, after the concrete had been placed for at least 60 hours, and be brought forward, through a form already in place, for further use.

The conduits were built in sections from 12 to 13½ ft. in length, and there was very little, if any, difference in the labor required to build a section, in from 8 to 10 hours, of any of the three sizes. One foreman and eighteen men on top of the trench mixed and handled the concrete and granolithic mortar, while one foreman, one carpenter, and seven men in the trench set the forms, placed and rammed

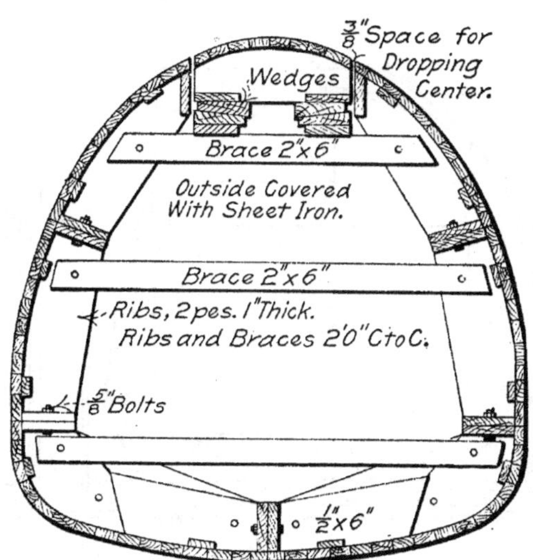

FIG. 307.—Form for 10-ft. Conduit, Torresdale Filters.

the concrete, etc., for one section in, generally, 8 hours. The 9-ft. conduit, as an average of the three sizes, contained 20 cu. yds. of concrete, 1,200 sq. ft. of expanded metal, and required 125 bags of cement to build a section 13½ ft. long.

The concrete was mixed rather wet and placed in six-inch layers. The one inch of granolithic face required on the inside of the conduit was mixed one part cement, one part granolithic grit, and one part sand.

In building a section the bulkhead (Fig. 308) is set up the length of the form in advance of the face of the conduit already built, and the concrete is deposited in this space, to within 1 in. of the invert bottom; the bottom section of the form—made in two pieces in order to be moved forward through a center already in place—is then set, the rear end being bolted to the face of the last form used, and the front end resting in the bulkhead. About 2 tons of pig iron is placed on the invert form to keep it from floating, and the granolithic mixture is then poured, through four spouts, at the corners, into the one-inch space between the concrete and the form. There was some doubt, at first, as to the granolithic mixture or grout flowing entirely under the form, but it did, and the result is a very smooth invert.

In building up the sides the grout is poured between the outside of the conduit forms and a sheet-metal board with pipe ribs on each side, the small rib of ¾-in. pipe forming the 1-in. space for the

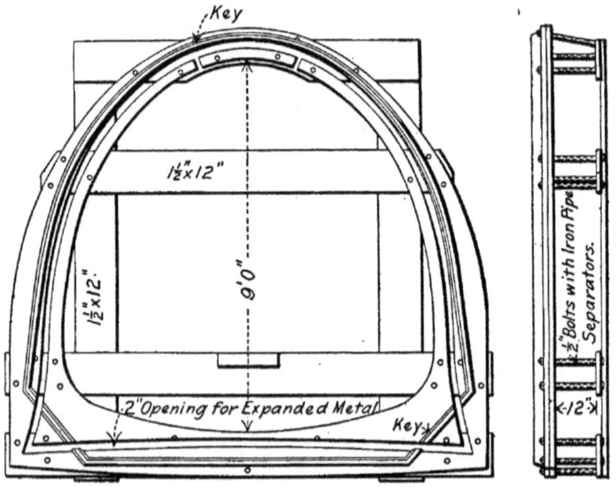

Fig. 308.—Bulkhead Form for End of Section, Torresdale Filters.

granolithic finish. The expanded metal was kept against the outside or large rib of 1½-in. pipe. Two widths of boards were used, and their length was slightly longer than half the length of the section. Two were used on each side, slightly lapping in the center, as follows: The metal board was placed with the small rib against the conduit form, and the larger rib kept the expanded metal 3 ins. from the inside of the conduit; a six-inch layer of concrete was placed between the metal board and the outside form or planks; the granolithic mixture was lowered in buckets, and poured in the inch space between the conduit form and the iron board to a depth of six inches. The board was then raised six inches and the concrete rammed, another layer of concrete placed, and so on. The granolithic face on the top section of the form was made of stiff mortar. The face left from the above work was smooth, entirely free from the appearance of stone, and did not require any plastering.

The expanded metal was cut on sheets corresponding to the length of the sides of the form and lapped 6 ins. in all directions; the bulkhead having a slot as shown permitting the expanded metal to project 6 ins. from the face of the concrete in order to tie two sections together, and also a rib to form a depression into which the concrete of the next section was rammed.

A large portion of these conduits have been built over a year and are free of all cracks, except that in a few instances a very fine crack has developed at the junction of two sections. The conduits will be under about 18 feet head of water when in operation, and on this account they were built in monolithic sections.

The cost of the above work, exclusive of excavation or any profit, but including forms, expanded metal, concrete material, and labor, was about $10.50 per cu. yd.

The excavation was mostly in sand or gravel containing water. The trench for the 10-ft. conduit leading from the filtered-water basin to shaft No. 1 was 34 ft. deep, and the bottom 6 or 8 ft. was in river silt, a dark mud, firm when first exposed, but mucked up badly when much disturbed. A slab of concrete 12 ins. thick and the width of the trench was built in advance and as a sub-foundation for the conduit between these points.

The trenches were drained through a 10-in. square box, made of 1½-in. boards, with a number of ¾-in. holes bored in the sides and top, laid below the sub-grade of the conduit and discharging into a sump from which the water was pumped. Coarse gravel was placed next to the box, and these drains worked very well on a grade of 2/10 ft. per 100 ft. for a length of 600 ft. In one case a similar drain was laid level for a distance of 500 ft. and kept clear by frequently working a wire cable and chain back and forth inside the box.

Manufacture of Concrete-steel Pipe.—In the preceding sections a number of instances have been mentioned of the use of separately cast lengths of reinforced concrete pipe for constructing aqueducts and sewers. The casting of these lengths of pipe is usually performed in special plants the apparatus and processes of which are controlled by patents, and it is, therefore, of general interest only to engineers. The leading forms of cast concrete-steel pipe are the Monier, the Bordenave, and the Bonna, and the methods of casting each of these forms is described in outline in the following paragraphs.

Monier Pipe.—Several methods are employed to cast Monier pipe. When the shell of the pipe being cast is of sufficient thickness to permit tamping between the reinforcing network and the walls of the molds, it is formed by inserting the network in the annular space between a cylindrical outside mold and a core or mandrel set on end and ramming the mortar in place in horizontal layers. For pipes with very thin shells this procedure is not practicable. The process adopted in this case is to employ a cylindrical core or mandrel only, and after placing the netting around and the proper distance away from it to plaster the mortar against the mold in successive thin coats until the required thickness of shell has been attained. A third process of manufacture which is practiced by the firm of Wayes & Co. of Berlin, Germany, is molding by machinery. The molding-machine employed consists essentially of a horizontal drum of the exact interior diameter of the pipe, and of rolls for spreading the mortar onto the drum. In operation mortar is spread onto the drum, which is rotated so as to pass under the rolls which spread the mortar uniformly over the cylindrical surface. The wire network is then wrapped around the drum on top of the mortar coat, and succeeding coats are added like the first until the required thickness of shell is attained.

Bordenave Pipe.—The process of manufacture employed in making Bordenave pipe is more truly a casting process than the processes described in the preceding paragraph. The mortar is made very thin and is simply poured into the annular space between a hollow cylinder and a cylindrical mandrel in which the reinforcing netting has been previously erected. The process is a simple one in itself, and interest attaches principally to the plant used for performing it on a large scale. The first operation is the construction of the lengths of reinforcement. The helical rod is first formed by a special binding-machine. This helix is then threaded onto a mandrel, where it is adjusted to the proper spacing and where the longitudinal rods are placed and attached. Marks on the mandrel aid in the spacing of the various members of the reinforcement. The placing of the molds and reinforcement and the casting of the pipe are performed by means of an elevated platform which turns on a circular track on the molding-floor. This platform is provided with apparatus for handling the molds. Its operation is as follows: The reinforcement is first set on end on the molding-floor at the rear of the platform, and the mandrel is lowered into it and centered. This mandrel is adjustable in diameter. The two halves of a hollow cylinder are next placed around the reinforcement, centered, and locked rigidly together and in place. The liquid mortar is then poured into the top of the mould and the platform moved ahead to the next casting position. The circular track is of such diameter that the platform makes its circuit in from two to three days, so that when the position of the first pipe cast is reached a second time that pipe is ready to be removed to make place for a new casting. The cast pipes are lifted by means of a traveling-crane which runs on a circular track concentric with and inside the molding track.

Bonna Pipe.—The process adopted for casting Bonna pipe is in all respects similar to that adopted in casting Bordenave pipe. Taking for illustration the small distributing sewer-pipe for the Parc Agricole d'Achères, the following was the general mode of procedure: The rings of the inner reinforcement were first erected and adjusted to space on a mandrel and then the longitudinal bars were attached to them. The next step was the construction of the interior steel tube, which was done by special machines for bending the sheets and forming the joints. The outside reinforcement was then erected exactly like the inside as described above. The two reinforcing skeletons with the steel tube between them were then set on end on wooden templates, a steel hollow cylinder mould was clamped outside of them and a collapsible steel mandrel was placed inside. The mold was then ready for pouring the liquid mortar. For placing the molds and pouring the mortar and for handling the cast pipe a traveling platform and crane much similar to those used for casting Bordenave pipe were employed.

CHAPTER XVII.—FACING AND FINISHING EXPOSED CONCRETE SURFACES.

THE difficulty of securing an even-grained surface of uniform color on concrete work is one of the most annoying which builders of such work have to overcome. Concrete work is subject to various sorts of surface imperfection, but the two most common imperfections are roughness or irregular surface texture and variability of color or discoloration. Either of these imperfections is capable of disfiguring an otherwise sightly structure, and the task of avoiding them is one which warrants serious attention from those undertaking work in reinforced concrete. Unfortunately practice has not settled upon a solution of the problem, hence its consideration here is rather a record of experience than a set of instructions which can be followed with the certainty that successful results will ensue.

Causes of Roughness and Discoloration.—There are several conditions which may result in a concrete surface of uneven texture and with mechanical roughnesses, such as projections, bulges, ridges, pits, bubble-holes, and scales. One of these is imperfections in the molds. The use of rough lagging of uneven thickness and with open cracks and allowing the forms to become distorted and warped are certain to leave their impress upon the plastic concrete in the form of ridges, tongues, and bulges. Failure to pack the concrete filling tightly and evenly against the mold will result in rough places. Lack of homogeneity in the concrete is another prolific cause of variation in the surface texture of concrete work. This lack of homogeneity may result from failure to mix the concrete materials thoroughly and evenly in the first place, or to the segregation of the coarse and fine parts of the mixture during its deposition and ramming into place. In both cases the result is a material of alternate coarse and fine texture. Dirt or cement adhering to the molds will leave pits in the concrete surface, and the pulling away of the concrete in spots when it adheres to the molds when they are removed will cause similar roughnesses.

Variations in the color of concrete surfaces probably result from a variety of causes. Some of these are obvious and others are difficult to determine with any exactness. Roughness or uneven surface texture is a common cause of variation in color, since the alternate rough and smooth parts weather differently and collect and hold dirt and soot in different degrees. Another cause of variation in color is the use of different cements in adjacent parts of the surface work. No two cements are of exactly the same shade of color, and the concrete made of them partakes of this variation. In a similar manner sand of different shades of color or of different degrees of cleanliness will cause a cloudy and streaky appearance in concrete. Dirt adhering to the molds will frequently stain the adjacent concrete surface.

Even when the smoothness of the surface is satisfactory, however, and when there is no criticism possible as to the kind and quality of the aggregates, their deposition and the cleanliness with which the work is done, concrete surfaces frequently vary in color and have a cloudy light and dark appearance. In many cases there seems to be good reason for attributing this to the leaching out of lime conpounds and their deposition in the form of an efflorescence on the concrete surface. The extent of this efflorescence varies; at times the deposit is so thin as merely to give a lighter shade to the places where it appears, but it will often form an encrustation of considerable body and thickness which may be readily scraped off as a white or yellowish-white powder. The nature of this discoloration and the preventive and remedial treatments which have been practiced in its cure are discussed more fully in a succeeding paragraph.

Construction of Forms.—Very slight imperfections in the face of the forms against which the concrete is molded are sufficient to leave an unsightly impression on the plastic mixture when it hardens. Even the grain of smoothly dressed timber will show on the surface of concrete which has been deposited with a mortar facing. It is very difficult to construct forms so that they will not leave slight impressions of this character, and it is generally better not to attempt the task in any but exceptional instances. In these a straight-grained, smoothly dressed timber, with its pores filled with soap or paraffine well rubbed in, or a rougher timber covered with sheet metal, can be used. Generally speaking, all has been done that is practicable so far as the forms are concerned when the face-lagging is kept true to surface and has close-fitting joints. Grain-marks and similar minor impressions of the forms can usually be eliminated by rubbing the surface or floating it with grout, at less cost, than by attempting to perfect the molds beyond a reasonable measure. In fact many engineers experienced in concrete work prefer not to attempt to secure particularly perfect finish in the forms, but to dress the entire surface by some style of tooling or rubbing process after the forms have been removed. The most apparent imperfection in concrete surfaces is usually the joint-marks of the lagging-boards. These may be due either to slight differences in the thickness of adjoining boards or to open joints. The remedy for the first cause is obvious, but it is not so easy to insure smooth, tight joints and keep them smooth and tight when the boards swell from the moisture absorbed from the wet concrete. One of the most successful forms of joint is that shown by the sketch Fig. 309. In this construction the wedge edge presses into the edge of the adjoining board without distorting or bulging the lagging. Pointing the joints with hard soap or putty, packing them with oakum and covering them with pasted strips of cloth, are other means which have been practiced for preventing joint-marks on the concrete. A method of eliminating grain-marks, which was used with success in constructing the piers of the Frazer River Bridge in British Columbia, consisted in covering the tightly laid matched lagging with gloss oil and then blowing sand into the oil with hand-bellows.

FIG. 309.—Joint for Lagging.

Mortar or Grout Facing.—One of the most frequently employed means for securing a smooth surface finish on concrete is to use a mortar or grout facing. This facing differs from plastering in being laid on as the concrete is deposited, thus forming a single piece with it. The thickness of mortar facing employed in practice varies from ½ in. to 3 ins., but the usual practice is to make it 1 in. or 1½ ins. thick. A facing as thick as 3 ins. is a rather unnecessary waste of mortar, while one which is much less than 1 in. thick is likely to be pierced by the stones in the concrete unless great care is taken in ramming the concrete filling behind the mortar facing. A mortar or grout facing shows the impress of small roughnesses on the mold more readily than does concrete, and particular care is necessary to secure a smooth surface in the mold when a mortar facing is adopted. The compositiou of the facing mortar is usually specified as 1 part of Portland cement to 2 or 3 parts of sand. These ingredients are mixed rather wet, since the paste must completely fill the facing-mold, but care must be had not to have so thin a paste that the stones from the concrete behind will be pushed through it during the subsequent filling and ramming.

The following method of placing mortar facing is practiced by the Illinois Central R.R. and has gained wide adoption during the last few years. A sheet-iron plate 6 or 8 ins. wide and about 6 ft. long has riveted across it on one side 1½-in. angles spaced about 2 ft. apart. One edge of this plate is provided with handles. This device is employed as a mold for the facing and is operated in the following manner: The plate is set up against the face of the form with its angle-ribs close against the timber and its handles upward. In this position of the plate there is between it and the form an open slot 1½ ins. wide. This slot is filled with mortar which is tamped thoroughly, and immediately afterwards the concrete backing is deposited behind the plate. When this has been done the plate is withdrawn by the handles and the backing and facing are rammed together to a close bond. The mortar facing is mixed in small batches as it is needed, and no delay is permitted in placing the concrete backing, the essential principle and purpose of the method being to secure as nearly as is possible the simultaneous construction of the backing of concrete and its facing of mortar.

Fig. 310 shows an excellent form of surface mold of the type just described. By varying the size of the angle-ribs any desired thickness of facing can be constructed, and the flare of the top edge facilitates the placing of the mortar, which is usually done with shovels. In lieu of a steel plate, use is sometimes made of a board provided with furring-strips on one side. This is a more unwieldy device than the one illustrated, and it is objectionable because of the large crevice left upon withdrawal into which the mortar facing is likely to slough and which is less easily closed and bonded by the final ramming. In constructing mortar facing with either iron or board molds perfect success is secured only at the expense of great care. The mortar must be mixed in small batches and only as needed, and it must be thoroughly rammed and churned into the facing-mold. The concrete backing must be deposited behind the mold without delay and firmly rammed against it, and finally the ramming together of the facing and backing must be thorough.

The following method of applying grout facing was employed with success in constructing the Atlantic Avenue subway for the Long Island R.R. in Brooklyn, N. Y. The concrete was deposited in 6-in. or 8-in. layers, and after ramming the concrete at the face was pushed back from the form about 1 in. with an ordinary gardener's spade and a thick grout of 1 part cement and 2 parts sand was poured into

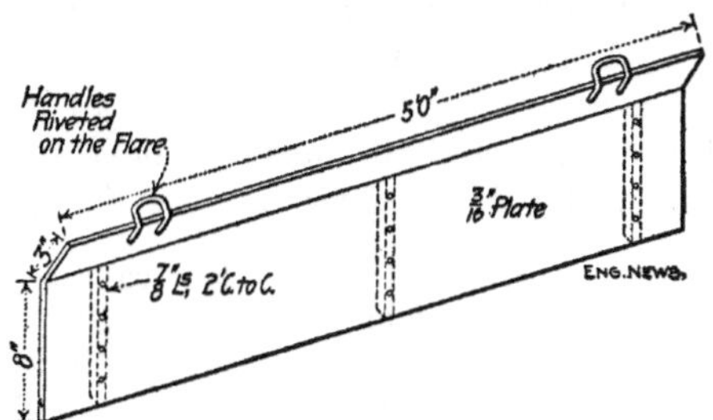

FIG. 310.—Mold for Mortar Facing.

the space. The forms used were tongued and grooved yellow pine painted with paraffine paint. In this work a good surface was invariably secured when the men did their work faithfully, but any carelessness on their part evidenced itself in a rough spot when the forms were removed. As an indication of the susceptibility of mortar facing in taking impressions from the forms, it may be noted that even with the dressed and paraffined lagging the grain of the wood was shown perfectly on the mortar facing.

Finishing Mortar Facing.—When mortar or grout facing is employed as described in the preceding paragraphs the slightest imperfections in the grain of smoothly dressed wood is clearly impressed on the plastic material. There will also be occasional rough spots, pittings, or bubble-holes even with the most careful construction. To get rid of these some method of surface finishing must be resorted to. A number of methods have been practiced. In recent concrete culvert work on the New York Central & Hudson River R.R. an excellent surface finish was obtained by the following procedure: The forms of 2-in. dressed and matched pine, after being put in place, were painted with a coat of thin soft soap, then as the layers of concrete were brought up the face was drawn back with a square-pointed shovel, the edges of which had been hammered flat. Mortar in the proportion of 1 part cement to 2 parts sand, mixed rather wet, was then poured in along the form and the layer rammed against it. Hard soap was used to fill openings left by joints of the lagging. When the forms were removed and while the concrete was yet "green," the surface was carefully rubbed with a circular motion, with pieces of white fire-brick or briquettes of 1 cement to 1 sand, made in molds about the size of a building-brick, handles being pressed in while soft. The surface was then dampened and painted with a coat of grout of 1 cement to 1 sifted sand, and this was closely followed by a final rubbing with a circular movement, using a wooden float. All edges were rounded with a Crafts edger, or with wood fillet, and the coping joints were struck with a Crafts jointer.

In the specifications for concrete presented by the special committee of the Engineering and Maintenance of Way Association the following requirement for finishing was adopted:

After the forms are removed, any small cavities or openings in the concrete shall be neatly filled with mortar if necessary. Any ridges due to cracks or joints in the lumber shall be rubbed down; the entire face shall then be washed with a thin grout of the consistency of whitewash, mixed in the proportion of 1 part of cement to 2 parts of sand. The wash should be applied with a brush.

In the extensive concrete construction of the Aurora, Elgin & Chicago R.R. the exposed surfaces were all finished according to the following specifications:

All walls when finished must present a smooth, uniform surface of cement mortar, and all disfigurements must be effaced, and if there are any open, porous places, they must be neatly filled with mortar of 1 cement and 2 sand, well rubbed in, which finishing must be done immediately upon the removal of the forms. Compensation for all labor and material required in such finishing, including the mortar facing when required, with the finishing of bridge seats and other parts, is included in the price per cubic yard for concrete work.

Mr. Edwin Thacher, in his general specifications for concrete-steel, requires the following surface finish:

For plain flat surfaces, the concrete may be rammed directly against the molds, and, after the molds have been removed, all exposed surfaces shall be floated to a smooth finish with semi-liquid mortar, composed of 1 part cement and 2 parts of fine, sharp sand, care being taken that no body of mortar is left on the face, sufficient only being used to fill the pores and give a smooth finish.

A very effective finish is obtained by etching the mortar facing with acid. The method consists of using a facing mortar composed of Portland cement and finely crushed stone, the kind of stone depending upon the appearance desired. Thus any shade of red or gray granite, sandstone, etc., can be obtained, and special effects can be obtained by the use of sand, pigments, etc., in the mixture. This mortar is composed of about 1 part Portland cement to 2 or 3 parts of the finely crushed stone. The exposed surfaces are then treated by chemical or mechanical means to remove the cement matrix at the face, leaving the granular particles of stone partly exposed. In general this is done by washing the surface with a weak acid solution, then with clean water, and finally with an alkaline solution to neutralize any effects of the acid. In the finished work it is difficult to detect that the material is not natural stone, except by close inspection. The stone is crushed to pass through a sieve of 10 to 30 meshes per square inch according to the character of finish desired, and enough water is used to make a soft plastic mixture.

Plastering.—Plastering as a method of finishing concrete surfaces deserves mention for the purpose alone of calling a warning against its adoption. It is practically impossible to apply mortar in thin layers to a concrete surface and make it adhere for any length of time, and when it once begins to scale off the result is a surface many times worse in looks than the unfinished concrete that it was intended to render more sightly.

Pebble Dash Facing.—An effective surface finish for certain classes of concrete work can be secured by using large rounded pebbles in place of the usual aggregate for the surface layer of concrete, and then, while the concrete is soft, removing the mortar between the pebbles by wire brushing until approximately half the pebbles are exposed. The following specification for this style of facing was employed in constructing a small concrete road-bridge in the National Park at Washington, D. C.:

The concrete, which will be in the exterior faces of the bridge and the parapet walls for a thickness of 18 ins., will be made of gravel and rounded stone varying in the concrete below the belting course between 1½ and 2 ins. in their smallest diameters. This gravel will be mixed in the concrete as aggregate instead of broken stone. The mixture will consist of 1 part Portland cement, 2 parts sand, and 5 parts of aggregate. The parapet walls will be made in a similar manner, with the aggregate composed of gravel not exceeding 1 in. in its smallest diameter. When the forms are removed the cement and sand must be brushed from around the face of the gravel with steel brushes, leaving approximately half of the gravel exposed.

In this work it was found by test that at the age of 12 hours the concrete was not sufficiently set to hold the pebbles from being torn out by the brushing, and that at the age of 36 hours it was too hard to permit the brushing, to remove a sufficient depth of mortar without undue labor. At 24 hours' age the brushing proved most successful.

Tooled Surfaces.—A method of finishing concrete surfaces which is preferred by many experienced engineers is to dress the concrete after it has hardened by means of hammers or pointed chisels. The process is exactly analogous to stone dressing, and any of the forms of finish employed for cut stone can be employed equally well for concrete. In connection with tooled surfaces it is common to mold the concrete to represent ashlar masonry by means of horizontal and vertical V-shaped depressions formed as shown by Fig. 311. This

FIG. 311.—Form for Molding Concrete to Resemble Masonry.

style of finish has been extensively employed by Mr. E. L. Ransome, who gives the following directions for securing it: In imitating rough-dressed work the mold is removed from the concrete while it is yet tender, and with small light picks the face is picked over with great rapidity, an ordinary workman finishing about 1,000 square feet per day. For imitations of finer-tooled work the concrete should be left to harden longer before being spalled or cut, and the work should be done with a chisel. Most natural stone and especially granite makes excellent material for the face, but ordinary gravel will do. Whatever is used, let it be uniform in color and of an even grade. When a very fine and close imitation of a natural stone is required take the same stone, crush it and mix it with cement colored to correspond. The finer the stone is crushed the nearer the resemblance will be upon close inspection; but for fine work it is generally sufficient to reduce the stone to the size of buckshot or fine gravel.

Masonry Facing.—A facing of masonry is often employed on reinforced-concrete arch bridges, and is a very satisfactory solution of the problem of surface finish for such structures. Masonry facing may be of any style of stonework which is used for true masonry arches, and coursed ashlar, random rubble, and boulder masonry facings have all been employed. Exactly the same care should be exercised in selecting stones and laying them up into arch ring and wall, cornice and parapet, as if the structure were entirely of masonry. Beyond this the most important feature to be observed is close bonding of the masonry facing to the concrete backing. To insure this there should be a liberal use of stretchers reaching well into the backing, and these can be supplemented with metal cramps to the advantage of the work in many instances. For facing the arch ring the stones should be cut to true voussoir shape, and laid quite as perfectly as if they were a part of a true voussoir arch ring. The soffit of the arch ring is not stone-faced. In place of stone a brick facing may be employed.

The following specifications for stone and brick facing, which were prepared by Mr. Edwin Thacher, M. Am. Soc. C. E., to control work conducted by him, give a fair idea of the requirements of high-class work of this character:

Stone Facing.—If stone facing is used, the ring stones, cornices, and faces of spandrels, piers, and abutments shall be of an approved quality of stone. The stone must be of a compact texture, free from loose seams, flaws, discolorations, or imperfections of any kind, and of such a character as will stand the action of the weather. The spandrel-walls will be backed with concrete, or rubble masonry, to the thickness required. The stone facing shall in all cases be securely bonded or clamped to the backing. All stone shall be rock-faced with the exception of cornices and string courses, which shall be sawed or bush-hammered. The ring stones shall be dressed to true radial lines, and laid in Portland cement mortar, with ¼-in. joints. All other stones shall be dressed to true beds and vertical joints. No joint shall exceed ½ in. in thickness and shall be laid to break joints at least 9 ins. with the course below. All joints shall be cleaned, wet, and neatly pointed. The faces of the walls shall be laid in true lines, and to the dimensions given on plans, and the corners shall have a chisel draft 1 in. wide carried up to the springing lines of the arch, or string course. All cornices, moldings, capitals, keystones, brackets, etc., shall be built into the work in the proper positions and shall be of the forms and dimensions shown on plans.

Brick Facing with Concrete Trimmings.—The arch rings, cornices, string courses, and quoins shall be concrete-faced as described above, the arch rings and quoins being marked and beveled to represent masonry. The piers, abutments, and spandrels shall be faced with vitrified brick as shown on plans. The brick facing shall be plain below the springing lines of the arches, and rock-faced above these lines.

All rock-faced brick shall be chipped by hand from true pitch lines. All brick-facing shall be bonded as shown on plans, at least one-fifth of the face of the wall being headers. The brick must be of the best quality of hard-burned paving brick, and must stand all tests as to durability and fitness required by the engineer in charge. The bricks must be regular in shape and practically uniform in size and color. They shall be free from lime and other impurities; shall be free from checks or fire cracks, and as nearly uniform in every respect as possible; shall be burned so as to secure the maximum hardness; so annealed as to reach the ultimate degree of toughness; and be thoroughly vitrified so as to make a homogeneous mass.

The backing shall be carried up simultaneously with the face work, and be thoroughly bonded with it.

The use of boulder facing will ordinarily be limited to structures of special character, and its success will depend very largely upon the care with which the stones are selected, their size, and their arrangement in the structure. In constructing a boulder-faced concrete arch at Washington, D. C., the following requirements were specified for the facing:

The term boulder here is meant to cover loose rock, which shall be hard, durable, and of a quality to be approved by the engineer, whose edges have become weathered or water-worn, or both, and are more or less rounded. It is the intention to obtain a decidedly rustic effect on the facing, and to that end extreme care must be taken in the selection of the stones, and only mechanics who show an aptitude for this class of work shall be employed. No tool marks or fresh fractures will be allowed on the showing faces.

The boulder face of such stone shall project at least 2 ins. beyond the neat lines of the bridge, and this projection shall not exceed 15 ins., nor shall it be greater than one-half the least horizontal dimension of the stone. All joints shall be scraped and brushed clear of mortar to the depth indicated by the engineer. The mortar shall consist of 1 part Portland cement and 2 parts sand. The backs of all boulders shall be plastered with a layer of mortar as specified, at least $\frac{1}{4}$ in. thick, immediately before ramming the concrete against them.

The arch-stones shall have a depth of between 3 and 4 ft., a width of not less than 18 ins., nor more than 36 ins.; all dimensions to be measured exclusive of the projections beyond the neat lines. The joints shall be dressed so as not to exceed $1\frac{1}{2}$ ins. at any point for at least two-thirds their depth and two-thirds their length, and as much more as the stones will admit. Each arch-stone shall be cramped to the adjacent steel girder by means of a wrought-iron cramp made from $\frac{3}{4}\times\frac{3}{8}$-in. bar, the cramps to reach at least 2 ins. into each boulder, to be well cemented into them, and securely cramped to the top of the girder. The outside girders shall be cramped to the adjacent girders by 10 wrought-iron cramps made from $\frac{3}{4}\times\frac{3}{8}$-in. bar (in construction we used $\frac{3}{4}\times\frac{1}{4}$-in., as it bent cold without fracture).

No dressing will be required on the stones used in abutments, spandrels, and wing walls of the work, but only well-shaped boulders, laid on their broadest bed, will be allowed. Dressing will be permitted on such stones as cannot be properly bedded without it. The parapet walls will be a continuation of the spandrel and wing walls. The boulder stone must reach entirely through the walls.

Cast Concrete Slab Veneer.—In constructing the arch bridge at Soissons, France, which is described on p. 206, the faces of the arch-ribs and the spandrel facing were formed of slabs of concrete-steel molded separately and set in place like stone veneer with the remainder of the concrete forming a backing. An essentially similar construction was employed in Chicago, Ill., in 1902, in constructing a number of recreation buildings in one of the city parks. In the last example mentioned the slabs were cast face down in wooden molds; the mode of procedure being as follows:

A layer of mortar composed of 1 part cement and 2 or 3 parts of finely crushed stone was first placed in the bottom of the mold to a depth of from $\frac{1}{2}$ in. to 1 in.; on this bed of mortar a 1–2–4 concrete, with $\frac{1}{2}$ to $\frac{3}{4}$ in. stone, was placed to the thickness desired and carefully rammed. After hardening, the blocks were removed from the molds and set aside to season until they were placed in the structure.

The construction of the slab veneer for the Soissons Bridge was as follows: For molding the arch-rib facing a smooth level platform or pavement of concrete was constructed on an adjacent level piece of ground. This molding platform was large enough to permit the arch-ribs to be delineated to full size on its surface. To prepare the mold the platform was covered smoothly with gunny cloth held down by battens, which also served to outline the extrados and intrados of the arch-rib. Radial strips of wood were then placed to divide the mold into voussoir-like sections. A thin bed of mortar was placed on the bottom of the mold and on this was laid four reinforcing-bars, one near and parallel to each edge of the voussoir being molded, so as to intersect at the corners. Under these bars at several points wire stirrups were looped with their fire ends projecting upward. The metal was then covered with a rich concrete of fine stone laid on the mortar-bed and compacted so that the total thickness was about 2 ins. When hardened, the product of the mold was a set of voussoir-shaped slabs with smooth faces and edges and a rough back with a number of projecting wires. In construction these facing slabs were set in place with mortar joints and backed with concrete. For the spandrel-wall facing the slabs were cast in rectangular molds in exactly the same manner. The engineers of the Soissons Bridge remark that the use of this cast concrete veneer enabled a considerable reduction of expense for forms and assured a surface finish of pleasing appearance.

Moldings and Ornamented Shapes.—The finishing of concrete structures in many instances comprehends the construction of moldings and ornamental shapes for cornices, corbels, medallions, keystones, and other architectural parts. These may either be molded in place by suitable construction of the stationary forms or they may be cast separately in portable molds and set in place in the structure as would be cut stone. Panels of simple form or plain cornice moldings can usually be molded in place without great trouble and expense, but in constructing corbels, complicated moldings, balusters, etc., particularly where one pattern is duplicated a number of times, time and expense will usually be saved by casting them separately or in sections, and afterwards erecting the separate pieces in the structure.

The casting of ornamental shapes in concrete may be accomplished either in sand-molds or in rigid molds of wood, metal, or plaster of Paris. Some very handsome work has been recently performed by sand-molding. The mode of procedure followed in making concrete castings in sand varies somewhat in practice, but it is substantially as follows: A pattern of the shape to be cast is first made in wood and to the exact size required, since no allowance for shrinkage is necessary. The pattern is then molded in sand in flasks exactly as is done in casting iron. The mixture used usually consists of cement and finely crushed stone of about the consistency of cream, and this is poured into the mold by means of a funnel and T pipe. The excess water in the mixture soaks into the sand and serves to keep the casting moist during setting. Generally the casting is left in the mold for three or four days, and is then removed and the projecting fins, if any, are cut off. The cast stone may be used immediately in the work, but it is preferable to let it season and harden for a fortnight or more before using. The product of these sand-molds has an unusually attractive surface texture. Sand-molding is particularly advantageous when balusters, corbels, medallions, and intricate moldings have to be cast, but for plain cornices and facing slabs it is generally as cheap and convenient to use wooden forms.

Efflorescence.—The leaching out of certain lime compounds and their deposition on the surfaces of concrete work are quite frequently the cause of the uneven color of such surfaces. In relation to this source of discoloration Mr. Clifford Richardson, Director of the New York Testing Laboratory, says:

It is primarily due to variations in the amount of water in the mortar of which the cement is composed. It will be readily understood that when any excess of water is used, segregation of the coarse and fine particles will take place, with a resulting difference in color. When a large amount of water is used the concrete is more porous and the very considerable percentage of free lime liberated from the Portland cement in the course of setting is more readily brought to the surface at such a point. . . . The amount of water in a concrete, the face of which is to be exposed, should be neither too small nor too large, but such a concrete should certainly not be dry or the exposed face will be honeycombed. . . . Where the greatest care is used as to the amount of water added to the mortar and to prevent its loss, and where separation of the mortar from the broken stone is carefully avoided in depositing the concrete and in ramming it, the exposed surface, after the removal of the molds, is fairly uniform in color. . . . A more uniform color will always be obtained when some puzzolanic material is ground in with the cement such as slag or tross. This hydrated silicious material combines with the lime which has been liberated and prevents it washing out on the surface. . . . Exactness in the amount of water used in the concrete, when the elimination of the stain caused by the free lime is considered desirable, the addition of some substance containing silica in an active form are the two steps to be taken to produce a concrete surface which should present a uniform color and a pleasing appearance.

The measures whose adoption are recommended in the quotation just made are designed to prevent the occurrence of efflorescence by adopting certain precautions in the materials and workmanship of the original construction. Their adoption, however, if it gives the success that Mr. Richardson anticipates, is obviously the way to get at the root of the trouble, but such action involves a degree of skill and watchfulness in constructing concrete work which is difficult of attainment under ordinary conditions of engineering construction, and which if attained will add materially to the cost of construction. They have the further objection that a special mixture of cement is required about which our information is not entirely certain. In default of preventive measures, which recommend themselves to general use, the engineer who encounters the trouble of efflorescence must overcome it by remedial measures. There are a number of these available. The most practical ones are the washing of the discolored surfaces by solution, which will remove the incrustation, or the removal of the original surface by dressing it down with hammers or tooling of some sort.

The manner of dressing down concrete surfaces to eliminate surface imperfections is discussed in a previous paragraph. The following account of the method of cleaning a concrete-steel bridge at Washington, D. C., gives instructive data as to this mode of procedure: The bridge in question had a mortar facing and after this was completed a heavy rain caused the entire north façade to become discolored by efflorescence. This discoloration was not uniform, but in streaks and blotches of a white color, which after weathering a short time turned into a dirty yellow. To clean the bridge trial was first made of water and wire brushes, but after a little work this method was considered impracticable owing chiefly to its cost, which was estimated at $2.40 per square yard. Washes of dilute hydrochloric acid, of dilute acetic acid, and of dilute oxalic acid were then tried in conjunction with ordinary scrubbing-brushes. The hydrochloric-acid wash proved the best, and the acetic-acid wash came next in efficiency. The wash finally adopted consisted of a solution of 1 part hydrochloric acid and 5 parts water. This was applied vigorously with scrubbing-brushes, water being constantly played on the work with a hose to prevent the penetration of the acid. One house-cleaner and five laborers were employed on the work, which cost 60 cents per square yard. This high cost was due largely to the difficulty of cleaning the balustrades; it was estimated that the cost of cleaning the spandrel- and wing-walls did not exceed 20 cents per square yard. The cleaning was thoroughly satisfactory. Some of the flour removed by the brushes was analyzed and found to be silicate of lime.

INDEX.

T.

W.

www.ingramcontent.com/pod-product-compliance
Lightning Source LLC
LaVergne TN
LVHW021939220826
846092LV00010B/1173

* 9 7 8 9 3 5 5 2 8 0 0 4 6 *